Emerging Infectious Diseases of the 21st Century

Series Editor: I. W. Fong

Professor of Medicine, University of Toronto
Head of Infectious Diseases, St. Michael's Hospital

For other volumes in this series, go to
http://www.springer.com/series/5903

I.W. Fong

Medico-Legal Issues in Infectious Diseases

Guide For Physicians

Springer

I.W. Fong
St. Michael's Hospital
University of Toronto
Room 4179 CC
30 Bond Street
Toronto, Ontario, Canada
M5B 1W8
fongi@smh.toronto.on.ca

ISBN 978-1-4419-8052-6 e-ISBN 978-1-4419-8053-3
DOI 10.1007/978-1-4419-8053-3
Springer New York Dordrecht Heidelberg London

Library of Congress Control Number: 2011922490

Printed on acid-free paper

Springer is part of Springer Science+Business Media (www.springer.com)

This book is dedicated to my children Michelle (Reyes) and Michael.

Preface

The current state of medical malpractice is of serious concern to physicians, medical organizations, and to many governments of developed nations. Many developed countries (including Canada and the United States) have seen a great increase in medical malpractice claims and a dramatic rise in premium rates over the past decade. Physicians face many challenges besides obtaining appropriate liability protection and practicing good cost-effective medicine, without the spectre of malpractice lawsuit. Unfortunately, although all practicing physicians and healthcare providers are vulnerable to medico-legal liabilities, this is a neglected and overlooked area in the education of medical student, residents and fellows, who will face the harsh realities of medical practice largely unprepared.

Physicians in developed nations have to a great degree, lost the respect and confidence of the citizens in their countries, and litigations for perceived wrong or harm occurs to a liberal degree for trivial or serious injuries. Besides seeking legal defense from their respective legal representatives, there are little or no guidelines or source of information in the medical body of literature. With this in mind, this volume is being written to help provide some guide to physicians dealing with a variety of various infections to avoid pitfalls in diagnosis and management that often predispose to litigations.

The author has many years of experience as a consultant to both lawyers for the defense and the plaintiff, with an average of 8–10 cases per year for the past 20 years. The text will give case scenarios in the format of problem-solving challenges that are faced by physicians in various practices and subspecialties such as family physicians, internists, emergency physicians and various surgical disciplines. The aim of this book is not to encourage defensive medical practice, but to help provide better, optimum care to patients and to be forthright and honest to our dear customers about our inevitable mistakes.

There are many books on jurisprudence and medicine, which deals with the law of the land (country, province, or state) and medicine, which are geared for the legal professions and students of law. A few books have provided some legal guidelines for physicians on general issues. This book will focus on clinical issues facing physicians in different settings (which can lead to malpractice), and the best approach to use to avoid litigations, and practice good medicine.

Acknowledgement

I am very grateful to Maria Isabel Suarez whose dedication and arduous assistance have made completion of this book possible. I am also indebted to Carolyn Ziegler for her assistance in performing literature searches, and to Debbie Reid for her administrative assistance.

Contents

Part I
Introduction

Chapter 1
General Principles

1.1 Introduction

To err is human and we all make mistakes, but this simple truth is not a satisfactory excuse for many of our affected patients, especially in developed countries. Despite the fact that many physicians will at sometime in their career encounter medical-malpractice litigation, we are ill prepared. It is, in fact, surprising that such an important topic is not included in the curriculum of all medical schools, and it is ignored to a large degree in post-doctoral training of residents and fellows. Physicians consider malpractice claims not only as a nuisance in their busy practice, but as unwanted events that result in anxiety, emotional pain, and possible loss of income and unnecessary expense.

Many industrialized nations have seen a dramatic and continued increase of malpractice litigations over the last two to three decades. A rise in medical malpractice claims and increase in settlements or awards (often termed "lawsuit lottery") have resulted in a dramatic rise in premium rates, and is considered at crisis level in the United States. Insolvency of both mutual medical defense organizations and commercial insurers is a challenge faced by many physicians and health care providers in obtaining appropriate liability protection. Moreover, an increasing number of physicians have abandoned or shunned high-risk practices associated with enormous medical malpractice insurance fees (i.e., obstetrics), resulting in great needs for medical coverage in some communities.

1.2 Global Trend in Medical Malpractice

The current state of medical malpractice protection is of great concern, not only to physicians and their families, but also to patients, many governments, and medical organizations in the developed world. The Organization for Economic Cooperation and Development (OECD) research indicates that in many countries, a supply 'crisis' for medical malpractice insurance is reducing the security and confidence of the citizens health care system. On the other hand, some consumer groups and

I.W. Fong, *Medico-Legal Issues in Infectious Diseases*, Emerging Infectious Diseases of the 21st Century, DOI 10.1007/978-1-4419-8053-3_1,

plaintiffs' attorneys view litigation as an indispensable form of protection against medical carelessness. Advocacy groups claim that physicians are the only ones who benefit from a decreased risk of being sued from lower malpractice premiums, and that increased risk of litigations and medical malpractice expenditure are a net-positive effect to the public, based on cost-benefit analysis.[1] This view is supported by the landmark 1999 Institute of Medicine (IOM) report that as many as 98,000 deaths in the US each year result from medical errors.[2] However, the IOM also found that 90% of these deaths were the result of failed systems and procedures, not negligence of physicians.

Critics of the current medical malpractice system in the US charge that frivolous litigation (claims that lack evidence of injury, substandard care or both) is common and costly, and calls for widespread tort reform. Malpractice law is part of tort, or personal injury law. To succeed in a tort lawsuit, the plaintiff must prove that the physician or hospital (defendant) owed a duty of care to the patient (plaintiff), that the defendant breached this duty by failing to adhere to the standard of care expected, and that this breach of duty caused an injury to the plaintiffs.[3]

In a recent retrospective review of 1,452 closed claims, trained physicians investigated the merits and outcomes of malpractice litigation in the US.[4] For 3% of the claims there were no verifiable medical injuries and 37% did not involve errors. Most of the claims that were not associated with errors (370 of 515 [72%]) or injuries (31 of 37 [73%]) did not result in compensation. Most of the claims that involved injuries due to errors (653 of 889 [73%]) received compensation. Payment of claims not involving errors occurred less frequently than nonpayment of claims associated with errors. The average payments for claims involving errors were significantly greater than those not involving errors ($521,560 vs $313,205, $p = 0.004$). Overall, unjustified claims (not involving errors) accounted for 13–16% of the systems monetary cost, but for every dollar spent on compensation 54 cents went to administrative expenses (including for lawyers, experts and courts).[4] Thus, the exorbitant costs of malpractice litigations are one of the main factors responsible for the high expenditure in this field.

There have been calls for sweeping reform of the medical liability system in the US and other countries, backed by organized medicine and the insurance industry and willingness of the government (Bush administration) to put caps on damages.[3] Opponents to widespread reform have claimed that the patient's right to safety by improving the quality of care is through litigation against hospitals and physicians. They point to the success of patient safety programs for anesthesiologists, which are motivated by liability suits and high rates of medical-malpractice insurance. During the last 25 years, the risk of death from anesthesia dropped from 1 in 5,000 to about 1 in 250,000, and the insurance for anesthesiologists (once the highest in medicine) is now among the lowest.[5] Furthermore, it is argued that the American Medical Association's own studies show no appreciable increase in malpractice premium over a 30-year period relative to physician practice expense and gross revenue, with no effect on physician net income.[1] The author's present additional data suggesting that medical malpractice insurance have no effect on the supply or location of physicians.

Government and legislative reforms in the US have been non-uniform and state-dependent, and have been in two stages. The first-generation reforms involve minor modifications designed to have immediate impact, and second-generation reforms include fundamental changes in dispute resolution.[1] The first-generation reform of caps for non-economic damages, for instance, include the California law enacted in 1975 to cap payments of $250,000, and in 2003 Texas lawmakers passed a $750,000 cap for noneconomic damages.[6] Despite the fact that the California state reform has been shown to reduce the frequency and severity of malpractice claims and to reduce malpractice premiums, it has been argued (by an economist and lawyer) that such caps are regressive, not indexed for inflation, unfair for persons with severe injuries as a result of medical negligence, and does not encourage improvement of the level of medical care.[1]

To improve both safety and the medical liability system, Clinton and Obama[7] proposed that the tort system must achieve four goals: reduce rates of preventable patient injuries, promote open communication between physicians and patients, ensure patient access to fair compensation for legitimate injuries, and reduce liability insurance premiums. To achieve these goals their main means of implementation is through the MEDIC (National Medical Error Disclosure and Compensation) program, based on the premise that the most important factor in people's decisions to file lawsuits is not negligence, but ineffective communication between patients and providers.[8] As noted by Studdert and Mello,[9] "lack of transparency in the current system is a destructively self-perpetuating phenomenon; physicians are reluctant to discuss injuries and errors out of fear of litigation, and patients sometimes sue to find out what happened." Injured patients frequently do not understand what happened to them, and the uncertainty with lack of empathy from physicians and a withholding of essential information, when an unexpected outcome occurs, is the main driving force in most instances of malpractice suits.[7]

Multiple factors are contributing to the increased medical liability costs in the US. The US malpractice system is a hodgepodge of medical liability insurance providers, most of which are state based. As of January 2006, the American Medical Association considered 21 states as being in a medical liability crisis, as the cost and/or availability of liability protection was adversely affecting the supply of physicians in high-risk specialites.[9] The per capita rate of medical litigation in the US is one of the highest in the world, and significantly higher than it is in Canada.

1.2.1 Liability System in Other Countries

1.2.1.1 Canada

The Canadian Medical Protective Association (CMPA) is a non-profit medical mutual defense association founded in 1901 and incorporated by a Parliament Act by 1913.[10] Its members include most physicians (66,000) across Canada; hospitals and other health care facilities utilize separate malpractice insurance agencies. The current medical liability system in Canada is a tort-based system, and had been undergoing significant adjustments, and trends in medical liability are changing. Based on data from 6 years (1999–2005) the CMPA estimates that the cost of the current Canadian liability system (including indemnities, legal and administrative costs) to be approximately $225 million per year.[10] In a study published in 2005, the average annual real growth of total malpractice claims in Canada during the period 1998–2001 was 20% (almost four times higher than in the US).[11] However, the number of new legal files opened annually on a national basis by the CMPA has shown a downward trend since 1998, and the cost of the medical liability system has remained relatively steady since 2001.[12] Although the Canadian liability model is based on three main tenets: patient safety and prevention, provider/facility accountability, and liability and accountability, there are a number of perceived faults. The Canadian model appears expensive in terms of operating or overhead costs; it may not support patient safety to the extent that no fault and other models might; and patients may not receive adequate compensation in a timely manner.[11]

1.2.1.2 France

The French medical liability system is complex and is composed of elements of both no-fault and fault approaches.[9] The no-fault system is when either no fault is declared by a regional commission (L' Office National d' Indemnization des Accidents Médicaux [L'ONIAM]), when the injury is the result of nosocomial infection, or the injuries resulting in an invalidity of at least 25% of the total damage or claim. It is the physicians' responsibility to demonstrate to the regional commission that the injury was not caused by the physicians' actions. The injured patient still has access to civil, criminal administrative and professional tribunals. Patients also have the right to refuse a compensation offer and seek judicial resolution for a higher award. Physicians in the public (no-fault) system have their liability insurance paid by their institutions, and those in the private system must pay their own premiums. Whilst the introduction of no-fault has diverted a large number of cases from the insured (private) system, it has not resulted in reduced premiums.[9] Furthermore, the increased cost of protection in the insured system has had a negative impact on the number of specialists in private practice.

1.2.1.3 New Zealand

The New Zealand liability model is supposed to be a no-fault system, but includes significant element of fault finding.[9] The health care system is a combination of public and private health care in which subjects pay for private care. The Accident Compensation Corporation (ACC) is a national insurance program that covers all bodily accidents or injuries caused by automobile, workplace, and medical treatment. Prior to 2005, the patient had to establish fault (as determined by the ACC) to receive compensation, unless the medical injury was a rare and severe complication. Despite the fact that 60% of the liability claims were rejected the per capita cost of the New Zealand system was 15% higher than that in Canada in 2003. Recent reforms in 2005 by legislation removed requirements for both the determination of fault and the rare and severe restriction; and treatment injuries of both serious and minor nature caused by health care treatment are now considered by the ACC.[12] It is estimated by the ACC that the number of claims will rise by 50%, and whether the added cost-burden will improve patient safety is yet to be determined.

1.2.1.4 Sweden

The Swedish health care system is a Universal public sector format as part of an extensive social welfare program, where individuals over age 20 pay small patient care fees, limited to a maximum of $150 per year.[9] Benefits provided to injured patients, through the medical liability system, form a part of this extensive system. Damages by medical provider (inferred fault) have to be deemed unintended and avoidable in comparison with standard of care from an experienced physician. Medical liability claims are approved in 45% and 10% of rejected claims are appealed.[9] The compensation system is also supported by a separate patient safety/risk management effort and by a separate accountability plan. The Swedish model appears to be relatively inexpensive, but since this represent only a part of the extensive expensive social support framework (common to Nordic countries), it really cannot be applied to other countries.

1.2.1.5 The United Kingdom

The medical liability system in the United Kingdom (UK) is also tort-based. Three medical defense societies provide medico-legal protection and advice to physicians in private practice.[9] The National Health Service (NHS) Trusts manage public hospitals and clinics and the NHS Litigation Authority (NHSLA) provide some malpractice protection for those in the NHS Trusts system. However, the NHSLA does not assist physicians by providing medico-legal assistance for members facing accountability enquiries by the regulatory body (the General Medical Council), and many physicians working in the NHS Trusts also join one of the three medical defense societies as well.[9] Although the medical liability system appears to working effectively, costs are rising from growing liabilities facing the government-run NHSLA.

1.2.1.6 Summary of Medical-Liability Systems

The international review completed by Secor Consulting (sponsored by the CMPA)[9] did not identify a single best model. Each system have their faults and advantages, and although the New Zealand and Swedish models are described as "no-fault," theirs are not pure no-fault systems and there is a significant amount of physician fault finding. The Swedish system appears to work well with their extensive social safety network, which if adopted by other countries would require extensive social system overhaul with huge budgets and much higher taxes.

1.3 Principles of Good Medical Practice to Avoid Litigation

Physicians need not practice defensive medicine by ordering unnecessary tests or referring patients needlessly for every condition to a specialist, in order to reduce the risk of litigation. In fact, unnecessary diagnostic tests and unwarranted therapy may have the reverse effect of greater risk of a malpractice suite.

The main tenets for reducing the risk of malpractice litigation can be categorized under the following: (1) good medical practice, (2) effective communication with patients, (3) proper and adequate notes and records, (4) informed consent, (5) accessibility and approachability, and (6) prompt attention.

Patients commonly complain that physicians do not take the time to adequately listen to their complaints. To address a patient's symptoms, a proper, adequately detailed history is essential, including relevant negative and positive functional inquiry. It goes without saying that previous illness and past medical history is part of a good history taking. The physical examination should be sufficiently detailed to assess the patients' symptoms, not only for the working or primary diagnosis but for possible other differential diagnoses.

Too often in medical malpractice cases, the entire focus has been on one primary diagnosis, with all the diagnostic, therapeutic and consultation efforts aimed in one direction. When the primary working diagnosis eventually is proven to be incorrect, recognition of the correct diagnosis is often too late, with catastrophic results. In the majority of significant medical illness, there should be two to three other differential diagnoses that should be considered and excluded. Thus, for a physician in any field (family practice, emergency medicine or other subspecialties) a list of possible other differential diagnosis should be considered and recorded. To practice good medicine in any field, continuous education by reading current literature, attending medical education courses or meetings is necessary to keep abreast of various medical conditions, and the ways to diagnose and treat these illnesses. In my experience, most malpractice litigation resulting from delayed diagnosis and treatment was due to physicians accepting an initial impression from a previous assessment or the patients' perceived self-diagnosis, without having a clear differential diagnosis in mind. Failure to make a proper diagnosis of any malady is usually due to failure to even think of the condition as a possible diagnosis, and this may lead to catastrophic outcome in serious illnesses that do not resolve spontaneously.

The skill in diagnosing a condition is judged by the benchmark of other physicians of the same specialty under the same conditions. The health care provider must conform to the standards and recognized procedures followed by members of the profession. The touchstone is based upon the conduct of the ordinarily careful and competent physician (under similar setting and qualification). However, the degree of care required to comply with the standard of care is dependent on the circumstances of the case.

In a recent retrospective review of 307-closed malpractice claims (in the ambulatory setting) alleged missed or delayed diagnosis was assessed and analyzed.[13] A total of 181 claims (59%) involved diagnosis errors that harmed patients. Most of these errors (59%) were associated with serious harm, and 30% (55 of 181) resulted in death. The most common breakdowns in the diagnostic process were failure to order an appropriate diagnostic test (55%), failure to create a proper follow-up plan (45%), failure to obtain an adequate history or perform an adequate examination (42%), and incorrect interpretation of diagnostic tests (39%). In this study, the leading factors that contributed to the errors were failures in judgment (79%), vigilance or memory (59%), knowledge (48%), patient-related factors (40%), and handoffs (20%).[13] The diagnostic errors frequently involved multiple process breakdowns, contributing factors and contributing physicians. There was a median of three process breakdowns per error, 54% of errors had three or more process breakdowns, and 29% had four or more. Moreover, in 43% of cases two or more clinicians contributed to the missed diagnosis, and in 16%, three or more physicians contributed.[13]

The legal considerations in the appropriateness of making a diagnosis are scrutinized under the following: (1) adequacy of the medical history, (2) appropriate and satisfactory examination of the patient, (3) conducting the appropriate medical tests and diagnostic investigations, (4) appropriateness (or lack of) in consulting or referring to other physicians (specialists), (5) medical judgment in making the diagnosis, (6) informing the patient of the diagnosis, and (7) making additional diagnosis if necessary. The liability of an erroneous diagnosis does not depend on the diagnosis alone, but on proof of breach of the standard of care required of the particular physician (health care provider) who caused the injury. A physician is only liable for a wrong diagnosis if it were overtly wrong as to constitute negligence. A health care provider is not considered negligent for failing to diagnose very rare conditions or in the early stages of a disease when it is difficult or almost impossible to diagnose. An honest error in exercise of judgment (even though other peer practitioners disagree with the judgment) is usually not considered negligence by the court.

As previously mentioned, ineffective communication by physicians with their patients is probably the single most important factor driving the impetus to sue. This is not limited to any one group of health care providers, but appears to be more common with very busy practices (i.e., emergency visits, busy office practice etc). As a group, physicians are not the best communicators, and we need to make a special effort to discuss issues in layperson language, without any judgmental or pedantic overtones. Physicians should take the time and effort, (otherwise make another appointment just for discussion) for any serious or perceived serious illness, to discuss the working diagnosis, differential diagnosis, planned investigations, and treatment.

Based on my experience of reviewing numerous medico-legal cases, the plaintiffs most common grievances against physicians lack of openness involve one or more of the following areas: (1) incomplete or absence of discussion on causation of symptoms, (2) failure to discuss prognosis and expected outcome, (3) incommunicable attitude of physicians after patients suffer from a complication or unexpected outcome after an intervention, (4) failure of the health care provider to give satisfactory explanation for delay of diagnosis and therapy that resulted in poor outcome, (5) inadequate discussion before development of adverse event of possible side-effects (from treatment or investigation), and the availability of alternate therapy.

Although patient safety is a priority in our modern health care system, 3–17% of hospital admissions result in an adverse event,[2,14–17] and almost 50% of these events are preventable. An adverse event can be related to drug toxicity, complications of surgical procedures or investigations, and is defined as unintended injury or complication caused by delivery of clinical care rather than the underlying condition. Some of these adverse events are intrinsic to the medications and therapy but certain groups of subjects have higher risk for various reasons, (i.e., genetic predisposition, underlying kidney or liver impairment etc.). Thus, it is imperative that the health care providers (physicians or pharmacists) avoid these therapies for high-risk patients. Other modifiable risk factors have been identified where preventable adverse events were mainly due to drug errors (40%) or poor clinical management (32%).[18] Elderly patients and those with communication problems (blindness, deafness, hard of hearing or language barrier) are at the highest risk of preventable adverse events (two to four times greater).[18] Subjects with poor reading skills and little education also have problems with verbal explanation and comprehension, as well as those with psychiatric disorders. Health care providers simply do not take the time and effort to ensure adequate comprehension by these persons of their conditions and medications. Simple means of communication to ensure patients understand their illness and therapy has been instituted in Iowa State by asking patients and families to answer these key questions:

(1) What is my main problem? (2) What do I need to do? (3) Why is it important for me to do this? (www.npsf.org.askme3). These basic questions should be adopted to ensure comprehension in all patients with communication problems and poor education, including the elderly.

Poor clinical management is the second most common factor associated with preventable adverse events,[18] and may be harder to fix. This is partly physician generated and partly a breakdown in our health-care system or structure. Simplification and standardization of care with performance measurements appears to be one solution that looks promising.[19] In 2003, the National Quality Forum (NQF) endorsed a set of 30 safe practices in the US that should be universally utilized in applicable clinical settings to reduce risk of harm to patients.[20] Patients expect to be informed promptly when any injury or adverse event occurs through medical care, especially those with serious consequences. However, frequently practitioners have not met these expectations, and until recently there were no guidelines for health-care providers as to when and how to disclose these errors. Previous reports have found that prompt and open disclosure policies had reduced the risk of malpractice litigation.[7,21] A few states have mandated the disclosure of certain adverse events to patients, and many states passed laws to protect health-care providers from litigation, to avoid using apologies for unanticipated outcomes from being used as evidence of fault in lawsuits.[22] The United Kingdom[23] and Australia[24] have also instituted disclosure programs.

In 2001 in the US, the Joint Commission on Accreditation of Healthcare Organizations (now the Joint Commission) issued nationwide disclosure standard, which is linked to the accreditation status of the hospital.[22] The standard did not mandate that unanticipated outcomes be admitted as being errors, nor specify the content of disclosure. By 2005, about 69% of healthcare organizations in the US had established disclosure practices, and in March 2006, the Full Disclosure Working Group of the Harvard Hospitals outlined a policy of full disclosure, taking responsibility, apologizing, and discussing steps to prevent recurrences.[22] Recommended practice guidelines by the NQF, for disclosing unanticipated outcomes to patients[20] included the following: (1) full disclosure to the patient by providing facts about the event, presence of error or system failure (if known), and results of event analysis, (2) expression of regret and formal apology for events caused by error or system failure, (3) institutional requirements should incorporate an integrated policy for disclosure, patient safety and risk management activities, including disclosure education and support system, and (4) institutions should provide emotional support for patients and families, as well as health-care workers, and use performance improvement tools to track and ensure disclosure.

In Canada, the CMPA guidelines for disclosure of adverse events or unexpected harm are also similar: (1) communicate the facts to the patient or family (to the extent they are known) in a gentle, non-rushed manner as soon as it is reasonable to do so, (2) discuss the options for dealing with the medical condition as it now exists, (3) express your feelings of concern, empathy and regret, as appropriate, (4) after investigation and all the facts are known, if the outcome is indisputably due to deficient care, the responsible health professionals may apologize and acknowledge responsibility.[25] However, the CMPA advises avoidance of the words fault or negligence, or reference to failing to meet the standard of care.

Detailed medical notes for initial assessment and progress are essential components of a good medical practice. Poor and inadequate records are common in many cases that come to litigation. Too frequently, there are only a few lines noted (often indecipherable) in the emergency department or office practice records, even for initial assessment or with new symptoms, which does not provide sufficient information on the physicians' thought process, or derivation of a differential diagnosis. Surprisingly, the nurses' records often provide better information about the patients' symptoms and current status than those of the physicians. It has been my experience, that there are frequent discrepancies between physicians' and nurses' notes, as if the attending physician was never aware of the situation or may not have read the supporting healthcare provider notes, nor reviewed any verbal report. These discrepancies and disharmony of records provide ammunition for the plaintiffs' lawyers, and is a poor reflection on the physician and health-care center. Another piece of information that is commonly missing from medical records, is the time of physicians assessment (including consultations), which may be important for time sensitive issues requiring prompt diagnosis and intervention.

There are certain legal aspects of the medical records that physicians should be familiar with: (1) medical records are considered legal documents, (2) medical records should not be tampered with or altered after the fact, (3) errors should be corrected promptly once detected, (4) after a record is made, errors should be corrected or edited only by using the proper methodology, (5) establish and implement policies for retention and destruction of medical records, (6) spoliation includes adding to an existing record at some later date, or the omission of significant medical information (fraudulent concealment), and (7) incorrect dating of the record or complete rewriting or retyping part of the medical record.[26]

In summary, timely and appropriate detailed documentation in the medical record is essential to good medical practice and neglect of this aspect is one of the most frequent breaches in patient care resulting in successful malpractice litigation. Carefully documented patient assessment, discussion, and interventions by detailed and accurate medical record are worth the time and effort as preventative measure to reduce litigation. Significant gaps in the medical record gives the impression of substandard care to a jury, whether or not it actually occurred.[27]

Lack of informed consent is sometimes an issue in malpractice litigation. A physician may be found negligent if he or she diagnoses or treats a patient without informed consent (or adequately informed consent) for a diagnostic or therapeutic procedure. Consent is a process, not a form, and we must provide adequate information with which the patient is to reach a decision regarding a diagnostic or treatment procedure, and must be given ample opportunity to discuss alternatives with the physician. Failure to give appropriate and informed consent may be considered departure from the recognized standard care.[28] Physicians should not just rely on an informed consent form to obtain adequate informed consent. Failure to discuss potential side effects of medication and alternative drugs beforehand may be considered a breach of informed consent to treatment if the patient develops a significant side effect. In these situations, a written consent is not required, but it is prudent for the physician to record that he or she has discussed the possible side effects with the patient. In discussion of informed consent and possible adverse effects, effective communication is crucial. The adage in legal circles: "Don't say it so that anyone can understand it; say it so no one can misunderstand it!" should be remembered by all physicians.

There are several golden rules worth remembering when dealing with informed consent process and refusal[29]:

1. The ultimate goal of informed consent process is not patient consent, but rather patient understanding of all circumstances pertinent to the decisions that must be made about care and treatment.
2. Patient autonomy outweighs the physicians' professional opinion and good intentions.
3. The patient consents help to protect the provider.
4. For consent to be meaningful, the patient should know what he or she is consenting to, and that person has to have the intellectual capacity to consent.
5. Treat refusal of recommended care like informed consent; good documentation would be recording it as an informed refusal.

6. Where the capacity of a patient is questioned, the physician should document his or her basis for this assessment.
7. Every single informing session for consent or refusal, should be approached with concern for the patients' well being, and reflect for them as individuals.

A common complaint of plaintiffs against physicians when an unexpected outcome occurs after surgical or medical therapy is that they were not adequately informed of the potential adverse events, or they would not have agreed to the treatment. The physicians usually have a signed consent form without any details of discussion, or they may have noted a discussion on possible outcome. Usually in these circumstances, the events were uncommon or rare, and it raises the issue of how detailed the information should be. It is not practical for physicians to list all possible side effects or adverse events to medications and interventions, but the most common ones are usually mentioned. Acceptance of this by the courts or jury may vary according to the circumstances, but in many cases, this is considered accepted standard. Physicians, however, should consider following the example of pharmaceuticals' commercial advertisements with a covering statement such as "and other rare and unforeseen events may occur."

Part of the problem of ineffective communication by healthcare providers as reported in many medico-legal transcripts, are their accessibility and approachability. Plaintiffs frequently report in litigation documents that they were unable to speak to their family physician or emergency physician etc, when phoning the physicians' office or emergency department to report new or worsening symptoms, and thus only received second-hand advice from the secretary, assistant, or nurse. When events go awry, there is usually no recall or documentation of these accounts to verify any verbal recommendation. In these situations, the physician (although busy) should make an attempt to speak to the patient or arrange a prompt appointment, and best to record any verbal discussion.

Some patients report feeling intimidated by their physician (usually a consultant or someone unfamiliar to the person), and thus report lack of rapport and feedback, and may complain that the physician does not address their symptoms and issues adequately. As a rule, physicians should always relate to patients as persons and treat them with respect. "Another word for an angry patient is a plaintiff."[27] Encounters with patients and families always should be transparent. Physicians should establish and maintain clear expectations of their patients, as compliance and behavior/alcohol/drugs may have an important impact on outcomes of treatment. A guarantee sometimes communicated to patients (medical or surgical treatment) is the promise to produce specific results, i.e., "We will fix your problem." This will often lead to malpractice litigations when the stated results are not attained, especially if associated with complications. It is best to give odds of successful outcome (therapeutic reassurance), as there are no warranties with medical or surgical treatment.[27] Physicians should take the time to have open discussion with their patients to provide information on the nature of their condition and options available for management, and to give patients the opportunity to have meaningful input into decisions about medical care.

Health-care providers should be accessible to their patients, especially when unpredictable events occur. At such times, patients and families may experience anxiety, grief, guilt, and anger and look to their physicians for support and answers. In such circumstances, honest, timely, and frank discussion with the patient and family will often alleviate concerns and prevent future difficulties. In deteriorating or complex situations, early consultation with a colleague can be helpful. In certain situations when the patient or family threatens legal action or complaint to authorities, the physician (still having a duty of care to the patient) would best arrange for the transfer of care to a colleague.

Prompt medical attention is a responsibility of both the patient (to seek early medical attention) and the healthcare providers (to provide timely assessment, diagnosis, and treatment). In the emergency department, this process is usually delegated to a triage nurse. However, timely management of several conditions can make a difference in the outcome in other ambulatory settings, such as in physicians' offices and clinics, where a triage system is usually non-existent. In these settings, a triage nurse is usually not necessary as the urgency of the conditions and outcomes are not dependent on rapid actions within minutes or hours. However, delays in diagnosis and treatment by days, weeks or months can make a difference in outcome in many illnesses not considered emergencies. Thus, delayed diagnoses (or missed) in the ambulatory setting are an important safety problem.

Delay in diagnosis and treatment can occur as a result of judgment errors or oversights during several steps, and often may involve more than one health-care provider. Errors at the initial assessment to consider the diagnosis or seriousness of the condition, delay in obtaining investigational results or appropriate consultation, misinterpretation of test results, delayed or inappropriate follow up to review results, and tardiness in instituting prompt appropriate therapy are the main problems. Diagnostic error is commonly multifactional in origin, typically involving both system-related and cognitive failing by the physician. The diagnostic acumen reflects the clinicians' knowledge, clinical skills, experience, and problem solving skills. Previous studies have found that system-related factors contributed to the diagnostic error in 65% of cases and physician cognitive factors in 74%.[30] Another common factor is delay in reviewing test results; this is probably more common in community practice than urban hospital practice. However, in a study of 262 physicians in 15 internal medicine practices associated with urban teaching hospitals, 83% of respondents reported at least one delay in reviewing test results during the previous 2 months.[31] Moreover, only 41% of physicians reported being satisfied with how they manage test results.

System failure includes lack of communication of test results and other clinical information, and several handoffs. Limited health care resources, such as availability of computerized scans (CT) or magnetic resonance imaging (MRI), are also a problem even in some developed countries (i.e. Canada), where the wait time for outpatient, non-emergency scans can be weeks to months. Failure to follow up on abnormal test results is a critical weakness in patient safety (especially in ambulatory care), and up to 33% of physicians do not regularly notify patients of abnormal test results.[32] A disturbing picture of widespread system failure in community practice has been reported by several studies, where 31% of women with abnormal mammograms do not receive timely appropriate follow-up care, and up to 33% of women with abnormal Papanicolauo (Pap) smears were lost to follow-up.[33,34]

Widespread nationwide or provincial high profile scandals involving shoddy laboratory medicine have recently been highlighted in Canada. Scandals involving erroneous or incomplete work by pathologists (resulting in missed or delayed diagnoses) have recently been investigated and are undergoing public inquiry.[35] These system failures as a result of significant laboratory errors involved four provinces (Newfoundland, New Brunswick, Ontario, and Manitoba) and prompted retesting in more than 60,000 cases. Incomplete or erroneous results may have contributed to inappropriate therapy and premature death in 100 women with breast cancer, and a wrong diagnosis of pancreatic cancer has resulted in unnecessary surgery and chemotherapy. The latter case has recently been settled out of court,[35] but class action lawsuits are expected to follow from the other mistakes.

1.4 Summary

To reduce the risk of malpractice litigation physicians in general need to practice good medicine, listen and talk to their patients. There are several golden rules that should be followed to be a respected and effective clinician:

1. Assiduity in history and physical examination are indispensable; too often short cuts lead to mistakes.
2. Continuous medical education and keeping abreast of current trends are essential to be a good clinician.
3. Think of a differential diagnosis and ways to exclude them; take a problem-solving approach.
4. Try to obtain prompt results and consultations for potentially serious illnesses.
5. Communicate effectively with patients and families. Be frank and honest, but be empathetic at the same time.
6. When in doubt, get help from consultants and when time sensitive conditions require urgent consultation, refer to the hospital emergency department.
7. Clear, detailed notes and records are a duty and not a burden.
8. Develop a standardized routine system to review results with automatic follow-up of patients and urgent re-appointments for serious abnormal test results.
9. Remember to treat patients as you would like to be treated, and be prompt and attentive.
10. Never guarantee results, and always provide options; patients should be part of the decision process.
11. Take time to explain possible outcomes, side effects and complications, and make a record of your discussion.
12. Phone advice can lead to a lawsuit, so it is best to assess a patient in person.

13. Phone consultations with other physicians or patients should be recorded in the patients file, or as a letter to the requesting physician with a paper or electronic trail.

References

1. Sloan, FA, Chepke LN (eds), (2008), Medical Malpractice. MIT Press, Cambridge, MA.
2. Kohn LT, Corrigan JM, Donaldson MS (eds), (2002). To err is human; building a safer health system. National Academy Press, Washington, DC.
3. Studdert, DM, Mello MM, Brennan TA, (2004). Medical Malpractice. N Engl J Med. 350: 283–292.
4. Studdert, DM, Mello MM, Gawande AA, Gandhi TK, Kachalia AB, Yoon C, Puopolo AL, Brennan TA, (2006), Claims, errors, and compensation payments in medical malpractice litigation. N Engl J Med. 354: 2024–2033.
5. Annas, GJ (2006). The patient's right to safety; improving the quality of care through litigation against hospitals. N Engl J Med. 354: 2063–2066.
6. Marcus, HR, (Texas Alliance for Patient Access) (2006). Medical malpractice and patient safety (letter to the editor). N Engl J Med. 355: 735.
7. Clinton, HR, Obama B, (2006). Making patient safety the centerpiece of medical liability reform. N Engl J Med. 354: 2205–2208.
8. Liebman, CB, Hyman CS, (2004). A mediation skills model to manage disclosure of errors and adverse events to patients. Health Affairs (Millwood); 23: 22–32.
9. Canadian Medical Protective Association, (2005). The medical liability environment outside of Canada. CMPA; http://www.cmpa-acmp.ca
10. Canadian Medical Protective Association, (2005). Medical liability in Canada: towards the right balance. CMPA; http://www.cmpa-acmp.ca
11. Anderson GF, Hussey PS, Waters HR, (2005). Healthcare spending in the United States and the rest of the industrialized world. Health Affairs; 24: 903–914.
12. Canadian Medical Protective Association, (2006). Medical liability – a physician primer. CMPA; http://www.cmpa-acmp.ca
13. Gandhi TK, Kachalia AB, Thomas EJ, Puopolo AL, Yoon C, Brennan TA, Studdert DM, (2006), Missed and delayed diagnoses in the ambulatory setting: a study of closed malpractice claims. Ann. Intern Med. 145: 488–496.
14. Baker GR, Norton PG, Flintoft V, Blais R, Brown A, Cox J, Etchells E, Ghali WA, Hébert P, Majumdar SR, O'Beirne M, Palacios-Derflingher L, Reid RJ, Sheps S, Tamblyn R, (2004), The Canadian Adverse Events Study: the incidence of adverse events among hospital patients in Canada. CMAJ; 170: 1678–1686.
15. Davis P, Lay-Yee R, Briant R, Ali W, Scott A, Schug S, (2002), Adverse events in New Zealand public hospitals I: occurrence and impact. N Z Med J 115; U271.
16. Vincent C, Neale G, Woloshynowych M, (2001), Adverse events in British hospitals: preliminary retrospective record review. BMJ; 322: 517–519.
17. Brennan TA, Leape LL, Laird NM, Hebert L, Localio AR, Lawthers AG, Newhouse JP, Weiler PC, Hiatt HH, (1991), Incidence of adverse events and negligence in hospitalized patients. Results of the Harvard Medical Practice Study I. N Engl J Med. 324: 370–376.
18. Bartlett G, Blais R, Tamblyn R, Clermont RJ, MacGibbon B, (2008), Impact of patient communication problems on the risk of preventable adverse events in acute care settings. CMAJ; 178: 1555–1562.
19. Frankel A, (2008), Health literacy and harm; who is at risk? What is the fix? CMAJ; 178: 1573–1574.
20. Safe practices for better healthcare (2007), Washington DC: National Quality Forum (accessed June 01, 2007; at http://www.qualityforum.org/projects/completed/safe_practices/.).

21. Kraman SS, Hamm G., (1999) Risk management: extreme honesty may be the best policy. Ann. Intern Med. 131: 963–967.
22. Gallagher TH, Studdert D, Levinson W, (2007), Disclosing harmful medical errors to patients. N Engl J Med. 356: 2713–2719.
23. Safer practice notice: being open when patients are harmed, (2005), London: National Patient Safety Agency (accessed June 1, 2007, at http://www.npsa.nhs.uk/site/mediadocuments/1314-SaferPracticeNotice.pdf.).
24. Australian Council for Safety and Quality in Healthcare (2003), Open disclosure standard: a national standard for open communication in public and private hospitals following an adverse event in healthcare- 2003 update (at http://www.safetyandquality.org/internet/safety/publishing.nsf./content/F87404B9B00D8E6CCA257160000F049/$File/OpenDisclosure-web.pdf.
25. Canadian Medical Protective Association, (2006), How to apologize when disclosing adverse events to patients. CMPA: http://www.cmpa-acmp.ca
26. Hoffmann, AC, Sanbar SS, (2007), Spoliation: record retention, destruction, and alteration. In: Sanbar SS (ed), The medical malpractice survival handbook, American College of Legal Medicine, Mosby, Elsevier Inc., Philadelphia, PA, p45–53.
27. Rich BA, (2007), Breach of the physician-patient relationship. In: Sanbar SS (ed), The Medical Malpractice Survival Handbook. American College of Legal Medicine, Mosby/Elsevier Inc, Philadelphia, PA, p133–143.
28. Sanbar SS (ed), (2007), Etiology of malpractice. In: The medical malpractice survival handbook, American College of Legal Medicine, Mosby/Elsevier Inc, Philadelphia, PA, p119–132.
29. Hill HF, (2007), Lack of informed consent and refusal. In: Sanbar SS (ed). The medical malpractice survival handbook, American College of Legal Medicine, Mosby/Elsevier Inc, Philadelphia, PA, p153–164.
30. Graber ML, Franklin N, Gordon, R, (2005), Diagnostic error in Internal Medicine. Arch. Intern Med. 165: 1493–1499.
31. Poon EG, Gandhi, T, Sequist, TD, Murff, HJ, Karson, AS, Bates, DW. (2004), "I wish I had seen this test result earlier!" Dissatisfaction with test result management systems in primary care. Arch. Intern Med. 164: 2223–2228.
32. Boohaker EA, Ward RE, Uman JE, McCarthy BD, (1996), Patient notification and follow-up of abnormal test results. A physician survey. Arch. Intern Med. 156: 327–331.
33. Haas JS, Cook EF, Puopolo AL, Burstin HR, Brennan TA, (2000), Differences in the quality of care for women with an abnormal mammogram or breast complaint. J. Gen. Intern Med. 15: 321–328.
34. Marcus AC, Crane LA, Kaplan CP, Reading AE, Savage E, Gunning J, Bernstein, B, Berek, JS, (1992), Improving adherence to screening follow-up among women with abnormal Pap smears: results from a large clinic based trial of three intervention strategies. Med. Care; 30: 216–230.
35. Lett, D, (2008), Canada lags in standardized protocols for medical labs (News). CMAJ; 179: 125–126.

Part II
Issues in the Emergency Department

Chapter 2
Complaints Related to the Head

2.1 Case 1: Irritability in a Young Child

A 16-month-old male child was taken to a community based emergency department by the mother at about 3 a.m. The child had been unwell the evening before with fever, vomiting, restlessness and irritability. The emergency physician evaluation 30 min later noted a temperature of 39°C, a pulse of 140/min, respiratory rate of 34/min, but there was no recorded blood pressure. According to the transcripts, the clinician felt the infant was not dehydrated, and the examinations of the ears, throat, chest, and abdomen were recorded as normal. A complete blood count was ordered, which revealed a hemoglobin of 91 g/dL, and a total normal white blood count (WBC) of about 9,000/μL. However, the differential count was not available until 3 h later. The child was treated with acetaminophen, which resulted in a reduction of the temperature later to 37.9°C, and he was also reassessed by the attending physician 2 h later (but not recorded in the chart), with apparently no worsening. The child was then discharged after 6 a.m., and the mother reassured that her son had only a "flu-like" illness. The leucocyte differential court subsequently was reported as showing 22% bands (normal 3–5% for a young child), but the family was never notified.

2.1.1 Comments

Fever, vomiting, and irritability are common with many young children and infants and often reflect self-limited viral infections. Thus, cases similar to the above represent common presentation to an emergency department. However, a serious bacterial infection has to be considered in the differential diagnosis to be excluded. There were several features of this case that should have alerted the physician to a serious illness and to perform a more detailed examination and investigation. Parents know their children best, and can often differentiate between a mild febrile illness and a more serious condition. Any parent who brings their child to an emergency department at 3 a.m. considers the child seriously ill. Although

I.W. Fong, *Medico-Legal Issues in Infectious Diseases*, Emerging Infectious Diseases of the 21st Century, DOI 10.1007/978-1-4419-8053-3_2,

respiratory and gastrointestinal infections are the most common diseases presenting in this manner in young children, the differential diagnosis should include pneumonia, urinary tract infections, appendicitis, bone-joint infections, meningitis-encephalitis and pneumococcal bacteremia (without focal findings). A normal (but incomplete) physical examination does not exclude any of the above differential diagnoses, especially in infants and young children. For instance, the sensitivity of a chest examination in detecting pneumonia is only about 50%, and young children often do not exhibit neck stiffness in meningitis. This latter sign was not even noted in the files, whether it was present or not.

The nurse's note in the emergency record indicated that the child was listless and restless with dry mucous membranes, both common signs of serious illness, including meningitis. Moreover, these signs persisted when the fever was reduced artificially. It is a common misconception that I have noticed in many malpractice cases, that clinicians falsely assess patients as not being very ill when the fever temporarily subsides after antipyretics, and patients are sent home based on this nonspecific therapeutic maneuver. Artificially reducing a high temperature is of no diagnostic or prognostic significance.

The clinical assessment should have included tests for meningeal irritation, lethargy, and toxicity. Moreover, the emergency physician note was scant and sketchy, with no details (mainly one or two words with corresponding tick marks). The terms "ill-looking" or "toxic-looking" are loosely applied and commonly used in medical jargon and charts. However, there have been guidelines established in 1993 to define toxic appearing infants and children.[1] Toxic is defined as a clinical picture consistent with the sepsis syndrome (i.e. lethargy, signs of poor perfusion, marked hypoventilation or hyperventilation, or cyanosis). Lethargy is defined as a level of consciousness characterized by poor or absent eye contact, or as failure of the child to recognize parents or to interact with persons or objects in the environment.[1] The probability of serious bacterial infections in toxic appearing young children range from 10% to 90%, depending on the criteria used to define toxic.

Others have used observation scales to identify serious illness in febrile children.[2] Six items were found to have significant and independent prediction of serious illness:

(1) Quality of cry, (2) reaction to parents, (3) state variation, (4) color, (5) state of hydration, and (6) response to social overtures. It has been well recognized for many decades that young children (especially <16 months of age) with fever and a serious illness may not manifest typical (classic) signs suggestive of that illness, e.g. the young child with meningitis will not infrequently fail to manifest a meningeal sign such as neck stiffness. Instead, the child may demonstrate "paradoxical irritability" – the neck is held stiffly and the child will cry when an attempt is made to flex the neck, even when cuddled by the mother.[3]

Discriminant function analysis of the six items used in the observational scales when used together revealed a specificity of 88% and sensitivity of 77% for serious illness.[2] Individual scores for each of the key six items were added to yield a total score for each patient (see Table 2.1). Only 2.7% of patient with a score of ≤ 10 had

Table 2.1 Observation items, scores and predictive values for severity of illness in infants (Adapted but modified from McCarthy et al.[2])

	Score I	3	5
Observation item:	Normal	Moderate impairment	Severe impairment
Quality of cry	No crying or strong cry	Whimpering or sobbing	Weak/moaning or high pitched
Reaction to parent stimulation	Not crying or cries briefly	Cries off and on	Continuous crying or little response
Wakeful state[‡]	Stays awake, or wakes easily	Eyes closed or awake only for prolonged stimulation	Asleep and will not arouse
Response to visual stimulation	Smiles or talk	Smile or alert only briefly	Listless, dull, expressionless, anxious, or unawake
Hydration	Skin, eyes normal, mucous membranes moist	Skin, eyes normal mouth slightly dry	Skin tented or doughy, eyes sunken, dry mucous membranes
Color	Pink	Extremities pale or acrocyanosis present	Ashen, pale, mottled or cyanotic

[‡]also called state variation

a serious illness, and 92.3% with a score of ≥16 had a serious illness.[2] The sensitivity of the six-item model for serious illness was increased to 92% when combined with history and physical examination.

Practice guideline for management of young children (3–36 months of age) with fever of 39°C or more and whose WBC count is ≥15,000/μL, recommend blood and urine cultures and a start on antibiotics until cultures are available (usually single dose ceftriaxone). However, up to 25% of children with bacteremia can have normal or lower leucocyte count early on, and increased bands (premature neutrophils) is a sign of acute bacterial infection. Hence, it is important to check the differential white blood count before discharge from the emergency, or recall the family to return once a high band count has been reported.

Fever without an obvious focus of infection is a common diagnostic dilemma, as up to 30% of febrile children have no localizing signs or symptoms. Besides obtaining a detailed history from the parents, it is important to observe the child without direct contact to determine the child's behavior and response to the parents, surroundings and social overtures, as these are important indicators of the seriousness of the child's condition. In febrile children, the higher the fever, the greater the risk of serious illness[3] and children under 3 months of age are particularly of concern because of their premature immune system. Although the majority of acute febrile children have a benign self-limited viral infection, serious infectious illnesses can be found in about 8% (pneumonia, meningitis, occult bacteremia, soft tissue, bone, joint, urinary tract or bacterial diarrheas).[3] Before the advent of routine conjugate pneumococcal vaccine (2000), occult bacteremia occurred in 1.5% of acute febrile children (90% due to *Streptococcus pneumoniae*).

Since 2002, the rate of occult bacteremia has declined to <1%.[4] Although 30–40% of *S. pneumoniae* bacteremia resolves spontaneously, the remainder would manifest evidence of pneumonia, meningitis, acute otitis media, and severe sepsis if left untreated.[4] The risk (prior to 2000) of occult bacteremia in children (3–36 months old) with a fever without a source and with a temperature of ≥39°C has been reported to be 3–11%, with a mean of 4.3%.[1] The overall risks of persistent fever, persistent bacteremia, and meningitis have been found to be 56%, 21% and 9%, respectively, in a review of 20 studies of the outcomes of bacteremia in febrile children.[1]

Most guidelines do not recommend routine chest radiograph for febrile children unless they have symptoms or signs of lower respiratory tract infection (e.g. tachypnoea, grunting respiration, cough, rales, or rhonchi). However, chest radiographs are recommended for toxic, ill looking children with temperatures ≥40°C, WBC count ≥15,000/μL or increased bands, or those found to have a positive blood culture.

When should a child have a lumbar puncture? A lumbar puncture for cerebrospinal fluid (CSF) analysis is indicated in any child with a possible diagnosis of sepsis or meningitis based on history, observational assessment, or physical examination. The clinical presentation of meningitis frequently starts with a non-specific viral-like illness with fever, poor appetite, mild headaches, upper respiratory tract symptoms, body aching and represent a viral upper respiratory tract infection which precedes or predisposes to bacterial meningitis.[5] Unlike most benign viral infections, the symptoms continue to progress over the next few days, into the meningitis phase. This stage represents onset of meningeal inflammation with persistent or increasing fever, protracted severe headaches and may be manifested by irritability and vomiting by the child. It may also be followed by listlessness, drowsiness, inability to eat or feed, stupor, and coma. The child may or may not manifest evidence of seizures. Sometimes, acute bacterial meningitis may be preceded by symptoms and signs of acute otitis media or pneumonia. The physical examination may or may not reveal nuchal rigidity and other signs of meningeal irritation; Kernig's and Brudzinki's signs are late manifestations which are often absent on presentation. In infants, there may be bulging fontanel, secondary to increased intracranial pressure, but the fontanel may also be sunken from dehydration secondary to vomiting and poor oral intake. "Paradoxical irritability" may be present instead of neck stiffness, and a persistently irritable, listless child especially with fever should warrant a lumbar puncture.

2.2 Progress Report

The ill child was taken the same day at noon, to an urban pediatric hospital emergency department. The nurses' note at 1:30 p.m. indicated that the child was very ill looking, pale, lethargic, irritable and with a stiff neck. The admission note indicated a history of vomiting, irritability, decreased oral intake, unfocused eyes, inability to stand up, grunting, and neck stiffness from the night before. The emergency physician noted the child was very unwell, with grunting respiration, back arched, neck hyperextended (opisthotonus), marked neck stiffness, positive Kernig's and Brudzinski's signs, bulging fontanel, red right tympanic membrane and crackles of the night anterior chest. A diagnosis of acute meningitis, right middle lobe pneumonia was made. A lumbar was not attempted as the child was too ill, and computerized tomography (CT) of the brain was ordered. Blood cultures were ordered, emergency intravenous fluids administered, ceftriaxone ordered (but was not administered until 4 p.m.), and vancomycin was given at 7 p.m. S. pneumoniae was recovered from the blood and the child survived, but was left with deafness, blindness, and probably developmental or mental impairment, and also required a permanent ventricular-peritoneal shunt. The family subsequently launched malpractice litigation against both hospitals and physicians involved.

2.2.1 Comments

When the child presented to the urban pediatric hospital emergency, he was recognized by the nurse to be gravely ill with obvious signs of advanced meningitis. Quite rightly, a lumbar puncture was delayed until a CT scan of the brain was done because of marked decrease in level of consciousness; a bulging fontanel alone is not an indication for CT scan or delay. Other indications for CT scan or putting off a lumbar puncture in a child include focal neurological signs, significant bradycardia, hypoventilation, impending respiratory failure, bleeding tendency (as in thrombocytopenia and disseminated intravascular coagulopathy [DIC]), and septic shock. In these situations, it is advisable to take blood cultures and start empiric antibiotics for meningitis as soon as possible.

2.2.2 Medico-legal Issues

The lawyer for the plaintiffs (parents of the child), have listed several areas of negligence and errors in the diagnosis and management of this case, which directly resulted in a poor outcome and fell below the standard of care.

1. The initial ER physician at the general hospital failed to take a proper history and examination, and was negligent in not performing a lumbar puncture, admitting the child to hospital, and starting appropriate antibiotics.
2. The health-care system at the first hospital and the medical staff should have known the significance of a high band leucocyte count, indicating acute bacterial infection, thus requiring further investigations (blood cultures, chest radiograph and a lumbar puncture) and antibiotics before discharge from the emergency department. Even if the results were not available before the child was sent

home, it was the duty of the hospital emergency staff to notify the parents once the results were available, to return immediately to the hospital. This error by the medical staff and the hospital health-care system directly led to a delay in the diagnosis and subsequent poor outcome.
3. The medical staff (including the attending ER physician) at the second (pediatric) hospital, fell below the standard of care in not initiating rapid investigations and treatment immediately after registration at 12 noon. It was obvious that the child was gravely ill, as noted by the triage nurse, but he was not assessed by the physician until an hour later, and the first dose of antibiotic was not given until 4 h later.
4. This delay in instituting therapy contributed to the poor outcome of the child. Furthermore, if antibiotics were instituted within an hour of registration in the emergency department along with dexamethasone, the outcome would more likely have been better.
5. The plaintiffs were seeking compensation for damages that have resulted in severe disability, poor prospects for future development, education, and employment, and he will likely need lifelong, constant support.

2.2.3 Comments of Medical Aspects

There are several aspects of this case worth reviewing for discussion. Firstly, there was a significant discrepancy between the first ER physician's (defendant) depositions in the transcripts of the appearance of the child (that the child was not ill looking) versus that of the parents. This is a frequent occurrence in many malpractice litigations that I have reviewed. Whom will the judge or jury believe? There is a high probably that a jury or judge would believe the plaintiffs' deposition rather than the clinician in this case (or similar cases). A major limitation of the defendant's (physician) evidence is the actual contents of the medical file, which is very sketchy with no details of history or examination, or differential diagnosis. This is a major limitation of the defense and it would favor the plaintiffs' evidence that the child was gravely ill from the onset. Moreover, the admission history at the pediatric (second) hospital clearly describes a very ill child from the night before admission, with symptoms and behavior that were highly suspicious for acute meningitis, even at that time. Thus, one of the lessons that clinicians should learn from this case is that adequately detailed notes of the history, physical examination, planned investigation, and differential diagnosis are essential components of a good defense. Moreover, juries consist of parents and grandparents and (although there is a tendency to sympathize with the plaintiffs when children are involved) they would recognize that a child had to be seriously ill for parents to bring him to a hospital emergency room either late at night or very early in morning (3 a.m.). They would also recognize that the clinician should have realized that the child would unlikely be suffering from a benign viral infection (flu-like illness) based on this fact alone.

Secondly, had the initial ER physician taken a proper history, examination and appropriate tests, a diagnosis of acute bacterial meningitis could have been made in the early morning and appropriate treatment instituted. The lawyer for the plaintiffs claim there was sufficient indication at the initial visit to warrant evaluation for treatment of meningitis or a serious bacterial infection, and that failure to do so contributed to an (avoidable) severe adverse outcome. In this case, the lawyers argue that bacterial meningitis was present at the initial visit, and delayed diagnosis and treatment increased the risk for neurological damage or death. In some case scenarios involving bacterial meningitis, a child with symptoms and signs of an infectious illness, the physician may be considered negligent in not diagnosing and treating a pre-meningitis condition (e.g. bacteremia). In that instance, the argument implies that bacteremia (and not bacterial meningitis) was present at the initial visit and early antibiotic therapy could have prevented bacterial seeding of the meninges or CSF.[6]

Medical malpractice litigations charging clinicians' negligence contributing to adverse outcome of patients (especially children) with bacterial meningitis are among the most common claims filed against emergency medicine physicians and pediatricians.[6] Long-term, neurological sequela occurs in about 30% of young infants, and 15–20% of older children, with a mortality of 5–10%.[7] These sequelae can result from cranial nerve dysfunction, or cortical brain damage resulting from cerebral vascular impairment (causing ataxia, paresis, spasticity, visual and hearing impairment), increased intracranial pressure from adhesions (causing chronic hydrocephalus), and parenchymal brain injury (causing cognitive deficits, seizure disorder, learning disabilities and behavior problems).[7,8] Bacterial meningitis was also recognized as one of the leading causes of acquired deafness in children,[9] before the advent of conjugate vaccines for *H. influenzae* and *S. pneumoniae*. Long-term studies have demonstrated differences in intellectual and cognitive function 12 years after meningitis in up to 30–47% of children.[10]

2.3 Medico-legal Discussion

The medical-legal issue concerning the first hospital and the ER physician is negligent care in missing the diagnosis of bacterial meningitis, implying medical care falls below the standard care of the majority of competent emergency medicine physicians in similar circumstances. Moreover, the outcome was adversely affected by care rendered or omitted (causation), and an earlier diagnosis and treatment would have prevented the dire consequences suffered by the child. An important medical issue is how long a delay in treatment of bacterial meningitis is acceptable, or would affect the outcome of a child. Although they are closely linked, an acceptable timing of antibiotics, in actuality, is a separate issue from the effect of delay in antibiotic treatment in outcome of bacterial meningitis. Most recent guidelines and review of the management of bacterial meningitis recommend that specific antibiotics be given as soon as possible, and that bacterial meningitis is considered a neurological emergency.[5,11,12] Expert opinion on the standard time from presentation in an emergency department to administration of antibiotics varies from a mean time of 0.93 h for emergency medicine physician to 1.45 h for an Infectious Disease Specialist; but in actuality, studies reveal the mean time is 2 h based on a review of 93 cases bacterial meningitis in two university-affiliated pediatric hospitals.[13]

The question of whether the standard of care always should be equated with "ordinary care" or some other benchmark? A judge or jury may not consider a 2 h delay justified for a gravely ill child highly suspicious of severe bacterial meningitis, and that to administer antibiotics "as soon as possible" would be compatible with administration of intravenous therapy within 30 min of assessment. Whereas, in a less severely ill child where diagnosis of meningitis is possible, but not very highly probable, a 2 h delay (after reviewing the CSF analysis) would be considered justifiable. As noted by Judge L. Hand (cited by the Illinois Supreme Court in *Darling vs Charleston Community Memorial Hospital*), "in most cases, reasonable prudence is in fact common prudence; but strictly it is never its measure."[13] Thus, the courts found that although custom is relevant in determining the standard of care, custom by itself is never conclusive. The court may reserve for itself the power to find a medical practice negligent, no matter how common the practice or trend, to protect the population against widespread derelict behavior of the medical industry.[13]

Separate from the accepted definition of "standard of care," with respect to inappropriate delay in administering antibiotics for bacterial meningitis (as well as defining a time frame for "antibiotics as soon as possible"), is the scientific evidence of the "delay" in treating meningitis on the effect of patient outcome. The medical literature has not provided any definite answers based on tier I data (double blind, randomized, controlled trials); as this would be considered unethical to randomize patients into groups receiving prompt antibiotics versus delayed antibiotics. However, accumulative observational studies over the years have provided sufficient information to recommend rapid institution of intravenous antibiotics in bacterial meningitis.

The limitations of older studies assessing the relationship of duration of patient's symptoms and their outcome, with respect to onset of antibiotics, were their inability to distinguish the duration of patients' symptoms (due to preceding viral infection) and duration or onset of meningitis. From clinical observation, the progression to bacterial meningitis is usually a tri-phasic process: (1) nonspecific viral prodrome (usually with upper respiratory tract symptoms), followed by (2) a bacteremic phase, resulting in (3) seeding to the meninges and CSF during the meningitic phase.[6] In a previous review of this topic, most of the studies ($N = 27$) suffered from major methodological limitations and most commonly retrospectively performed, and only five studies specifically defined symptoms assessed, and seven were prospective cohort studies.[6] It would appear from this previous review that most studies did not differentiate between pre-meningitic nonspecific symptoms from those related to onset of meningitis (not an easy task), thus their findings were not valid. Even in one prospective study of 286 children using duration of illness (which was duration of fever in most cases), it was found the duration of illness (which likely included the prodromal viral infection) was inversely correlated with the outcome.[14] The authors postulated that those with a shorter course of illness had more fulminant disease, but it also likely that those with a shorter course had abrupt meningitis without the prodromal phase (pre-meningitic), whereas patients who had a longer course had a prodromal phase with fever.

In a more recent study of 288 children with meningitis fever interval before diagnosis, prior antibiotic treatment, and clinical outcome were assessed.[15] *S. pneumoniae* infection was associated with the longest duration of fever interval prior to the diagnosis of meningitis, the highest frequency of contact with a physician before hospitalization, and the highest rate of morbidity and mortality. There was an association between antibiotic treatment received at prior clinician assessment (possibly the bacteremic phase) and reduced rate of meningitis-related complications (odds ratio (OR), 0.14, $p = 0.02$).[15]

In adult community-acquired bacterial meningitis ($N = 269$), adverse clinical outcome was more common for patients in whom the prognostic stages advanced from low risk or intermediate risk at arrival in the emergency department to high risk before administration of antibiotics.[16] Three baseline clinical features (hypotension, altered mental status, and seizures) were independently associated with adverse clinical outcome and were used to create a prognostic model. In this study, the delay in antibiotic therapy after arrival in the emergency was associated with adverse outcome mainly when the patients' condition advanced to the highest stage of prognostic severity before the initial antibiotic dose was given. Thus, clinicians should aim to administer antibiotics very soon after arrival in the emergency department before progression of disease.

Three relatively recent retrospective studies in adults with bacterial meningitis also found that delay in antibiotic administration was associated with increased mortality and adverse outcome (total of 403 cases),[17–19] In one of these studies, the mortality rate for patients who received antibiotics in the emergency department (1:08 h $\pm$ 13 min meantime) was 7.9% versus those who received antibiotics as inpatient (meantime 6 ± 9 h) was 29%.[17] In another study, timing of antimicrobial therapy (as defined by consciousness level) was a major determinant of survival and neurological outcome; and the first dose of antibiotics should be administered before the consciousness deteriorates to Glasgow coma scale lower than 10 (normal conscious 15, stupor 4–9, and coma 3).[18]

In a more recent prospective observational study of 156 adults with pneumococcal meningitis, delay in antibiotic administration >3 h was the strongest predictor of mortality and outcome (OR, 14.12, $p < 0.0004$).[20] Less predictive of mortality were acute physiological score II (OR 1.12, $p = 0.002$), and isolation of a penicillin non-susceptible strain (OR, 6.83, $p = <0.0004$).[20]

Although results of retrospective and even prospective observational studies cannot conclusively prove that delay in antibiotic administration causes adverse outcome, the best current data supports this contention. This paradigm is supported by our understanding of the pathophysiology in bacterial meningitis, as the longer the time for initiating treatment lengthens the greater the risk of ongoing infection/

inflammation and can result in progressive vasculitis, cerebral edema, increased intracranial pressure, inadequate perfusion of tissues, and brain or neuronal damage.[6]

Therefore, it could be strongly argued by the plaintiffs' lawyer that if the child were treated for bacterial meningitis early in the morning of presentation at the first emergency department, when the child was in a lower risk source, the prognosis and outcome would have been much better.

2.3.1 Issues at the Pediatric Emergency Department

Three medical-legal issues have been raised about the care at the urban pediatric hospital emergency department. When the child was assessed by the triage nurse, he was recognized to be gravely ill, yet the clinician's assessment occurred between 45 and 60 min later. According to guidelines,[21] emergent conditions (potential threat to life, limb or function), require rapid medical intervention and physician assessment should occur $\leq$15 min. Thus, a delay of 45–60 min represents a decline below accepted standard of care.

Even though it was recognized that the child had severe signs of meningitis and impaired level of consciousness, the first dose of intravenous antibiotic was not given until 4 h after registration. Although this may be considered below accepted standard of care, the effect of the delay of antibiotic administration would be less certain (as the child would then have a high risk score with stupor) on the outcome.

A third issue was the omission of administering dexamethasone, before or with, the first dose of antibiotics. Would dexamethasone have affected or improved the adverse outcome or neurological sequelae? Since the early 1990s, dexamethasone has been recommended as an adjunctive initial therapy for bacterial childhood meningitis. Previous randomized controlled trials in childhood *Hemophilius influenzae* meningitis had shown that dexamethasone (0.15 mg/kg every 6 h) for 4 days, significantly reduced the risk of eighth nerve deafness.[22] As a result of this and other studies, adjunctive dexamethasone was recommended as adjunctive treatment for children with *H. influenzae* meningitis and was sanctioned by the committee on Infectious Disease of the American Academic of Pediatrics.[23] A subsequent meta-analysis of 11 randomized controlled trials (RCT) of dexamethasone in childhood meningitis was reported in 1997.[24] It confirmed the benefit of dexamethasone in *H. influenzae* meningitis, and indicated that for pneumococcal meningitis, early dexamethasone administration suggested a benefit in reducing any neurological deficit (OR, 0.23, 95% CI, 0.04–1.05), but the sample size was too small to prove a significant difference.

In a large, non-blinded, randomized control trial in 429 children and adults (ages 3 months to 60 years of age but more than two-thirds were less than 13 years) in Egypt, the mortality and permanent neurological sequelae were reduced in dexamethasone treated patients with pneumococcal meningitis (13.5% vs 40.7%, $p < 0.002$).[25] In a more recent RCT in 598 children with bacterial meningitis in

Malawi (Africa), 338 (40%) due to *S. pneumoniae*, dexamethasone for 2 days did not improve the outcome or sequelae.[26] Failure of steroids to improve the outcome in Malawi has been attributed to multiple factors, including high incidence of human immunodeficiency virus (HIV) infection (24%), late presentation in very advanced stages of disease, and underlying high prevalence of malnutrition.[27]

Large multicenter RCTs in Europe and Vietnam have confirmed that dexamethasone (10 mg every 6h) for 4 days in adults with pneumococcal meningitis significantly reduces unfavorable outcome (by 40%, $p = 0.03$) or death (by 52%, $p = 0.04$),[28,29] but again, had no benefit in adults in Malawi[30] (probably for the same reason as in children as 90% of the adults were HIV infected). Although dexamethasone has not been definitely proven to improve outcome in children with pneumococcal meningitis (in developed countries) the evidence currently supports its use. Biologically, there is no evidence that the pathogenesis of bacterial meningitis in children is any different from adults, and therefore steroids should benefit any age group. Pediatric guidelines for pneumococcal meningitis state "adjunctive therapy may be considered after weighing the potential benefits and possible risk."[31]

Further support for the use of steroids in childhood pneumococcal meningitis has recently been reported from a population-based study in Australia.[32] In a retrospective cohort of 122 cases (aged 0–14 years), early use of corticosteroids protected against death or severe morbidity (OR, 0.21, 95% CI, 0.05–0.77). Delayed diagnosis and treatment was also associated with increased morbidity (severe disability), OR 3.4, 95% (I 1.03–11.4) but not with mortality.[32]

Based on the cumulative data from various clinical trials and our understanding of the biologic mechanisms, a strong argument can be made scientifically, to recommend dexamethasone adjunctive therapy in all patients with severe bacterial meningitis (especially *S. pneumoniae*), irrespective of age. Previous RCTs[28] found the greatest benefit of corticosteroids in the highest risk score patients with severe disease. Thus, in this present case scenario, there is a reasonable medico-legal argument that administration of dexamethasone before or with the first dose antibiotic, more likely than not would have lessened the disability suffered by the child. However, the residual neurological disability even after corticosteroids, may still not have improved his functional capability substantially.

2.4 Case 2: Fever, Earache, and Headache in a Young Adult

A 23-year-old female attended a small community hospital emergency department at 4 a.m. with symptoms of headache, neck pain, vomiting, and low-grade fever. She was assessed the day before with left earache, sore throat, and fever, and treated for acute otitis media with oral Amoxil 250 mg every 8 h. The ER physician noted an oral temperature of 37.7°C, a pulse of 132/min and blood pressure (BP) of 124/72 mmHg. Further examination revealed a red tympanic membrane in the left ear and pain with neck flexion, but no definite nuchal-rigidity (neck-stiffness).

Blood tests, including blood cultures were taken and intravenous fluids, an anti-emetic, and analgesics were administered. The ER physician subsequently consulted the internist on call to discuss the need for a lumbar puncture to rule out bacterial meningitis. He was advised to wait until assessment by the internist later that morning before any further procedure.

The patient was assessed 4 h later that morning by the internist who noted mild erythema of the left tympanic membrane, no decrease in level of consciousness and no evidence of neck stiffness. A diagnosis of viral infection was considered most likely. Neither antibiotic, nor lumbar puncture was recommended by the consultant. However, the patient was admitted for observation, and further blood tests and a chest-radiograph were ordered. Eight hours after arrival in the emergency department, the patient was reassessed in the medical floor by the internist. She had then a temperature of 37.7° and continued to have headaches, nausea, and vomiting. Her leucocyte count was mildly elevated at 12,500 cells/μL with predominant neutrophils (85%) and the chest-radiograph was found to be normal. No evidence of repeat physical examination was performed, and the patient was treated only for symptomatic relief.

2.4.1 Comments

Unlike children, adults are better in verbalizing their symptoms and can follow directions that are helpful in eliciting specific signs. Yet meningitis continues to be a diagnostic challenge in early cases and a source of adverse outcome and medico-legal risk. Early recognition, diagnosis, and specific therapy are necessary to obtain optimal results, and minimalize the risk of death and permanent disability. However, even in previously healthy adult patients who survive bacterial meningitis, up to 18% may suffer from long-term sequelae including dizziness, excessive fatigue, and gait ataxia.[33]

Although lumbar puncture is a relatively safe procedure, it is an invasive one that frequently results in severe, post-procedure headaches. Yet it is the best diagnostic tool along with CSF analysis and culture to confirm the diagnosis and assess the etiology for specific therapy. However, because headaches and fever are common symptoms in many febrile illnesses, it would not be practical to perform lumbar puncture in all patients with these symptoms. The physical examination is used to improve the accuracy and precision of the above symptoms to select patients for lumbar puncture. A previous review had assessed ten studies for analysis to determine various signs used to make the clinical diagnosis of meningitis.[34] However, all but one of these studies was a retrospective chart review series. In this review history of headaches, surprisingly, the group only had a pooled sensitivity of 50%, nausea and vomiting 30% and neck pain 28% (none of which are very specific). The presence of the classic triad of signs for meningitis of fever, neck stiffness and a change in mental status (or headaches) had a pooled sensitivity of only 46%. However, the study found that the absence of all three signs independently effectively eliminates meningitis (sensitivity, 99–100% for the

Table 2.2 Sensitivities of clinical features of meningitis

Clinical features	Retrospective review[34] pooled sensitivities	Prospective[35] sensitivities
Headache	50%	87%
Fever	85%	77%
Nausea/Vomiting	30%	74% (nausea)
Neck stiffness	70%	83%
Altered mental status	67%	69%
Fever, neck stiffness and altered mental status	46%	44%
Jolt accentuation of headache	97%	NA
Any 2: headaches, fever, neck stiffness, altered mental status	NA	95%

NA not assessed

presence of one of these findings).[34] Documented fever itself, had a pooled sensitivity of 85% but a specificity of only 45% for the diagnosis of meningitis. In a recent large prospective study of 696 episodes of community-acquired acute bacterial meningitis, 95% of patients had at least two of the four hallmark symptoms and signs: headaches, fever, neck stiffness, and altered mental status.[35] Neck stiffness, which is the most frequent sign of meningeal irritation or inflammation, had a pooled sensitivity of 70% in the review,[33] and was present in 83% in the prospective study.[35] Among patients with fever and headache, jolt accentuation of headaches was found to be a useful sign in the review,[34] with a sensitivity of 100% but specificity of 54%. Indicating that a negative jolt accentuation (the patient rotates his or her head horizontally, 2–3 rotations per second, and worsening of the headache represents a positive sign) effectively excludes acute meningitis at that time. However, this sign was only assessed in one study in 97 episodes,[36] thus its value has not been verified. See Table 2.2 for sensitivities of symptoms and signs for meningitis, comparison between retrospective pooled date and large prospective data.

2.4.2 Course in Hospital

While in hospital, the patient continued to have symptoms of headache, nausea, and fever and was unable to stay awake by 7 p.m., with a temperature of 39°C. There was no evidence of any clinician reassessment since noon that day. At 9:30 p.m., she was unresponsive to verbal commands and the ER physician on call was called to reassess the patient. She was transferred to a tertiary care hospital at midnight, then suffered cardiac arrest and died. The first dose of antibiotics and dexamethasone was administered in the tertiary care emergency department. Autopsy confirmed that the patient has acute pneumococcal meningitis and acute otitis media, with cerebral edema.

2.4.3 Medico-legal Issues

The lawyer for the plaintiff (husband of patient) charged negligence in medical care against the physicians, nurses, and the community hospital that ultimately led to death of his clients' wife. He sought compensation for pain and suffering of the family, loss of a mothers' care for their child, and loss of future income. The lists of blunders were as follows:

- The initial ER physician should have performed a lumbar puncture soon after arrival in the emergency department, since he considered the diagnosis of meningitis. All ER physicians should be capable of performing this procedure, thus making the diagnosis and instituting early treatment. If this procedure were performed, then the patient would have survived and probably be fully functional. Thus, his actions fell below the standard of care.
- The consulting internist was negligent in not recommending a lumbar puncture, either before his clinical assessment, or afterwards. He should have recognized that absence of neck stiffness does not exclude meningitis, and his recommendation directly led to the catastrophic outcome of the plaintiffs' wife. He should also have recognized that increased leucocyte count was indicative of an acute bacterial infection rather than a viral illness. Moreover, once the patient was admitted for observation, he should have performed repeated examinations to look for signs of meningitis (altered mental status, neck stiffness, and jolt accentuation of headaches). Even if these maneuvers had been performed later in the day, the diagnosis would have been obvious, and appropriate antibiotic/steroid treatment would have saved her life.
- The nurse attending the patient on the medical ward was negligent in not notifying the attending or physician on call to reassess the patient when her condition (especially mental status) deteriorated.
- Furthermore, the ER physician on call that assessed the patient at 9:30 p.m. should have immediately started intravenous antibiotics and dexamethasone, as he recognized she was gravely ill. These interventions (which were only started after midnight at the tertiary care emergency department) may have prevented her death. At the initial examination (from the reported transcripts) of the second defendant (internist), he claimed to have performed an equivalent test to the jolt accentuation, by moving the patients' head from side to side, and found no worsening of her symptoms (not noted in the medical records). Thus, he was of the medical opinion she did not have meningitis at the time of his assessment. It not known whether this maneuver is equivalent to the jolt accentuation of headache, which is an active movement performed by an alert subject. Most experts are of the opinion that if the jolt accentuation were negative, that the lumbar puncture would not be necessary. However, if a patient is admitted for observation, then examinations should be repeated at regular intervals. Furthermore, since the jolt accentuation of headaches has not been subjected to rigorous assessment in large trials, lumbar puncture should be performed if there is suspicion of meningitis.

2.5 Conclusion

The diagnosis and prompt treatment of acute bacterial meningitis are still being missed and delayed, resulting in malpractice litigation for adverse outcome and death. These events continue to occur in both children and adults. The lessons we need to learn as physicians from these cases are diverse.

1. Meningitis should be considered in all children with fever, headaches, nausea/vomiting, irritability and restlessness.
2. The examination should begin with observing the child's behavior and response to parents, environment, and social contacts before physical contact, to assess severity of illness.
3. Neck stiffness and other signs of meningeal irritation can be absent in both children and adults.
4. Jolt accentuation of headaches in alert adults appears to be a very useful sign if negative, to avoid, or delay lumbar puncture. However, patients with negative jolt accentuation of headaches should be closely followed for reassessment and re-examination. Lumbar puncture is the only accurate procedure with CSF analysis to confirm or exclude meningitis, and patients should be given this option even with a negative jolt accentuation of headaches.
5. Prompt institution of appropriate antibiotics and dexamethasone are important to reduce complications and mortality. For gravely ill patients, immediate treatment ($\leq$30 min) should be the aim, and for less ill subjects, treatments and confirmation of the diagnosis within 2 h appears to be appropriate.
6. CT scan of the brain is not necessary for most cases with symptoms suggestive of meningitis, unless there are certain conditions present (immunosuppression, bleeding tendency, septic shock, impending respiratory failure, papilledema, focal neurological signs, recent onset seizures, or known or suspected brain mass or tumor). Patients with suspected meningitis requiring a CT scan should have at least two sets of blood cultures taken, and then started on antibiotics plus dexamethasone until results of CSF analysis and cultures become available.

References

1. Baraff LJ, Bass JW, Fleisher GR, Klein JO, Mc Cracken Jr. GH, Powell KR, Schriger DL, (1993), Practice guideline for the management of infants and children 0 to 36 months of age with fever without source. Ann, Emerg. Med. 22: 1198–1210.
2. McCarthy PL, Sharpe MR, Spiesel SZ, Dolan TF, Forsyth BW, De Witt TG, Fink HD, Baron MA, Cicchetti DV, (1982), Observation scales to identify serious illness in febrile children. Pediatrics: 70: 802–809.
3. McCarthy PL, (2007), Evaluation of the sick child in the office and clinic (chapter 60). In: Kliegman RG, Behrman RE, Jensen HB, Stanton BF (eds); Nelson textbook of Pediatrics, 18th Edition. Saunders/Elsevier, Philadelphia, PA, p 363–366.

4. Powell KR, (2007), Fever without focus. In; Kliegman RG, Behrman RE, Jensen HB, Stanton BF (eds), Nelson Textbook of Pediatrics, 18th Edition. Saunders/Elsevier, Philadelphia, PA, p1087–1093.
5. Prober CG, (2007), Central nervous system infections. In: Kliegman RG, Behrman RE, Jensen HB, Stanton BF (eds), Nelson Textbook of Pediatrics, 18th Edition, Saunders/Elsevier, Philadelphia, PA, p 2512–2524.
6. Bonadio WA, (1997), Medical-legal considerations related to symptoms duration and patient outcome after bacterial meningitis. Am. J. Emerg. Med. 15; 420–423.
7. Klein J, Feigin R, Mc Cracken G, (1993), Report of the Task Force on Diagnosis and Management of meningitis. Pediatr. Infect. Dis J. 328: 42-47.
8. Taylor H, Mills E, Ciampi A, du Berger R, Watters GV, Gold R, MacDonald N, Michaels RH. (1990), The sequelae of *Hemophilus influenzae* meningitis in school-age children. N. Engl. J. Med. 323:1657–1663.
9. Wolff A, Brown S, (1987), Demographics of meningitis-induced hearing impairment: implications for immunization of children against *Hemophilus influenzae* type B. Am. Ann. Deaf. 132: 26–30.
10. Grimwood K, Anderson P, Anderson V, Tan L, Nolan TE, (2000), Twelve year outcomes following bacterial meningitis: further evidence for persisting effects. Arch. Dis. Child. 83:111–116.
11. Tunkel AR, Hartman BJ, Kaplan SL, Kaufman BA, Roos KL, Scheld WM, Whitley RJ, (2004), Practice guidelines for the management of bacterial meningitis. Clin. Infect. Dis. 39: 1267–1284.
12. Tunkel AR, Scheld WM, (1995), Acute bacterial meningitis. Lancet; 346: 1675–1680.
13. Meadow WL, Lantos J, Tanz RR, Mendez D, Unger R, Wallskog P, (1993), Ought 'standard care' be the 'standard of care'? A study of the time to administration of antibiotics in children with meningitis. Am. J. Dis. Child. 147: 40–44.
14. Kilpi T, Anttila M, Kallio MJ, Peltola H, (1991), Severity of childhood bacterial meningitis and duration of illness before diagnosis. Lancet; 338: 406–410.
15. Bonsu BK, Harper MB, (2001), Fever interval before diagnosis, prior antibiotic treatment, and clinical outcome for young children with bacterial meningitis. Clin. Infect. Dis. 32: 566–572.
16. Aronin SI, Peduzzi P, Quagliarello VJ, (1998), Community-acquired bacterial meningitis: risk stratification for adverse clinical outcome and effect of antibiotic timing. Ann. Intern. Med. 129: 862–869.
17. Miner JR, Heegaard W, Mapes A, Biros M, (2001), Presentation, time to antibiotics, and mortality of patients with bacterial meningitis at an urban county medical center. J. Emerg. Med. 21: 387–392
18. Lu C-H, Huang CR, Chang WN, Chang C-J, Cheng BC, Lee PY, Lin M-W, Chang H-W, (2002), Community-acquired bacterial meningitis in adults: the epidemiology, timing of appropriate antimicrobial therapy, and prognostic factors. Clin. Neurol. Neurosurg. 104: 352–358.
19. Proulx N, Fréchette D, Toye B, Chan J, Kravcik S, (2005), Delays in the administration of antibiotics are associated with mortality from adult acute bacterial meningitis. Q. J. Med. 98: 291–298.
20. Auburtin M, Wolff M, Charpentier J, Varon E, Le Tulzo Y, Girault C, Mohammedi I, Renard B, Mourvillier B, Bruneel F, Ricard JD, Timsit JF, (2006), Detrimental role of delayed antibiotic administration and penicillin-non-susceptible strains in adult intensive care unit patients with pneumococcal meningitis: The PNEUMOREA prospective multicenter study. Crit. Care. Med. 34: 2758–2765.
21. Beveridge R, Clarke B, Janes L, Savage N, Thompson J, Dodd G, Murray M, Jordan CN, Warren D, Vadeboncoeur A, (1998), Implementation guidelines for the Canadian Emergency Department Triage & Acuity Scale CTAS. Version: CTAS 16. Doc Dec.16, 1998.
22. Lebel MH, Freij BJ, Syrogiannopoulos GA, Chrane DF, Hoyt MJ, Stewart SM, Kennard BD, Olsen KD, McCracken GH Jr, (1988), Dexamethasone therapy for bacterial meningitis. Results of two double blind, placebo-controlled trials. N. Engl. J. Med. 319: 964–971.

23. American Academic of Pediatrics committee on Infectious Disease, (1990), Dexamethasone therapy for bacterial meningitis in infants and children. Pediatrics; 86: 130–133.
24. McIntyre PB, Berkey CS, King SM, Schaad UB, Kilpi T, Kanra GY, Odio Perez CM, (1997), Dexamethasone as adjunctive therapy in bacterial meningitis. A meta-analysis of randomized clinical trials since 1988. JAMA; 278: 925–931.
25. Girgis NI, Farid Z, Mikhail IA, Farrag I, Sultan Y, Kilpatrick ME, (1989), Dexamethasone treatment for bacterial meningitis in children and adults. Pediatr. Infect. Dis. J. 8: 848–851.
26. Molyneux EM, Walsh AL, Forsyth H, Tembo M, Mwenechanya J, Kayira K, Bwanaisa L, Njobvu A, Rogerson S, Malenga G, (2002), Dexamethasone treatment in childhood bacterial meningitis in Malawi: a randomized controlled trial. Lancet. 360: 211–218.
27. McCracken GH Jr, (2002), Rich nations, poor nations, and bacterial meningitis. Lancet; 360:183 (editorial).
28. de Gans J, van de Beek D, for the European Dexamethasone in Adulthood bacterial meningitis study investigators, (2002), Dexamethasone in adults with bacterial meningitis. N Engl J Med; 347:1549–1556.
29. Nguyen TH, Tran TH, Thwaites G, Ly VC, Dinh XS, Ho Dang TN, Dang QT, Nguyen DP, Nguyen HP, To SD, Nguyen VC, Nguyen MD, Campbell J, Schultsz C, Parry C, Torok ME, White N, Nguyen TC, Tran TH, Stepniewska K, Farrar JJ, (2007), Dexamethasone in Vietnamese adolescents and adults with bacterial meningitis. N. Engl. J. Med. 357: 2431–2440.
30. Scarborough M, Gordon SB, Whitty CJ, French N, Njalale Y, Chitani A, Peto TE, Lalloo DG, Zijlstra EE, (2007), Corticosteroids for bacterial meningitis in adults in sub-Saharan Africa. N. Engl. J. Med. 357: 2441–2450.
31. American Academic of Pediatrics, (2003), Pneumococcal infections. In: Pickering LK (ed). Red book: 2003 report of the committee on Infectious Disease. 26th Edition, ELK Grove Village, II, American Academics of Pediatrics, p490–500.
32. McIntyre PB, Macintyre CR, Gilmour R, Wang H, (2005), A population based study of the impact of corticosteroid therapy and delayed diagnosis on the outcome of childhood pneumococcal meningitis. Arch. Dis. Child. 90: 391–396.
33. Bohr V, Paulson OB, Rasmussen N, (1984), Pneumococcal meningitis: late neurological sequelae and features of prognostic impact. Arch. Neurol. 41:1045–1049.
34. Attia J, Hatala R, Cook DJ, Wong JG, (1999), The rational clinical examination. Does this adult patient have acute meningitis? JAMA. 282: 175–181.
35. van de Beek D, de Gans J, Spanjaard L, Weisfelt M, Reitsma JB, Vermeulen M, (2004), Clinical features and prognostic factors in adults with bacterial meningitis. N. Engl. J. Med. 351:1849–1859.
36. Uchihara T, Tsukagoshi H, (1991), Jolt accentuation of headache: the most sensitive sign of CSF pleocytosis. Headache.31: 167–171.

Chapter 3
Conditions of the Skin/Soft Tissue in the ER

3.1 Case 1: Musculoskeletal Pain Without Trauma

A 29-year-old female was presented to the emergency department of a suburban general hospital at 11 p.m. with severe pain of her left thigh. The patient had a medical history of asthma since childhood, otherwise she was well. The day before, she had an argument with her live-in boyfriend that precipitated an attack of asthma, and she gave herself an epiPen (epinephrine) intramuscular injection in the left thigh without cleansing the skin beforehand. Apparently, the epiPen injector had been in her handbag for the past 2 years and the cap was off when it was used. The pain in her left thigh had been progressing over the previous 12 h, but there was no history of fever or chills.

The ER physician noted the patient was lying in obvious pain and the vital signs were normal except for a tachycardia of 120/min. Examination of the left thigh revealed slight swelling with very minimal erythema near the ejection site. She was given an oral analgesic and reassessed 1 h later, with no significant change in her condition. The patient was discharged that night, advised to see her family physician the next day, and told to use analgesics (a non-steroidal anti-inflammatory agent [NSAID]) for the pain. The patient had to be assisted by her boyfriend to a taxi as she could not bear any weight on the left lower limb.

During the night, the patient was feeling worse with increasing excruciating pain and sweating. The boyfriend called the emergency department twice, 1 a.m. and 2 a.m., reporting that the patient could not sleep and had severe pain which was unrelieved by the acetaminophen 600 mg/codeine 60 mg. He was not able to speak to the physician, but he was told by the attending nurse that the clinician's advice was to take more analgesics, that nothing serious was wrong with his girlfriend, but she should try to see her family physician later that morning.

Apparently, the patient's condition worsened during the early morning, because 911 was called and she was taken to the hospital emergency department by ambulance in respiratory failure (cyanosed), shock, stupor, and a grossly swollen left thigh with some mild erythema and crepitus extending to her hip and lower left abdomen. Resuscitation was attempted, but the patient went into cardiac arrest and died at 6 a.m. that same morning.

I.W. Fong, *Medico-Legal Issues in Infectious Diseases*, Emerging Infectious Diseases of the 21st Century, DOI 10.1007/978-1-4419-8053-3_3,

3.1.1 Medico-legal Issues

The attorney representing the family and boyfriend of the patient (plaintiffs) filed suit, claiming negligence of the hospital's emergency department health-care providers.

1. The plaintiffs charge that the ER physician was negligent in not recognizing the seriousness of the patient's condition, no investigations were performed to make a diagnosis, and no surgeon was consulted to assess her condition.
2. Moreover, the patient was evidently in severe pain, and the ER physician was negligent for not admitting her for observation and management, even if he were aware of the proper diagnosis.
3. The emergency department attending nurse and physician were grossly negligent in not advising the boyfriend to bring the patient back to the emergency department when he called to report that her condition had worsened.
4. If the patient were admitted in the first instance, or advised to return for reassessment, it is more likely that the seriousness of her condition would have been recognized earlier, appropriate treatment could have been instituted, and her life could have been saved.

The lawyers for the defendants (ER physician and the hospital), claim that they were not negligent, and that appropriate and proper history and physical examinations were performed, but did not indicate any serious underlying condition. Thus, there were no indications to admit the patient, to perform any investigations, nor refer her to a surgeon. Furthermore, the patient suffered from a rare and unusual condition, and at the time she presented, the signs were not evident for any clinician to recognize and make the proper diagnosis. Moreover, the onus was on the patient and her boyfriend to return for reassessment at the emergency department if her condition worsened.

From the transcript of the examination of the discovery of the ER physician, it became evident that the clinician thought that the patient was "histrionic" and was seeking narcotics, as he could find no clinical evidence to explain her severe pain. He also denied or could not recall advising the nurse about the phone calls made by the boyfriend.

3.1.2 Autopsy

At autopsy, there was evidence of profuse gas in the tissues (subcutaneously and in the muscles) of the entire left thigh, extending to the hip, buttock and left lower abdominal wall. Macroscopically and microscopically, there was extensive necrotizing fasciitis and gangrene of the muscle of the same areas ("gas gangrene"), with large gram-positive rods seen on gram-stain. Cultures of the blood taken in the emergency department during the second visit, and tissues from autopsy grew

Clostridium perfringens. Final autopsy report: The patient died from overwhelming septic shock from clostridial necrotizing fasciitis and myonecrosis induced by epiPen injection of epinephrine intramuscularly.

3.1.3 Medical Issues

Clostridial necrotizing fasciitis and myonecrosis (gas gangrene) are extremely rare in developed countries, but physicians need to be aware of the clinical manifestations in order to think of the diagnosis, otherwise they will not be able to diagnose these conditions until it is too late. Clostridial fasciitis and gas gangrene are most commonly seen following war wounds and crush injuries (farm mishaps and motor vehicle accidents), as devitalized tissue and compromise of blood supply predisposed to proliferation of this ubiquitous anaerobe. It is now rarely ever seen, except in patients with peripheral vascular disease, following abdominal gunshot injuries, or post-colon surgery, as the majority of these infections are from mixed organisms. Clostridial infection of an amputation stump may also occur or rare occasions, and may be masked by the tightly bound surgical dressings or casts that are often used to shape or mold the stump for prosthesis fitting.

Spontaneous clostridial crepitant cellulitis, fasciitis, or myonecrosis may occur without any obvious injuries or surgery. These may fall into two major groups, those occurring after subcutaneous or intramuscular injections and those with no preceding injections. There is a well-established association between black-tar heroin "skin-popping" or soft-tissue injections and clostridial, crepitant fasciitis-cellulitis or myonecrosis.[1–3] It is believed that the black-tar heroin becomes contaminated with *Clostridia* spores when mixed with adulterants (i.e. methamphelamine, strychnine, dyed paper) or diluted with water.[2] The clostridial spores can survive heating or boiling, which is typically done before use of the black-tar heroin. Repeated injections of black-tar heroin into soft tissues (usually when there is loss of venous access) results in tissue ischemia and necrosis, predisposing to germination, growth and elaboration of clostridial toxin (as in wound botulisms).[4]

Injection of cocaine into soft tissue is another potential cause of clostridial infection as it is a potent vasoconstrictor that can lead to local tissue ischemia[5] and predisposal to anaerobic infection (personal experience). Whereas *Clostridium perfringens* is the predominant cause of infections after injuries and surgery, in drug abusers, the soft tissue infections are more frequently due to *Clostridium sordelli* or *Clostridium novyi*.

Rare but fatal cases of clostridial myonecrosis have also been reported with repeated intramuscular injections with anti-inflammatory substances (in non-drug abusers).[6] More disturbing are several reports of gas gangrene following subcutaneous epinephrine (adrenaline) or intramuscular injections.[7–14]

Clostridial gas gangrene after epinephrine injection had been recognized at the beginning of the twentieth century, and was more frequently reported before the 1960s. In 1936, Mabin[15] reviewed 84 cases of gas gangrene following intramuscular injection and one-third of these were associated with injections of epinephrine. In the same year, another review by Touraine[16] found that epinephrine was the most frequently implicated drug. Although most reports involved intramuscular injections in the buttock, cases have also been described after injections in the thigh and rarely in the deltoid muscle.[14] Most of the reports of clostridial myonecrosis occurring after epinephrine injections were due to *C. perfringens* (previously *C. welchii*). Earlier reports in the first half of the twentieth century noted a fulminant course after injection with death usually within 48 h, and with a mortality rate of 94%.[7] In a review of the English literature, there have been only ten cases of clostridial gas gangrene after epinephrine injection reported since 1960 (see Table 3.1). The underlying conditions requiring epinephrine injections (mainly used as epinephrine-in-oil) were asthmatic attack in eight, and severe urticaria in two cases. Nearly all the cases received intramuscular injections (six in the buttock, three in the thigh and one in the deltoid) and one subcutaneous injection over the deltoid muscle. The incubation periods from injection to onset of first symptoms (mainly severe pain) varied from within 4–72 h. Although a high mortality and rapid progression to circulatory collapse were typical, radical surgery, appropriate antibiotics and supportive therapy resulted in four (40%) survivals.

Table 3.1 Clostridial gas gangrene after epinephrine since 1960

	Cases/Ref	Condition	Presentation	Injection site	Outcome
1	63 year-male (Koons et al.[7])	Asthma	Pain/weakness after 14 h	Buttock, IM	Died
2	43 year-male	Urticaria	Pain, swelling in 12 h	Buttock, IM	Survived
3	52 year-male	Asthma	Severe pain, swelling crepitus in 13 h	Buttock, IM	Died
4	48 year-male (Marshall and Sims[8])	Asthma	Pain, swelling by 15 h	Deltoid, IM	Died
5	22 year-male (Harvey and Purnell[9])	Status asthmaticus	Pain in 18 h, Circulatory collapse by 48 h	Buttock, IM	Died
6	18 year-female (Gaylis[10])	Severe urticaria	Pain, fever Swelling within 72 h	Thigh, IM	Died
7	Male (Maguire and Langley[11])	Asthma	Pain after 30 h, swelling	Thigh, IM	Survived
8	64 year-male (Van Hook and Vandievelde[12])	Asthma	Severe pain in 6 h Surgery by 30 h	Buttock, IM	Died
9	13 year-female (Teo and Balasurbramaniam[13])	Asthma	Pain within 12 h, in shock within 72 h	Buttock, IM	Survived
10	32 year-male (Hallagan et al.[14])	Asthma	Pain in 4 h, swelling, chills, crepitus within 12 h, hypotension by 24 h	Deltoid, subcutaneous	Survived

Clostridial organisms are frequently found on the buttocks, and on skin where alcohol preparation fails to eliminate the spores (and increases their dispersal), so iodine preparations have been recommended.[12] However, the organism is ubiquitous and can be found in dust, and contaminated needles and syringes were a problem before disposables became the standard. Epinephrine causes vasoconstriction of blood vessels of the muscle or subcutaneous tissue creating an anaerobic environment suitable for the vegetation and growth of *clostridia*. In the present case being discussed, the clostridial spore could have contaminated the epiPen injector needle in the patient's handbag, or could have colonized in her thigh.

Spontaneous clostridial myonecrosis with no previous injections is a very rare condition, but well documented in the literature. Most of these cases are secondary to *Clostridium septicum* and 70–80% of cases are associated with underlying malignancies, particularly carcinoma of the colon and relapsing leukemias.[17–21] The pathogenesis is believed to be secondary to hematogenous seeding from mucosal ulceration overlying a malignancy, or mucosal damage and granulocytopenia secondary to chemotherapy. *C. septicum* is relatively aerotolerant and can initiate infection in the absence of tissue damage and ischemia, but diabetes and leucopenia are risk factors.[20] In a more recent case series of *C. septicum* infection, 50% were found to have a primary malignancy, 75% of which were colonic and 40% of those were cecal.[21] Occasionally *C. septicum* myonecrosis has been associated with diverticulitis, intestinal infarction, enterocolitis, volvulus and instrumentation,[20] and following intramuscular injection.[22]

Other unusual causes of spontaneous clostridial necrotizing fasciitis and gangrene include *Clostridium tertium* in a patient with alcoholic liver disease, receiving chemotherapy for lymphoma[23]; and a recent case of *Clostridium difficile* gangrene after trauma from a motor-vehicle accident have been reported.[24] Although abdominal wall clostridial (*C. perfringens*) myonecrosis was a known rare complication of open cholecytectomy,[25] it can also occur after laparoscopic cholecystectomy.[26] These episodes likely represent direct inoculation of the surgical site by *clostridia*, as the organism has been isolated in 10–20% of all diseased gall bladders removed at surgery.[27] However, the development of gas gangrene at a remote site (buttock) occurred unexpectedly in one case after laparoscopic cholecystectomy, with no prior subcutaneous or intramuscular injection (case under review for litigation purposes, unreported). The proposed mechanism in this latter case is hematogenous seeding from the gallbladder during the procedure to the buttock muscles, possibly predisposed by relative ischemia from the pressure-effect of lying supine on the operating table for more than an hour.

Clostridial gas gangrene (especially secondary to *C. perfringens*) is the most fulminant necrotizing infection that affects humans. Infection can become established in tissues within 6–8 h, with destruction of healthy muscle or soft tissue and progression by several centimeters per hour, despite appropriate antibiotics. Although injured tissue and ischemia are more susceptible to infection, healthy tissue can less commonly be affected as well. Early studies before 1950 in experimental animals demonstrated that epinephrine could drastically reduce the infective dose of *C. perfringens* by 10,000–100,000-fold.[28] Shock and organ failure are present in 50% of patients. Even with modern intensive care, the mortality rate is about 40%,[29] and radical amputation is often the best life-saving treatment.

Institution of appropriate combined medical-surgical therapy early in the onset of illness is critical for preservation of tissues (limbs etc.) or life. Thus, familiarity with early manifestations is important to recognize and consider the diagnosis. Early manifestation of clostridial fasciitis or myonecrosis in most cases is a sudden or gradual onset of severe pain, with minimal to moderate soft tissue swelling and without much evidence of inflammation (erythema or increased warmth). In fact, histopathology usually demonstrates sparse inflammatory reactions, and mainly shows accumulation of neutrophils within small blood vessels near the border between healthy and necrotic tissue. Furthermore, at surgery, these infected wounds usually show no pus, but instead have a "dishwater" discharge. These observations suggest that destruction of tissue in clostrial gas gangrene is initially a local ischemic process (causing sudden pain from acute arterial thrombosis), and with time, expands regionally to involve the surrounding tissues or entire limb. Experimental studies have implicated toxin-mediated microvascular events that result in the tissue destruction.[30,31] Previous studies have identified phospholipase C (PLC) and theta (θ) toxin (a thiol-activated cytolysin) as the principal toxins involved in the pathogenesis. Tissue destruction is caused by decreased muscle blood flow induced by PLC stimulated platelets and neutrophils aggregation.[31] The θ-toxin contributes to the pathogenesis by direct destruction of the host's inflammatory cells and tissues, and by promoting dysregulated neutrophils/endothelial cell interaction.[32]

To aid in the early recognition of gas gangrene Altemeier and Fullen[33] have listed eight important points.

1. The incubation period can be short, as early as 6 h, but averages 53 h.
2. Persistent, severe pain is the earliest and most important symptom, usually progressively increasing even after medical treatment.
3. Tachycardia out of proportion to the degree of elevated temperature, and marked increase in the pulse rate may herald severe circulatory collapse.
4. Early in the course, the blood pressure is usually normal, but as the disease progresses, hypotension, septic shock, and anuria commonly develop.
5. Fever or elevated temperature is not consistently present in the early stages (often normal), and subnormal temperature with marked tachycardia indicates a grave prognosis.
6. The patient's general appearance usually shows a peculiar grey color, unlike the expected facial flush seen with other pyogenic infections.
7. Altered mental status such as stupor, delirium, prostration, and coma are very late manifestations of overwhelming infection, while early changes often consist of apathy and indifference (the patient unaware of the seriousness of the condition).
8. In the early stages, the local lesion can be normal in appearance, demonstrate edema, or be white, shiny, and tense. In the late stages, it becomes a dusky bronze color, then progresses to fluid-filled blisters with a dark-red, purplish appearance and may show evidence of crepitus.

Previous case series of clostridial soft tissue infection have found laboratory results of limited value in the early diagnosis. The leucocyte count is usually normal to slightly increased in the early stages, and the hemoglobin or hematocrit is usually decreased in the later stage.[33,34] Plain radiograph may show gas in the soft tissues before palpable crepitation (usually as linear streaks along the muscle and facial planes), but in suspicious cases the radiograph should be repeated at 2–4 h intervals if not initially diagnostic.[33] Computerized tomography (CT) and magnetic resonance imaging (MRI) should be more sensitive than routine radiography. Direct aspirate for immediate gram stain may be very helpful to demonstrate large gram-positive bacilli without much neutrophils infiltration. An elevation of serum creative phosphokinase (CK) should be seen early in the course of illness, but this has not been rigorously assessed in studies.

The success of treatment of clostridial fasciitis and myonecrosis largely depend on early diagnosis. A delay of 24 h often results in high fatality and 48 h delay after onset of symptoms results in 100% mortality.[34] Current practice of the management has been guided largely by retrospective studies and animal experiments. One of the best results reported in civilian population of gas gangrene was a series of 54 cases with a mortality of only 14.8%.[33] Primary treatment was early aggressive surgical decompression, debridement, or amputation, combined with circulatory support, large doses of intravenous penicillin and tetracycline, and polyvalent clostridial anti-toxin. Tetracycline was used based on experimental data to inhibit toxin production by inhibition of protein synthesis.[35] Most centers nowadays will use metronidazole or clindamycin in combination with penicillin instead, to achieve the same aim.[34,36] However, antitoxin therapy is no longer used, as subsequent studies found no therapeutic benefit.[34] Controversy exists about the use of hyperbaric oxygen, which is not available in most centers. Some guidelines recommend hyperbaric oxygen after surgery if readily available,[34] but it should be noted that Altemeier and Fullen[33] achieved a survival rate of 85.2% in patients with gas gangrene without using hyperbaric oxygen.

3.1.4 Comments

Primary clostridial soft tissue infection is a rare but a life threatening medical emergency. Frequently, when gas gangrene occurs, a suit is filed against the physician for negligence.[37] This often occurs because of the sudden onset and rapid progression with unexpected death or loss of limb. Most liability suits involving patients who develop gas gangrene are related to charges of negligence due to missed diagnosis or delay in appropriate treatment. As in, all liability cases, the plaintiff must prove that the gangrene resulted from negligence on the part of the physician, or the adverse outcome was a result of failure to recognize or treat in a timely fashion.

Unavoidable results stemming from the serious condition of the patient impose no legal liability on a physician. In this particular case initially presented, the ER physician clearly could not have made the correct diagnosis at the first presentation to the emergency department. However, it can be argued that the physician should have advised the patient to return to the emergency department for reassessment if she experienced any worsening. Moreover, when the boyfriend called the Emergency department later in the early morning, this advice should have been given. The plaintiff's lawyer would need to prove in court that if that patient were reassessed before the final visit, an appropriate diagnosis could be made to institute lifesaving surgery and medical therapy that would have altered the outcome. The outcome would then depend on the presence of circulatory collapse, and rapidity with which radical surgery could be performed (within 1–2 h). If a physician deals with a condition promptly by using the applicable standard of due care, he or she is not held legally responsible if the patient cannot be cured.[37]

3.2 Case 2: Soft Tissue Pain and Swelling with Minimal Trauma

A 34-year-old male, previously well, arrived at an outlying small suburban hospital emergency department at 7 a.m. His main complaint was of severe pain and swelling of the right shoulder and arm, with fever and chills for 2 days. He was assessed at another hospital emergency department 48 h before, with symptoms of fever (38.8°C), chills, pain, and decreased mobility of his right forearm, sore throat and a cough. There was a history of the patient using a wrench to remove an overhead bolt the day before. He was sent home on acetaminophen without investigation and with a diagnosis of having influenza or viral infection plus muscle strain.

At the second emergency visit the patient had a temperature of 34.9°C, blood pressure of 102/61 mmgHg and pulse of 114/min. There was marked swelling of the entire right arm, a small abrasion at the shoulder tip, some erythema around the elbow, a blister in the antecubital fossa, absent radical pulse and bluish discoloration of the right hand. The working diagnosis at the time was possible compartment syndrome or axillary vein thrombosis. A venogram was performed which was interpreted as either occlusion of the axillary vein or external compression. An orthopedic surgeon was consulted who felt the condition did not require surgery but should be treated as venous thrombosis. Blood tests performed that morning became available 1.5 h later reported a leucocyte count of 9,700/mm^3 *(the differential white blood count was reported 2 h later) with increased bands and a left shift; an elevated glucose of 12.8 mmol/L (normal 3.0–7.0); an elevated creatinine of 254 µmol/L (normal <120); and creatine kinase of 1582 µ/L (normal <110). Four hours after admission, his temperature was noted to be 38.4°C and his blood pressure was 92/60 mmHg. An internist was consulted for further care 6 h after arrival in the emergency department. At that time, there was massive swelling of the patient's right shoulder, axilla, upper arm, forearm, hand, and fingers. Cultures were requested, but the considered diagnosis was still axillary vein thrombosis, and it was treated with intravenous saline and heparin.*

3.2.1 Comments on Medical Aspect

There are very few conditions that can present with acute pain, swelling, and fever of a limb, and infections should be among the preeminent differential diagnosis. Although axillary vein thrombosis can produce swelling of the arm and hand, it is usually seen in hospitals from intravenous catheters, intravenous drug abuse, hypercoagulable state, and occasionally from trauma. Spontaneously axillary vein thrombosis has rarely occurred after strenuous activity (such as rock climbing, pole vaulting, shot-putting, boxing, javelin throwing etc.) presenting with swelling and muscle pain (dull aching), venous distension, and either none or mild tenderness on palpation, and normal shoulder movements.[38] Primary upper limb deep vein thrombosis with no predisposition is a very rare disorder.[39] Severe pain is usually not a feature of this condition, but a dull ache may be present and fever is usually absent unless it is secondary to septic thrombophlebitis. Thus, primary axillary vein thrombosis should not have been seriously considered to explain the patient's symptoms and manifestations.

Compartment syndrome typically presents with swelling and severe pain on movement or palpation (but mild discomfort at rest), and numbness and tingling of the extremity are frequent with hyperesthesia of the web space.[39] Usually, the distal pulses are palpable and the overlying skin and soft tissue are very tense. Pain is reproduced by passive stretching of the digits (stretch pain).[40] Fever is not usually a feature of this disease, but could occur later with tissue necrosis. Myoglobulinuria with secondary renal impairment and high creatinine kinase are rare complications seen mainly from a large muscle compartment involvement in the lower limb. Compartment syndrome of the upper extremity is most commonly due to compression injuries, direct trauma with or without fracture, crushing of the upper arm muscles occasionally from prolonged unconsciousness (pressure effect). It is rarely from shoulder dislocation, avulsion of the triceps muscle, spontaneous hemorrhage from minor trauma (bleeding disorders such as hemophilia, or anticoagulation), pneumatic tourniquet, arteriography, and infection.[41] This patient had no predisposing factor for compartment syndrome other than infection, which should have been the primary concern and diagnosis. Although an increased leucocyte count may be seen with compartment syndrome, a significant left shift with increased bands would be unusual.

The three main infections that should be considered in patients presenting with pain, swelling and fever of a limb are cellulitis, fasciitis and myositis (pyogenic or with myonecrosis). Rarely, localized soft tissue abscess, septic arthritis, or bursitis can present with diffuse swelling of a limb secondary to compression or impairment of venous or lymphatic drainage (with or without evidence of thrombosis or obstruction). The case under discussion was very unlikely to have had cellulitis, as this is should be evident and easily diagnosed by the erythema or redness of the skin. Moreover, the degree of pain is usually mild and present mainly on movement or palpation. Thus, the two remaining conditions would be necrotizing fasciitis and pyogenic myositis. These two conditions can be present together as an extension of the infection.

Necrotizing fasciitis (inflammation with necrosis of the subcutaneous fascia) is uncommon, but not extremely rare. It is one of the most common infections to be misdiagnosed, and delays in diagnosis and treatment have led to litigations because of adverse outcome (personal experience). It is important that physicians (particularly emergency physicians, general practitioners, and internists) be familiar with recognizing the early stages of the disease. Similar to gas gangrene, necrotizing fasciitis can progress rapidly, and if not recognized and treated promptly can lead to loss of life or limb. Spontaneous, noncrepitant (no gas), fasciitis is most commonly due to *Streptococcus pyogenes*, and less frequently, secondary to *Staphylococcus aureus* (including methicillin resistant strains, MRSA), or other bacteria (*Vibrio vulnificus*). Crepitant necrotizing fasciitis is usually secondary to mixed organisms (including anaerobes, coliforms and *Streptococci*), especially in diabetes, vascular insufficiency, and the presence of decubitus ulcers, and following abdominal injuries or surgery. Primary clostridial necrotizing fasciitis is much less frequent as a cause of crepitant necrotizing fasciitis, and primary coliform (alone) fasciitis is distinctly rare.

The patient in this case scenario would best fit the diagnosis of streptococcal necrotizing fasciitis for several reasons. It is the most frequent cause in healthy subjects with no predisposition for the other types of necrotizing fasciitis i.e. diabetes, ischemia, drug abuse, crush injury etc. Patients commonly have no break in the skin, but sometimes have abrasions, lacerations, prior surgery, sore throat, chickenpox, or exposure to a household member with invasive streptococcal infection.

Streptococcal necrotizing fasciitis is commonly misdiagnosed for cellulitis, muscle strain, ankle sprain, gout, and ordinary wound infection, especially in the early stages, and occasionally as deep vein thrombosis. The classic appearance of necrotizing fasciitis such as violaceous bullae, reddish-purple discoloration of the skin, woody induration of the soft tissue (with or without areas of flocculence),[42] are late manifestations usually when there is circulatory collapse or septic shock, or when the disease carries a high mortality. A few salient features of necrotizing fasciitis in the early stages would be localized severe pain and tenderness out of proportion to the appearance of the skin (often pain at rest), or swelling of the soft tissue or limb (without erythema or minor redness until the disease progresses). Patients can initially present with nausea, vomiting, and diarrhea (which can be seen in any severe sepsis), so symptoms are often mistaken for gastroenteritis. These non-specific symptoms may be related to circulating toxins and massive release of cytokines. Systemic toxicity with rapid increase in area of involvement, and persistence or progressive pain, then follows later with changes as described above. Frequently, in the initial stages, there often are questions of drug-seeking (narcotics), or low-pain threshold, as there is a disassociation between symptoms and clinical findings. Unlike typical cellulitis, the erythema or redness may not be prominent as the infection appears to originate in the fascial plane or the subcutaneous tissue in some cases. In a report of 15 streptococcal necrotizing fasciitis, influenza-like and gastrointestinal symptoms were common, and the most consistent clinical clue was unrelenting pain out of proportion to the physical findings.[43] Moreover, local erythema and edema was only present at first presentation in three to four patients (20–27%).

The laboratory tests that are useful in help differencing necrotizing fasciitis from more benign conditions include marked elevation of the leucocyte count or marked left shift (with or before leucocytosis) often above 16,000 cells/mm,[3] with toxic granulation. These features are very uncommon in uncomplicated cellulitis, deep vein thrombosis, or early stages of compartment syndrome. The CK is frequently elevated in necrotizing fasciitis, pyogenic myositis, or gas gangrene, but not in cellulitis, or venous thrombosis. It can be increased with the compartment syndrome later, if untreated due to secondary muscle necrosis. Secondary organ dysfunction is also commonly seen with necrotizing fasciitis and clostridial myonecrosis: such as renal impairment, liver disturbance, and ARDS (acute respiratory distress syndrome), which are all manifestations of severe sepsis. Low serum albumen and low serum calcium are common non-specific findings.

Plain radiograph of the site or limb should be performed to exclude crepitant fasciitis or myositis. Ultrasonography is useful to exclude pyogenic myositis with abscess (mostly commonly due to *S. aureus*), which in children may demonstrate as distortion or thickening of the fascia with fluid accumulation.[42] CT scan is better than ultrasonography at defining the extent of the disease and may be more specific than MRI.[44] See Table 3.2 for summary of features of conditions mimicking necrotizing fasciitis.

3.2.2 Course in Hospital

The patient was transferred to the medical floor after 3:30 p.m. At 5 p.m., his vital signs were recorded as temperature of 39°C, blood pressure 95/60 mmHg, pulse of 152/min and respiratory rate of 28/min. He was visited by his family physician about an hour later and found to have a temperature of 40.3°C and systolic blood pressure of 70 mmHg. A diagnosis of necrotizing fasciitis was then made and he was transferred to the intensive care unit, where broad-spectrum antibiotics, vasopressors and oxygen were instituted (about 12 h after admission to the emergency department). He was transferred to a tertiary urban center in septic shock, and underwent radical debridement of the shoulder and arm early the next morning (about 20 h after presentation to the emergency department). He suffered from an intraoperative cardiac arrest and died. Previous blood cultures and debrided tissue grew S. pyogenes.

3.2.3 Comments on Medical Issues

Management of necrotizing fasciitis should be considered a medical-surgical emergency. Supportive care to maintain adequate circulation, tissue perfusion, and oxygenation are essential, as for all patients with severe sepsis. At least two sets of blood cultures and aseptic tissue aspirate for immediate gram and cultures should be attempted before starting antibiotics. In situations where streptococcal necrotizing

Table 3.2 Features of necrotizing fasciitis compared to other conditions

Early	Fasciitis	Compartment syndrome	DVT	Cellulitis	Pyogenic myositis
Pain/tenderness	Severe at rest and palpation	Moderate on movement/ palpation or stretching of muscle	Rare, dull ache or heaviness	Mild to moderate on movement/palpation	Severe, worst with movement
Skin changes	Mild erythema or normal	Tense, shiny	Normal or mild erythema, distended veins	Moderate to marked redness	Normal or mild erythema
Swelling	Mild to moderate, localized	Marked swelling, diffuse	None to moderate	Mild to moderate	Marked, localized bulging
Progression	Rapid, toxic	Moderate, numbness, hyperesthesia	Slow to none, but prone to pulmonary emboli, extension to proximal veins	Variable	Moderate to rapid, toxic
Fever	Usual high	Absent mostly, but can occur	Absent mostly, but can occur	Moderate to high	High
Leucocytosis	Very high, left shift	Usually absent but can occur	Absent	Mild to moderate	Very high, left shift
CK	Increase early	Increase late	Normal	Normal	High early
Renal impairment	Present late	Present occasionally late	Absent	Absent	Present late or absent
Lungs	ARDS late sepsis	Normal	Normal or emboli	Normal	ARDS late sepsis, depending on microorganism

Abbreviations: *DVT* deep vein thrombosis, *ARDS* acute respiratory distress syndrome, *CK* creatine phosphokinase fasciitis is very likely, high doses of intravenous penicillin and clindamycin are started immediately after cultures are obtained. Although *S. pyogenes* is uniformly penicillin sensitive, clindamycin is used to inhibit the toxins and super-antigens that are produced by the organisms, which largely account for the toxicity, shock, and tissue necrosis.[45] The evidence to support this paradigm is based on observational studies in animal and in vitro experiments. These studies indicate that clindamycin is more effective than penicillin in streptococcal myositis or necrotizing fasciitis,[46,47] but about 5% of group A *Streptococcus* are clindamycin resistant. Prompt aggressive surgery to debride necrotic tissue and open fascial planes for drainage, is a key component of appropriate management to preserve life and limb. This surgical procedure should be done immediately once the diagnosis is made, and can be performed in any rural or primary care hospital with a general surgeon. Initial surgery starting with an explorative incision and inspection of the fascia and muscles to confirm the diagnosis is preferable to delaying the surgery pending imaging (CT scan), or referral to a tertiary care center. Macroscopic clues to necrotizing fasciitis on incision include fat necrosis ("dishwater pus"), thrombosis of subcutaneous veins (called spider-web veins), and deeper areas of fascial necrosis. The surgeon should remove involved fascia, overlying skin, underlying muscle, until all devitalized tissue has been removed and healthy bleeding tissue is seen. Once initial surgical debridement is performed, the patient can be transferred to a tertiary care center for further plastic surgery if necessary. Repeat inspection and further debridement in the operating room should be performed within 24 h, and sometimes daily for the next several days.

If the diagnosis is in question after the initial explorative incision, frozen section and rapid histo-pathological assessment of biopsies may be needed to confirm the diagnosis. The value of intravenous gamma globulin (IVIG) in streptococcal fasciitis is still controversial. Initial case-control studies suggest a benefit of IVIG in reducing mortality[48] but a subsequent European multicenter randomized trial did not confirm a survival benefit,[49] but the sample size was probably too small. The putative effect of IVIG was to bind and neutralize circulating toxins or super-antigens. Although there was no proven survival benefit in the randomized trial: there was significant improvement in secondary end points, such as decrease in sepsis related organ failure at days 2–3 and increase in plasma neutralizing activity against super-antigens.[49]

3.2.4 Medico-legal Issues

The plaintiff's lawyer had filed litigations against all the physicians and the initial hospital involved in the care and management from the time of admission in the emergency department, until the time of transfer to the tertiary care hospital. The charges were as follows: (1) the emergency physician, the consulting orthopedic surgeon, and the internist were negligent in not diagnosing and treating for necrotizing fasciitis soon after admission, (2) the emergency physician on call, who assessed the patient when the diagnosis was made, should have initiated immediate specific treatment for necrotizing fasciitis and consulted a local surgeon for emergency surgery, rather than arranging transfer to another hospital, (3) failure to diagnose and initiate appropriate medical/surgical therapy earlier was the direct cause of the adverse outcome, (4) earlier diagnosis and optimal treatment, which was never considered or implemented until it was too late, would have prevented the patient's demise.

Lawyers for the defendants countered that the physicians were not negligent in missing the diagnosis earlier, as this is a rare condition and that their differential diagnosis, investigation, and management were appropriate and met the standard of medical care. Furthermore, they argued that even if the diagnosis and treatment were made and implemented earlier, the outcome would likely be the same, as this condition carries a high mortality.

3.2.5 Comments and Medico-legal Discussion

What are the facts of this case versus conjecture? Clear and established facts are: (1) the patient died of severe sepsis/septic shock as a consequence of streptococcal necrotizing fasciitis, (2) there was a delay in the diagnosis from the first initial assessment in the emergency department of $\geq$10 h, (3) there was a delay in the optimal therapy (including radical surgery) of about 20 h after admission to the emergency department, (4) the patient had the typical clinical manifestations and laboratory results consistent with the diagnosis of necrotizing fasciitis. Thus, the case presented is by no means atypical for this condition. Why then was this diagnosis not considered (to be excluded) in the initial differential diagnosis? This is difficult to explain but is a common occurrence in medico-legal malpractice cases, where the diagnosis is missed and treatment delayed due to errors committed by three to four health care practitioners. Too often, the consulting physicians are accepting of the initial impression of the first physician, without generating an unbiased independent opinion and diagnosis. This is tantamount to "tunnel vision" by physicians and may represent a problem in their earlier clinical medical training by not questioning their colleagues' opinions.

Although there are no randomized controlled trials (tier I data) to prove that early medical-surgical therapy improves the outcome of necrotizing fasciitis, this is the paradigm for all guidelines for the treatment of these infections. Thus, defendants can argue that it is pure conjecture. There is increasing evidence from large observational studies that early institution of appropriate antibiotics (even by 1–2 h) can significantly improve the outcome in severe sepsis.[50] Early institution of antibiotics and surgery would also very likely halt the progression of disease, from being simple necrotizing fasciitis to necrotizing fasciitis complicated by streptococcal toxic shock syndrome (or septic shock). Analysis of severe invasive streptococcal infections in the United States, 2000–2004, of 5,400 cases found the case fatality rate for necrotizing fasciitis (without shock) was 24% but for those with streptococcal toxic syndrome it was 36%.[51]

Prompt early and complete debridement should be undertaken as soon as possible.[52] Observational studies (mainly retrospective) have shown that early and complete debridement can affect the final outcome in patients with streptococcal necrotizing fasciitis.[53–56] When comparing earlier and complete, versus delayed or incomplete, debridement mortality is always lower with early aggressive approach. Occasionally amputation of a limb is necessary to preserve the life of the patient. In severely ill patients and at special times (weekends and nights), or in community hospitals with limited imaging facilities, it is preferable to do explorative surgery to confirm the diagnosis and perform radical debridement at the same time (if diagnosis confirmed), rather than waiting for imaging studies (CT), or transferring to a tertiary care center, as the time lost may adversely affect the outcome.

Any health care center with a general surgeon and an operating room can perform the primary debridement. The patient's surgical therapy should not be delayed in order to transfer to a tertiary care center. Limited surgical exploration (with a small incision) can be used as a diagnostic procedure and would be more sensitive and specific than any imaging modatities.[52] Thus, the best current data of the disease management strongly support the need for rapid initiation of appropriate antibiotics and immediate surgery in order to limit tissue damage and to preserve limb and life.

In the case under discussion, there is good scientific medical evidence that if appropriate antibiotics were instituted earlier, the patient would more likely have survived, especially if this was done before development of hypotension and accompanied by emergency surgical radical debridement. Later that evening, when the patient was evidently in septic or toxic shock, resuscitation with crystalloids, use of vasopressors in the intensive care unit, or possible emergency surgery might have had a good chance of saving the patients' life. Amputation of the limb might still have been necessary.

Another issue raised by the defendants' lawyers is that streptococcal necrotizing fasciitis is a very rare disease, and the physicians involved cannot be expected to recognize rare conditions they may never have seen before. This is a valid point that needs further discussion, but the issue will likely have to be settled by the courts. The term "rare" is defined by Collins English dictionary as "not widely known, or uncommon or unusual," but there is no medical or scientific definition is terms of incidence or prevalence in the community. The fact that a physician has never seen a case before is not necessarily an excuse or defense for not recognizing the disease. In fact, physicians should have knowledge through their continued medical education and training, or by reading medical literature about uncommon or rare conditions and should be familiar with their manifestations, ways of making diagnoses, and treatment. Moreover, the term "rare disease" is relative, rather than an absolute definition. For example, Lassa fever and Ebola fever are extremely rare in North America or developed countries, and a misdiagnosis in these extremely unusual rare conditions would be acceptable.

However, in a recent epidemiological study in the United States, invasive group A streptococcal infection incidence was reported as 3.5 cases per 100,000 persons[52] (which was likely underestimated). The mid-1996 invasive group A streptococcal infections became a reportable disease in the State of Florida, and over a 4-year period, 257 patients were hospitalized with this disease and 45 (18%) has necrotizing fasciitis.[57] In Ontario, Canada, another surveillance (also likely incomplete) of invasive group A streptococcal infection (which was not a reportable disease) noted 2,351 cases in the community, and 291 cases were hospital-acquired between 1992 and 2000.[58] Of these 142 (6.0%) patients with community-acquired invasive streptococcal infection who required intensive care, and 8.1% of nosocomial invasive streptococcal infection developed necrotizing fasciitis and streptococcal toxic shock syndrome, respectively. Moreover, ever since 1994, streptococcal necrotizing fasciitis (labeled "flesh eating disease") has received much attention in the popular press in Canada. Thus, although necrotizing fasciitis is uncommon, it could be argued that most physicians, especially emergency specialists, family physicians, internists, Infectious Diseases specialists, general surgeons, plastic and orthopedic surgeons, should be quite familiar with this condition. It should be noted, that where the diagnosis of streptococcal necrotizing fasciitis has been missed at the initial outpatient evaluation, it has resulted in a higher mortality than the average (8 of 15 [53%]).[43] Luckily for physicians, unlike the judiciary where "ignorance of the law is no excuse," ignorance of a disease can be used as defense, depending on the expected standards of other physicians practicing in the community under similar conditions.

3.2.6 Final Comments

What can we learn from these two cases presented in this chapter? Although both gas gangrene and streptococcal necrotizing fasciitis are rare and uncommon, front-line physicians (emergency physicians, general practitioners, general surgeons) should and are expected to be familiar with their clinical manifestations and be cognizant of the diagnosis and treatment. These conditions have a rapid course, and misdiagnosis that results in delayed treatment for a few hours may affect the outcome, resulting in death or loss of limb.

It is very important that physicians take patient's symptoms of severe pain seriously, and be aware that in these two conditions, it is common to not find much evidence of inflammation on examination early in the course. Clinicians also need to remember that absence of detectable fever or leucocytosis does not exclude a serious infectious disease, and one should always check the differential white blood count for a shift to the left even when the total leucocyte count is normal. In both clostridial myonecrosis and streptococcal necrotizing fasciitis, a serum CK should be elevated in the early stages, so a normal value would make the diagnosis unlikely. Although the conditions discussed in this chapter are considered uncommon or rare, a high proportion of these cases result in medico-legal malpractice suits.

A jury or court may regard physicians' misdiagnosis of certain rare conditions as acceptable, such as rabies or Jacob-Creutzfeldt disease, when no effective treatment is available, but necrotizing fasciitis and gas gangrene are easily treatable diseases in the early stages.

References

1. Gordon RJ, Lowy FD, (2005), Bacterial infections in drug users. N Engl J Med. 353: 1945–1954.
2. Kimura AC, Higa JI, Levin RM, Simpson G, Vargas Y, Vugia DJ, (2004), Outbreak of necrotizing fasciitis due to *Clostriduim sordelli* among black-tar heroin users. Clin. Infect. Dis. 38: e87–e91.
3. Dunbar NM, Harruff RC, (2007), Necrotizing fasciitis: manifestations, microbiology, and connection with black tar heroin. J Forensic Sci. 52: 920–923.
4. Werner SB, Passaro D, McGee J, Schechter R, Vugia DJ, (2000), Wound botulism in California, 1951-1998: recent epidemic in heroin injectors. Clin Infect Dis. 31:1018–1024.
5. Hoeger PH, Haupt G, Hoelzle E, (1996), Acute multifocal skin necrosis: synergism between invasive streptococcal infection and cocaine-induced tissue ischemia? Acta Derm Venereol. 76: 239–241.
6. Eckmann C, Kujath P, Shekarriz H, Staubach KH, (1997), *Clostridium* myonecrosis as a sequelae of intramuscular injections–description of 3 fatal outcomes. Langenbecks Arch Chirugie; 114 (suppl.): 553–555.
7. Koons TA, Boyden GM, Albuquerque NM, (1961), Gas gangrene from parental injections. JAMA; 175: 140–141.
8. Marshall V, Sims P, (1960), Gas gangrene after the injection of adrenaline-in-oil with a report of three cases. Med. J. Aust. 2: 653–656.
9. Harvey PW, Purnell GV, (1968), Fatal case of gas gangrene associated with intramuscular injections. Brit. Med. J. 1: 744–746.
10. Gaylis H, (1968), Gas gangrene and intramuscular injection. Brit. Med. J. 3: 59–60.
11. Maguire WB, Langley NF, (1967), Gas gangrene following an adrenaline-in-oil injection into the left thigh with survival. Med. J. Aust. 1: 973–975.
12. Van Hook R, Vandevelde AG, (1975), Gas gangrene after intramuscular injection of epinephrine: report of fatal case. Ann. Intern. Med. 83: 669–670.
13. Teo WS, Balasubramaniam P, (1983), Gas gangrene after intramuscular injection of adrenaline. Clin. Orthop. Relat. Res. 174: 206–207.
14. Hallagan LF, Scott JL, Horowitz BC, Feied CF, (1992), Clostridial myonecrosis resulting from subcutaneous epinephrine suspension injection. Ann. Emerg. Med. 4: 434–436.
15. Mabin G, (1936), La gangrene gazeuse après injections medicamenteuses. Thèse de Paris.
16. Touraine A, (1936), La gangrene gazeuse après injections medicamenteuses. Presse Med. 44: 674–677.
17. Koransky JR, Stargel MD, Dowell VR Jr, (1979), *Clostridium septicum* bacteremia, its clinical significance. Am. J. Med. 66: 63–66.
18. Collier PE, Diamond DL, Young JC, (1983), Nontraumatic *Clostridium septicum* gangrenous myonecrosis. Dis. Colon. Rectum. 26: 703–704.
19. Katlic MR, Derkac WM, Coleman WS, (1981), *Clostridium septicum* infection and malignancy. Ann. Surg. 193: 361–364.
20. Stevens DL, Musher DM, Watson DA, Eddy H, Hamill RJ, Gyorkey F, Rosen H, Mader JT, (1990), Spontaneous, non-traumatic gangrene due to *Clostridium septicum*. Rev. Infect. Dis. 12: 286–296.
21. Bretzke ML, Bubrick MP, Hitchcock CR, (1988), Diffuse spreading *Clostridium septicum* infection, malignant disease and immune suppression. Surg. Gynecol. Obstet. 166: 197–199.

22. Thomas MG, (1990), *Clostridium septicum* gas gangrene following intramuscular infection from an influenza vaccine booster. Br J. Clin. Pract. 44: 709–710.
23. Ray P, Das A, Singh K, Bhansali A, Yadav TD, (2003), *Clostridium tertium* in necrotizing fasciitis and gangrene. Emerg. Infect. Dis. 9:1347–1348.
24. Bhargava A, Sen P, Swaminathan A, Ogbolu C, Chechko S, Stone F, (2000), Rapidly progressive necrotizing fasciitis and gangrene due to *Clostridium difficile*: case report. Clin. Infect. Dis. 30: 954–955.
25. Willis AT, (1974), Some infections due to anaerobic spore forming bacilli. In: Balows A, Deehaan RM, Dowell VR Jr, Gaze L (eds), Anaerobic bacteria: role in disease. Charles C. Thomas Publisher, Springfield, Illinois, p195–217.
26. Mittermair RP, Schoberberger W, Hasibeder W, Allerberger F, Peer R, Bonatti H, (2002), Necrotizing fasciitis with *Clostridium perfringens* after laparoscopic cholecystectomy. Surg. Endoscopy; 16: 716.
27. Lorber B, (2000), Gas gangrene and other *Clostridium*-associated disease. In: Mandell GL, Bennett JE, Dolin R (eds), Principles and Practice of infectious Diseases, 5th Edition, Churchill Livingstone, Philadelphia, p 2549–2561.
28. Anonymous, (1961), Gas gangrene from adrenaline (editorial). Brit. Med. J. 730–731.
29. Stevens DL, Bryant AE, (1999), The pathogenesis of shock and tissue injury in clostridial gas gangrene. In: Alouf JE, Freer JH (eds). The comprehensive sourcebook of bacterial protein toxins. 2nd Edition, Academic Press, London; 623–636.
30. Bryant AE, Chen RY, Nagata Y, Wang Y, Lee CH, Finegold S, Guth PH, Stevens DL, (2000), Clostridial gas gangrene. I. Cellular and molecular mechanisms of microvascular dysfunction induced by exotoxins of *Clostridium perfringens*. J. Infect. Dis. 182: 799–807.
31. Bryant AE, Chen RY, Nagata Y, Wang Y, Lee CH, Finegold S, Guth PH, Stevens DL, (2000), Clostridial gas gangrene. II. Phospholipase C-induced activation of platelet gpIIbIIIa mediates vascular occlusion and myonecrosis in *Clostridium perfringens* gas gangrene. J. Infect. Dis. 182: 808–815.
32. Bryant AE, Bergstrom R, Zimmerman GA, Salyer JL, Hill HR, Tweten RK, Sato H, Stevens DL, (1993), *Clostridium perfringens* invasiveness is enhanced by effects of theta toxin upon PMNL structure and function: the roles of leukocytotoxicity and expression of CD11/CD18 adherence glycoprotein. FEMS Immunol. Med. Microbiol. 7: 321–336.
33. Altemeier WA, Fullen WD, (1971), Prevention and treatment of gas gangrene. JAMA; 217: 806–813.
34. Cline KA, Turnbull TL, (1985), Clostridial myonecrosis. Ann. Emerg. Med. 14: 459–466.
35. Altemeier WA, McMurrin JA, ALT LP, (1950), Chloromycetin and aureomycin in experimental gas gangrene. Surgery; 28: 621–631.
36. Stevens DL, (2004), Skin and soft tissue infections: cellulitis, fasciitis, myonecrosis. Consultant; 44: 694–700.
37. Holder AR, (1972), Liability for gas gangrene. JAMA; 221: 1083–1084.
38. Louis J, (1999), Axillary vein thrombosis mimicking muscular strain. J. Accid. Emerg. Med. 16: 233–234.
39. Sajid MS, Ahmed N, Desai M, Baker D, Hamilton G, (2007), Upper limb deep vein thrombosis: a literature review to streamline the protocol for management. Acta Haematol. 118: 10–18.
40. Amendola A, Twaddle BC, (1998), Compartment syndrome. In: Browner BD, Jupitor JB, Levine AM, Trafton PG (eds); Skeletal Trauma, 2nd Edition, WB Saunders Company, Philadelphia; p365–389.
41. Jobe MT, (2003), Compartment syndromes and Volkmann Contracture. In: Canale ST (ed), Campbell's Operative Orthopedics, 10th Edition, Mosby, St. Louis, p3739–3750.
42. Eron LJ, (2009), In the clinic: cellulitis and soft tissue infection. Ann. Intern. Med. 150: ITc 1–14.
43. Bisno AL, Cockerill FR III, Bermudez CT, (2000), The initial outpatient-physician encounter in group A streptococcal necrotizing fasciitis. Clin. Infect Dis. 31: 607–608.

44. Arsian A, Pierre-Jerome C, Barthne A, (2000), Necrotizing fasciitis: unreliable MRI findings in the pre-operative diagnosis. Eur. J. Radiol. 36: 139–143.
45. Stevens DL, Bisno AL, Chambers HF, Everett ED, Dellinger P, Goldstein EJ, Gorbach SL, Hirschmann JV, Kaplan EL, Montoya JG, Wade JC, (2005), Infectious Diseases Society of America. Practice guidelines for the diagnosis and management of skin and soft-tissue infections. Clin. Infect. Dis. 41: 1373–1406.
46. Stevens DL, Bryant-Gibbons AE, Bergstrom R, Winn V, (1988), The Eagle effect revisited: efficacy of clindamycin, erythromycin, and penicillin in the treatment of streptococcal myositis. J Infect Dis. 158: 23–28.
47. Zimbelman J, Palmer A, Todd J, (1999), Improved outcome of clindamycin compared with beta-lactam antibiotic treatment for invasive *Streptococcus pyogenes* infection. Pediatr. Infect. Dis J. 18: 1096–1010.
48. Kaul R, McGeer A, Norrby-Teglund A, Kotb M, Schwartz B, O'Rourke K, Talbot J, Low DE; The Canadian Streptococcal Study Group, (1999), Intravenous immunoglobulin therapy for streptococcal toxic shock syndrome–a comparative observational study. Clin. Infect. Dis. 28: 800–807.
49. Darenberg J, Ihendyane N, Sjölin J, Aufwerber E, Haidl S, Follin P, Andersson J, Norrby-Teglund A; Streptlg Study Group, (2003), Intravenous immunoglobulin G therapy in streptococcal toxic shock syndrome: a European randomized, double-blind, placebo-controlled trial. Clin. Infect. Dis. 37: 333–340.
50. Kumar A, Roberts D, Wood KE, Light B, Parrillo JE, Sharma S, Suppes R, Feinstein D, Zanotti S, Taiberg L, Gurka D, Kumar A, Cheang M, (2006), Duration of hypotension before initiation of effective antimicrobial therapy is the critical determinant of survival in human septic shock. Crit. Care. Med. 34: 1589–1596.
51. O'Loughlin RE, Roberson A, Cieslak PR, Lynfield R, Gershman K, Craig A, Albanese BA, Farley MM, Barrett NL, Spina NL, Beall B, Harrison LH, Reingold A, Van Beneden C; Active Bacterial Core Surveillance Team, (2007), The epidemiology of invasive group A streptococcal infection and potential vaccine implications: United States, 2000-2004. Clin. Infect. Dis. 45: 853–862.
52. Anaya DA, Dellinger EP, (2007), Necrotizing soft-tissue infection: diagnosis and management. Clin Infect Dis. 44: 705–710.
53. McHenry CR, Piotrowski JJ, Petrinic D, Malangoni MA, (1995), Determinants of mortality for necrotizing soft-tissue infections. Ann. Surg. 221: 558–563.
54. Bilton BD, Zibari GB, McMillan RW, Aultman DF, Dunn G, McDonald JC, (1998), Aggressive surgical management of necrotizing fasciitis serves to decrease mortality: a retrospective study. Am. Surg. 64: 397–400.
55. Lille ST, Sato TT, Engrav LH, Foy H, Jurkovich GJ, (1996), Necrotizing soft tissue infections: obstacles in diagnosis. J Am Coll Surg. 182: 7–11.
56. Majeski J, Majeski E, (1997), Necrotizing fasciitis: improved survival with early recognition by tissue biopsy and aggressive surgical treatment. South Med J. 90: 1065–1068.
57. Mulla ZD, Leaverton PE, Wiersma ST, (2003), Invasive group A streptococcal infections in Florida. South. Med. J. 96: 968–973.
58. Daneman N, McGeer A, Low DE, Tyrrell G, Simor AE, McArthur M, Schwartz B, Jessamine P, Croxford R, Green KA; Ontario Group A Streptococcal Study Group, (2005), Hospital-acquired invasive group a streptococcal infections in Ontario, Canada, 1992-2000. Clin. Infect. Dis. 41: 334–342.

Chapter 4
Litigation in Infections of Obstetrics and Gynecology

4.1 Case 1: Fever in a Pregnant Female

A 28-year-old female Native American Indian, in her third trimester of pregnancy (34 weeks), presented to an isolated, stand-alone medical center serving the local community (Indian Reservation) at 6 p.m. on a Sunday evening. This center serves the dual purpose of medical clinic and emergency facility. Available teleconsultation and transportation to a tertiary care center via air ambulance were accessible 24 h/day. Normally, air ambulance transfer to a distant tertiary care hospital can be accomplished within 3–4 h after notification by phone.

The patient presented with symptoms of fever, chills, and backache of 24 h duration, with more recent nausea, vomiting, and dizziness of 2 h duration. There was no significant past medical illness or allergies. Physical examination by the attending physician noted a blood pressure of 90/60 mmHg, pulse of 120/min, temperature of 39.5°C, and evidence of left costovertebral angle tenderness. Urinalysis showed many leucocytes and positive nitrates and a diagnosis of acute pyelonephritis was made. The patient was kept overnight in the medical center, and she was started on intravenous saline and gentamicin 60 mg every 8 h, after obtaining urine and blood cultures. The patient's weight was estimated to be about 65 kg. Two hours later the patient was stable with a blood pressure 100/70 pulse 120/min and temperature of 38.5°C.

The incoming physician starting the next shift at 7 a.m. the following morning was notified about the patient. At 9 a.m., when the new physician got around to the patient, she was found to be diaphoretic, toxic looking, with a systolic blood pressure of 70, pulse 150 min, weak in volume, cold cyanosed periphery, and respiratory rate of 40 min. There was no record of monitored vital signs in the preceding 12–14 h. An immediate arrangement for air-ambulance and transfer to the hospital was implemented. Increased volume of intravenous saline was given and another dose of gentamicin at 100 mg. The patient arrived at the tertiary care hospital within 3 h, and was found to be in septic shock. She subsequently lost the baby, and developed bilateral gangrene of the lower legs, which required amputation. She survived the surgery. Initial urine and blood cultures grew Eschericia coli fully sensitive to the usual antibiotics.

I.W. Fong, *Medico-Legal Issues in Infectious Diseases*, Emerging Infectious Diseases of the 21st Century, DOI 10.1007/978-1-4419-8053-3_4,

4.1.1 Medico-legal Issues

The lawyer for the plaintiff filed litigation against the two physicians involved in the case and the medical center. Against the first physician, the claims were of negligence for not arranging transfer to a hospital from the night before, as the patient had a complicated urinary tract infection in pregnancy. Moreover, when the patient was seen, the dosage of gentamicin used was inadequate to treat sepsis, and there was no monitoring of the vital signs throughout the night.

With respect to the second physician, he was charged negligent for the delay in assessing an ill patient. He knew the patient was admitted to the medical center, but did not assess her condition until 2 h after his arrival at the medical facility. The actions and decisions of both physicians, as claimed by the plaintiff, resulted in fetal loss, and indirectly, ischemic gangrene of the lower limbs that resulted in amputation. Compensation was thus being sought for injury to the mother and fetus, and life-long disability of the mother, which will affect her functioning and employability.

Liability claims against the medical center (run by the Provincial Government) was for failure to provide adequate prenatal care, as the employees or healthcare workers did not perform routine midstream urine (MSU) for culture in pregnancy. Furthermore, during the overnight admission, the nurse on duty should have been checking the patient's vital signs at regular intervals.

The defendants' legal defense responded that the physicians' actions were not negligent and they acted in good faith and appropriately, considering the limited resources available to them. Furthermore, even if a transfer to the tertiary care hospital were made earlier, it probably would not have affected the outcome. Moreover, the second physician was not aware that the patient was very ill and he was occupied seeing emergency patients waiting soon after his arrival at the facility.

4.1.2 Medical Issues

Urinary tract infections are the most common medical complications of pregnancy. These may be asymptomatic (asymptomatic bacteriuria of pregnancy) or symptomatic with cystitis or acute pyelonephritis. Studies performed in 1960 identified persistent asymptomatic bacteriuria in 6% of prenatal patients, and 40% of these patients would develop acute pyelonephritis when the bacteriuria was not eliminated.[1,2] Neonatal death rates and prematurity were noted to be two to three times greater in untreated bacteriuric women compared to those when the bacteriuria was eliminated.[3,4] However, the association between bacteriuria and prematurity and low birth weight infants has been controversial.

It has long been recognized that symptomatic urinary tract infection is more common in pregnant than non-pregnant women. Several factors are believed to play a role. The upper ureter and renal pelvices become dilated, resulting in a physiologic hydronephrosis of pregnancy. This is a result of effects of progesterone on muscle tone and peristalsis on the urinary collecting system, and the mechanical obstruction by the enlarged uterus. There is also decreased bladder tone, increased capacity, and incomplete emptying of the bladder, all factors that predispose to vesicoureteric reflux. The hypokinetic collecting system reduces urine flow and predispose to ascending infection from the bladder to the kidneys.[5]

Acute pyelonephritis is one of the most serious complications in pregnancy and is a threat to maternal and fetal well being. Several studies have confirmed the association between acute pyelonephritis and the increased risk of premature delivery.[6–10] The association of acute pyelonephritis and preterm delivery was known from the pre-antibiotic era with prematurity rates of up to 20–50%.[5] The proposed mechanisms include oxytoxic-like effect of endotoxin on the myometrium causing uterine contractions; high fever by pyrogens may increase myometrial activity, uteric contractions may result in reflex myometrial contractions, and the endotoxin may cross the placenta and produce fetal effects resulting in premature labor. Asymptomatic bacteriuria alone probably has no harmful effects unless it leads to symptomatic pyelonephritis.

The incidence of persistent asymptomatic bacteriuria in pregnancy varies from 2% to 7% and depends on parity, race, and socio-economic status. The highest incidence of asymptomatic bacteriuria is in African-Americans with sickle cell trait, and the lowest incidence is in affluent white women of low parity.[11] It is usually the same group of non-pregnant who has asymptomatic bacteriuria, but is prone to episodes of intermittent cystitis. The patient in the present case report would fall into the higher risk group for persistent bacteriuria.

Acute pyelonephritis during pregnancy occurs more frequently in the second trimester, and multi-parity and young ages are associated risk factors.[12] It can also occur during the third trimester and early post-partum, but there does not appear to be a significantly increased risk in the first trimester. Bacteremia occurs in 15–20% of women with acute pyelonephritis, and the sepsis syndrome is common.[11] Kidney infection was the second most common reason for non-delivery admission, 4 per 100 in one study of 833,264 hospitalizations for pregnancy complications.[13] Admissions for preterm labor (33%) were the most common reason for hospitalization, followed by genito-urinary infections (16%). In another survey of 46,179 pregnancies, 3.5% required antepartum admission for acute pyelonephritis.[14] Urosepsis also is the leading cause of septic shock in pregnancy, predominantly due to *E. coli*.[15] Over a 2-year audit at the Parkland Hospital Obstetrics intensive care unit, 12% of the admissions were for sepsis syndrome caused by acute pyelonephritis.[16] The overall incidence of acute pyelonephritis in pregnancy is about 1–2.5%, with an estimated recurrence rate of 10–18% during the same gestation.[5] Asymptomatic bacteriuria is the major predisposing factor and others include obstructive uropathy, neurologic disease, renal calculi, and the need for catheterization. Screening and treatment of asymptomatic bacteriuria decreases the risk of pyelonephritis significantly in pregnant patients.[17]

In a prospective study over a 4.5-year period, over 2% of 24,000 pregnant women were admitted for acute pyelonephritis.[18] Chills, fever, and back pain were the most common complaint in 656 women, while lower urinary tract symptoms, nausea, and vomiting were very common. Documented fever was present in 96% and costovertebral angle tenderness was positive in 97% (27% were bilateral).[18] In most (67%) of these women, pyelonephritis occurred during the last two trimesters; but 8% were diagnosed intra-partum and 19% occurred post-partum. Of 501 women with antepartum pyelonephritis, the pregnancy outcome was spontaneous abortion in 5, stillbirth in 4, neonatal death in 2 (for a prenatal mortality of 12 per 100), and 133 (23%) developed recurrent acute renal infection.[18]

A more recent longitudinal study over 2 years from the same center (Parkland Hospital, Dallas) reported on 440 cases of acute antepartum pyelonephritis (1.4% incidence).[12] Acute pyelonephritis occurred most often in the second trimester (53%) with *E. coli* accounting for 70% and group B *Streptococcus* in 10%. Complications included anemia (23%), septicemia (17%), transient renal dysfunction (2%), pulmonary insufficiency (7%), and 43 (10%) required admission to the intensive care unit (indicating severe sepsis syndrome).[12]

In a retrospective review of 18 patients with septic shock during pregnancy, the causes were acute pyelonephritis (33%), chorioamnionitis (16%), post-partum endometritis (11%), toxic shock syndrome (11%), and miscellaneous conditions (27%).[15] The incidence of septic shock was 1 per 8,338 deliveries, with a mortality of 28% (N = 5), and of nine women delivering while septic, only two babies survived with significant morbidity. However, three patients delivered 15–20 weeks after the episode of septic shock had uncomplicated deliveries. Of the 14 pregnancies that reached viability (≥24 weeks gestation), perinatal mortality was high (29%).[15]

4.1.3 Comments

Obstetric patients with acute pyelonephritis require hospitalization for intravenous fluids (saline), initial parenteral antibiotics, and close monitoring. Although most women will do well, the perinatal morbidity and mortality is significant. The majority of patients will respond within 48 h of treatment, and failure to do so requires ultrasonography to exclude obstruction and renal calculi. Transient renal impairment is commonly seen (about 20%) and usually improves with rehydration. The exact incidence of septic shock in pyelonephritis of pregnancy is unclear, but it appears to be very low.

Empiric therapy for acute pyelonephritis in pregnancy usually consists of intravenous ampicillin (for *streptococcus* and enterococci coverage) and gentamicin (for coliform coverage) until identification and susceptibility of the recovered microorganism is available.[5] The dose of gentamicin recommended has been 3–5 mg/kg/day, previously as three divided doses, but since the 1990s is commonly given as a single daily dose. Gentamicin levels (pre-and post-therapy) used to be recommended to assess toxic concentrations when multiple daily doses were in vogue. In pregnancy, there was an increase in vascular volume and the volume of distribution with enhanced glomerular filtration rate, resulting in a high prevalence of sub-therapeutic levels. With the advent of single dose of aminoglycosides, sub-therapeutic levels are no longer a problem as 3–5 mg/kg is given once daily. In this instance, the trough level only is assessed, and should be 0 to <2 μg/mL. In animal models, single daily doses of aminoglycosides are associated with less toxicity and may be more effective, but clinical studies have been less convincing (probably from small sample sizes).[19] Sub-therapeutic trough levels of aminoglycosides given for a few hours once a high peak level is achieved is probably not therapeutically important, as the aminoglycosides have a prolonged post-antibiotic effect on coliforms (inhibition of the bacteria even after the antibiotic is removed) and the rate of killing is concentration dependent. It should be noted that in patients with renal impairment, the initial dose would be same as for normal renal function, but the subsequent dose or dosing interval would be adjusted according to the trough level and estimated gentamicin clearance.

Thus, in the present case under discussion, there is valid argument by the plaintiffs' attorney that the first physician should have arranged for an air-ambulance transfer to the acute care hospital the night before, and that close monitoring and aggressive treatment for severe sepsis would have prevented the adverse outcome. Furthermore, the initial dose of gentamicin should have been at least 3 mg/kg ($\simeq$200 mg), and that failure to give the appropriate dose resulted in progression of the sepsis syndrome.

The charge of negligence against the second physician for the delay in assessing the patient until 2 h after his arrival is considered a break down in the health care system and the manner of patient handover. It is the duty of an incoming physician just starting his or her shift to assess the most seriously ill patient first, either in an emergency department, freestanding medical center, or hospital. However, whether a 2 h delay in appropriate aggressive treatment for sepsis would have made a difference in the outcome would be difficult to ascertain. This would largely depend on the duration of severe hypotension and poor peripheral tissue perfusion, and the response to aggressive fluid resuscitation at the remote medical facility.

The next issue brought forth by the plaintiff's lawyer was the fetal loss and liability of the physicians. It was stated that had the initial physician arranged transfer from the night before, progression to septic shock would have been prevented and the fetus or baby would likely have survived. Based on current data, it does appear that the prevalence of fetal loss is significantly less in pregnant women with uncomplicated pyelonephritis than in those with septic shock.

4.2 Case 2: Fever Post-cesarean Section

A 29-year-old woman presented to the emergency department of an urban hospital in early active labor at 8 a.m. one morning. She was 38 weeks pregnant and this was her first pregnancy. The patient noted premature rupture of her membranes (PROM) from the late evening before (14 h). Her past medical history was insignificant and the prenatal vaginal culture for group B Streptococcus was negative.

On admission, her vital signs (including temperature) were normal and the leaking amniotic fluid was clear. Since her cervix was inadequately dilated, an oxytoxin drip was started for induction of labor, an internal fetal electrode was inserted, and the condition of her cervix was assessed by repeated vaginal examinations (4). After 12 h, the cervix was still not fully dilated to facilitate delivery, and an emergency lower segment cesarean (C) section was performed, about 24 h after onset of PROM. A healthy baby was delivered (7 lb) with no operative complication, and the patient was discharged home 2 days later.

Three days after discharge from hospital, the patient was readmitted through the emergency department, with symptoms of fever, sweats, and a painful incision wound since the day before. The emergency physician assessment at 6 a.m. noted a temperature of 36.7°C, pulse of 120/min, blood pressure of 120/70 and normal respiratory rate and oxygen saturation. She was noted to be ill looking, and in severe pain with nausea and vomiting, plus there was mild erythema, and marked tenderness of the incisional wound and surrounding areas of the lower abdomen. A diagnosis of wound infection was made, blood cultures were obtained, and intravenous antibiotic (cefazolin) was started.

She was assessed that morning by the obstetrical service and found to be hypotensive with a blood pressure of 70/60 mmHg, but she responded to intravenous saline. Her temperature then was 38.6°C, and the wound appeared erythematous on the right side of the incision but there was marked tenderness, and firm induration around the entire surrounding areas of the wound. An Infectious Disease consultation was obtained 24 h later as her temperature was still 39°C, and the area of redness and tenderness had spread to involve the right lower abdomen and upper thigh. The patient was noted to be tachypneaic (respiratory rate 40/min), blood pressure 100/70 and the pulse 110/min. The previous complete blood count was 17,800 cells/uL, with a left shift and normal creatine. The antibiotics were then changed to intravenous clindamycin and gentamicin and a computerized tomography of the lower abdomen was performed, which showed gas in the soft tissue, with edema of the lower abdominal wall and right upper thigh. She underwent radical surgery for extensive debridement of the lower abdomen, right upper thigh, and flank, extending to the back 58 h after admission. Debrided tissues revealed necrotizing fasciitis and cultures grew mixed organisms consisting of coliforms and anaerobes. The patient eventually recovered, but required skin grafts and was left with scars and disfigurement of her lower abdomen, flank, and right thigh. She spent over a month in hospital.

4.2.1 Medico-legal Issues

The plaintiffs (patient and husband) filed litigations suits against the obstetrician, consultants, and the hospital. Compensation was sought for damages resulting in prolonged hospitalization, disfigurement, and mental anguish. The claims were negligence of the obstetrician, his assistant, and the anesthesiologist in not giving appropriately required prophylactic antibiotics to prevent this severe adverse outcome. In addition, the plaintiffs' lawyer claims that if appropriate antibiotic prophylaxis were given before the cesarean section, the patient would not have suffered from the post-operative infection.

Further claims of negligence on the part of the attending team (obstetric/gynecology service) and the consultants, were the delay in making the proper diagnosis of necrotizing fasciitis. This caused delay in surgery, which led to progression of the infection, resulting in extensive tissue damage. It was further stated, that if the correct diagnosis and proper treatment were implemented within 24 h, the degree of tissue damage, subsequent pain, suffering and disfigurement would have been much less.

The defendants lawyers' countered that the physicians and hospital treated the patient appropriately and timely, and that rare infections and complications can occur in cesarean sections, due to no fault of the healthcare professionals. Furthermore, the defendants should not be held accountable as their management during the two admissions met the standard of care.

4.2.2 Medical Issues

Prior to labor and rupture of the membranes (ROM), the amniotic cavity is usually sterile. The cervical mucus and intact placental membranes provide a physical, chemical, and microbiological barrier for bacterial entry. Once ROM occurs with labor, there is a potential for microorganisms colonizing in the lower genital tract to ascend and infect the amniotic cavity. The quantity of bacteria recovered from the amniotic cavity in some patients increases with duration of ROM before delivery. Thus, patients with prolonged rupture of the membranes (PROM) before delivery (defined as $\geq$12 h by some specialist and $\geq$18 h by others), would increase the risk of peripartum infections. The onset of labor with uterine contractions may facilitate the ascension of bacteria into the uterine cavity by a massaging effect.

Bacteria can gain access to intrauterine tissues in pregnancy by three mechanisms: (1) transplacental transfer of maternal systemic infection (rare, more common with viral infection), (2) retrograde flow of infection from the peritoneal cavity via the fallopian tubes (rare, possible with low grade pelvic infection) and (3) ascending infection from the vagina via the cervix–considered the most common.[20]

More recently is has been postulated (with some supporting evidence), that preterm labor (before 37 weeks gestation) is commonly precipitated by subclinical infection in the amniotic cavity with intact membranes.[20–22] It is believed that 40% of preterm labors are induced by intrauterine infection (without clinical manifestations of infection). Colonization of the vagina by certain microorganisms (*Gardinerella vaginalis, fusobacterium, Mycoplasma hominis, Ureaplasma urealyticum* etc.) may ascend from the vagina and colonize the decidua and possibly the fetal membranes, and may then enter the amniotic sack. However, the case under discussion, by definition, did not have preterm labor or delivery.

The patient's (case 2) pregnancy, however, was complicated by puerperal infection or sepsis. Puerperal fever is defined as $\geq$38°C on any two of the first 10 days post-partum, exclusive of the first 24 h.[21] The most common causes of puerperal fever are genital infections, urinary tract infection, wound infection, and less commonly pneumonia, atelectasis, deep vein thrombosis, and breast engorgement (the latter lasts under 24 h and usually <39°C). Although the genital infection of the uterus used to be subdivided as endometritis (inflammation of the endometrium), endomyometritis (inflammation of the myometrium), and endoparametritis (inflammation of soft tissues surrounding the uterus), varying degrees of all three layers of tissue are usually involved. Thus, recently the term "metritis" with pelvic cellulitis is more in vogue.

The most important predisposing factor for puerperal genital infection is the route of delivery. In vaginal delivery the average rate of metritis is 1.3%, with a higher rate of 6% for high-risk cases (prolonged rupture of membranes and labor, multiple cervical examinations, internal fetal monitoring), and up to 13% for presence of intrapartum chorioamnionitis.[21] All women undergoing C-section are considered high risk and routine antibiotic prophylaxis is now recommended. Before adoption of routine prophylaxis for C-section, the incidence of uterine infection was dependent on social economic status; 13% in affluent (especially white) women and from 27% to 50% in indigent women.[21] In women with high-risk delivery (prolonged rupture of membranes/labor, multiple vaginal/cervical examinations, internal fetal monitoring, cephalopelvic disproportion), serious pelvic infections after C-section (without antibiotic prophylaxis) occurred in up to 90%.[21]

Other risk factors for puerperal infection after C-section include race (African-American), bacterial colonization of the vagina with certain organisms (group B and A *Streptococcus*, bacterial vaginosis, *Chlamydia trachomatis, Gardineralla vaginalis, and Mycoplasma hominis*), young age and nulliparity, obesity and multifetal gestation.[21]

Wound infection post-C-section typically occurs 4–7 days postoperatively, but certain infections with virulent bacteria (group A *Streptococcus* and clostridial species) can appear within 2 days of surgery. The pathogenesis of wound infections are mainly of two sources, inoculation of the wound from the skin (*Staphylococcus aureus*), or endogenous inoculation from ascension from the vagina (mixed infection, group B *Streptococcus*, etc). Group A *Streptococcus* can be either externally from skin colonization via oropharynx, or from vaginal or rectal (rare) colonization. Rarely can the microorganisms be introduced from external sources at surgery, such as the environment of the operating room or the surgical team.

Cesarean delivery provides direct access to the wound from microorganisms that ascend to the uterine cavity, and most of the post C-section wound infections are believed to be of cervico-vaginal origin. Hence, post C-section wound infections commonly consists of mixed organisms (as in case 2), such as *Enterobacteriaceae* species, *Streptococcus* species, and anaerobes, and rarely genital mycosplasmas. *S. aureus* infections, which account for about 25% of the wound infections, usually arise from the patient's skin in chronic nasal colonizers. It should be noted that wound infection post-C-section is frequently complicated by endometritis or parametritis, and all serious wound infections with fever should be investigated for this complication with ultrasonography or CT scan.

The management of post-C-section wound infection includes opening of the wound down to the fascia, drainage of pus, and debridement of any necrotic or devitalized tissue.[23] Parental antibiotics are usually started to cover expected microorganisms until culture and susceptibility are available. Complete integrity of the fascial suture line should be assessed, and a more serious necrotizing fasciitis can be excluded by early surgical intervention. Usually the wound is liberally irrigated by sterile saline and packed daily with sterile parking gauze, or more often as needed.[23]

Necrotizing fasciitis is a rare rapidly progressive infectious complication of C-section, characterized by extensive necrosis of subcutaneous tissue and other adjacent surrounding soft tissue. Most cases of necrotizing fasciitis are secondary to mixed synergistic infection with coliforms, streptococci, and anaerobes (including bacteroides species and clostridia species). Occasionally, group A *Streptococcus* can be the cause of puerperal necrotizing fasciitis and be introduced externally from the skin or throat, or endogenously from the vagina or anal carriage. Post-C-section necrotizing fasciitis is associated with diabetes mellitus, obesity malnutrition, intravenous drug abuse, and hypertension.[23]

A retrospective review of 23 women admitted to an obstetric and gynecology service with necrotizing fasciitis in a single hospital over 14 years was recently reported.[24] Six women (26%) were puerperal complications, three of whom were associated with C-section and three from episiotomy infections. As noted in this report and other series, necrotizing fasciitis, whether in the abdomen or vulvar area, are usually polymicrobial.[24–27] Obesity was a major predisposition, occurring in 86.9% of the entire cohort of 23 women. Severe pain on presentation was a common manifestation in 83.3% of puerperal complications and 70.6% of the non-puerperal patients. Common findings on physical examination include marked degree of subcutaneous edema with varying degrees of skin discoloration, rare presence of overlying anesthesia (one patient), and presence of subcutaneous gas by radiography in eight (34.8%) patients.[24] The recognition and early diagnosis of necrotizing fasciitis post-C-section or delivery is critical because delay in a few hours in intervention can be fatal.[23] A characteristic feature of most cases that leads to delay in diagnosis is the benign-appearing wound and skin, but patients were systemically toxic or ill looking, often with persistent fever, marked leucocytosis and spreading inflammation of the surrounding tissues. Although radiographic studies,[28,29] ultrasonography, CT scan,[30] or magnetic resonance imaging (MRI)[31] have been used to assist in differentiating cellulitis or ordinary wound infection from necrotizing fasciitis, or bedside biopsy for frozen section,[32] there has been no prospective evaluation to assess differential merits and effect on outcome. In a small case control study of 21 necrotizing fasciitis and paired match controls, multivariate analysis found that a white blood count $>14 \times 10^9$/L, serum sodium <135 mmol/L, and blood uren nitrogen >15 mg/dL separated those with necrotizing fasciitis from those with non-necrotizing infection.[28] This study needs to be repeated in prospective larger trials and include other investigations, such as creatine phosphokinase, venous lactate, and easily obtainable portable imaging (ultrasonography).

Necrotizing fasciitis of any cause carries a high morbidity and significant mortality. The fundamental principles of therapy include prompt administration of broad-spectrum antibiotics, and early immediate surgery with radical debridement of necrotic and devitalized tissue, until normal bleeding tissue is visible.[23] Nutritional support and correction of fluid and electrolyte balance, anemia, and renal impairment are of major importance. The need for immediate aggressive surgery for necrotizing fasciitis has been recognized since the 1950s[33] and reinforced by more recent guidelines.[34] Initial observation showed that the average time from onset of disease to diagnosis and treatment was 4 days for those who lived, and 7 days for those who died.[35] Other subsequent studies found 48 h duration was a more significant time frame, after which the mortality rate was 75%.[27] The critical time for radical surgery after admission to hospital for manifestations of necrotizing fasciitis appears to less than 12 h, as significant morbidity exists if surgery is delayed for >12 h.[36]

4.2.3 Comments on Medico-legal Issues

There is increasing evidence that C-section carries a greater risk to the mother and baby than vaginal delivery. However, in the case presented, C-section was necessary and clinically indicated. The plaintiffs' main accusation of negligence was failure of the obstetrical service and/or anesthesiologist to provide prophylactic antibiotic before or during the C-section.

Current guidelines recommend antibiotic prophylaxis and 1–2 g of intravenous cefazolin after cord clamping, to prevent infection for elective or non-elective (emergency) C-sections.[37] Previous guidelines by the American College of Obstetricians and Gynecologists (ACOG) had recommended in 2003[38] that all high-risk patients undergoing cesarean delivery be given antibiotic prophylaxis (level A evidence). Although the evidence was inconclusive for low-risk patients undergoing C-section, use of antibiotic prophylaxis was also recommended (level C evidence). In a Cochrane review of the topic,[39] 81 studies were analyzed for elective C-section (N = 2,037) and non-elective C-section (N = 2,132). There was a finding of reduction of endometritis by two-thirds to three quarters, and a significant reduction of wound infections (36% for non-elective and 73% for elective C-section).[39] The policy of routine antibiotic prophylaxis for all C-section was also supported by other reviews.[40,41] Thus, a single dose of safe, inexpensive antibiotic is very affective in preventing endometritis and surgical wound infection after C-section. It is most likely that in case 2, the infection ascended from the vagina to the uterine cavity (with endometritis), then spreading to the wound and surrounding tissue causing necrotizing fasciitis. Thus, if prophylactic antibiotics were given, it is more likely than not that the serious puerperal infection would have been prevented. It should be noted that the patient under discussion fulfilled the criteria for high-risk

C-section, which is associated with post-operative infectious complications of 70–90% without antibiotic prophylaxis.

Another area of grievance expressed by the plaintiffs against the defendants was the delay in arriving at the correct diagnosis and implementing appropriate surgery. This delay resulted in spread of the infection causing extensive tissue damage and disfigurement. The patient had increased risk factor for ascending infection from the vagina (prolonged rupture of membranes and labor, multiple vaginal examinations, internal fetal monitoring etc.) and she also presented with the typical manifestation of wound necrotizing fasciitis (severe pain and tenderness, high fever and leucocytosis, with little skin erythema, and marked wound swelling and induration). Thus, the index of suspicion should have been high for this complication.

The defense put forward by the defendants lawyers argue that necrotizing fasciitis is a very rare condition and that none of the physicians attending the patient had personal experience managing such a case post-C-section. If we examine the management course of the patient after the second admission, it is evident that the nature of the infection could have been detected earlier just by following standard guidelines for an infected wound[23] – i.e. surgical opening of the wound for debridement and drainage soon after admission would have led to the proper diagnosis. It is not clear why this simple intervention was not performed soon after admission by the obstetrics and gynecology service.

Did the initial choice of antibiotics play a role in progression of the infection? Although gentamicin and clindamycin (started 24 h after admission) are considered the gold standard for endometritis and synergistic fasciitis,[41] other broad-spectrum monotherapy (piperacillin-tazobactum, ampicillin-sulbactum, ertapenem etc.) would be equally effective. Although cefazolin was initially started on admission, which is a reasonable choice for monomicrobial wound infection, it is doubtful whether earlier broad-spectrum therapy would have changed the course without adequate surgery.

4.3 Case 3: Laparoscopy for Pelvic Adhesions

The family physician (FP) of a 33-year-old female requested a gynecological consultation for the patient's symptoms of chronic lower abdominal pain and painful intercourse. The patient had a previous cesarean section a few years before, and a more recent laparoscopic tubal ligation. Based on the previous laparoscopic findings, the gynecologist attributed the young woman's symptoms to multiple adhesions involving the bladder, intestines, uterus, and lower abdominal wall. Thus, he recommended lysis of the adhesions by laparoscopy to be performed in the small town community hospital. The patient had no significant past medical illness and a pre-operative clinical assessment by the FP a week before the planned procedure was normal.

Three days before the surgery, the young woman called the FP's office complaining of sore throat, difficulty swallowing, and mild cough from trying to clear throat secretions. Apparently, the physician's secretary informed the FP who advised that an office visit was not necessary and no medications were needed before the surgery, but that throat lozenges could be used for symptomatic relief. On the day of surgery, the routine examination performed by the anesthesiologist was reported to be normal. The surgical procedure was uncomplicated and no preoperative antibiotic was administered.

Two days postoperatively, the patient presented to the hospital emergency department with symptoms of severe abdominal pain, vomiting, and fever. She was found to be hypotensive with tachycardia and febrile (38.7°C) with signs of acute peritonitis. Cultures of the peritoneal fluid grew Streptococcus pyogenes, but the blood cultures were negative. Intravenous fluids and broad-spectrum antibiotics were initiated and an emergency laparotomy was performed. Generalized peritonitis with serosanguinous fluid was found, but no evidence of perforated intestines or uterus. She was postoperatively transferred for further management at the ICU of a tertiary care university teaching hospital. Her course was complicated by respiratory failure (due to ARDS), renal failure, liver disturbance, and heart failure, secondary to toxic cardiomyopathy. The patient survived the ordeal, but she required mechanical ventilation for 3 weeks, prolonged hospitalization, and convalescence. Six months later, she was doing fairly well with no residual kidney or heart failure, but she still had problems with dyspareunia and poor sex drive.

4.3.1 Medico-legal Issues

The patient and her husband subsequently instigated medico-legal actions for medical negligence against the FP, gynecologist, and anesthesiologist. Charges against the FP included failure to advise or clinically assess the plaintiff for her sore throat before the surgery, as this directly affected her complications of streptococcal group A peritonitis. Furthermore, the FP should have notified the gynecologist and anesthesiologist, or advised the patient to inform them of her symptoms.

The gynecologist and anesthesiologist were blamed for not taking a clinical history for any intercurrent illness before the surgery and preoperative assessment. Their failure to obtain a history of sore throat represented substandard care, and failure to postpone the elective procedure until resolution of her upper respiratory tract infection was medical negligence that caused a near catastrophic outcome.

Defense counsel for the physicians argued that there was no proof that the patient's sore throat was related to the infectious complication. Moreover, medical expert witness for the defense indicated that most cases of sore throat are due to viruses, and that the anesthesiologist found no evidence of pharyngitis or tonsillitis to suggest a streptococcal infection. In addition, the defense experts argued that the group A *Streptococcus* more likely originated from the colonization of her skin or vagina. Furthermore, it was contended that there are no guidelines for surgeons or anesthesiologists to cancel surgery for patients with mild upper respiratory tract infections.

4.3.2 Medical Aspect

One of the critical issues in this case is the source or origin of the group A *Streptococcus*. Although a throat culture was never done to prove or establish the origin, this is not necessary in civil lawsuits, and if the plaintiff's lawyer can show greater probability than other sources, then this may be accepted by the courts.

Humans are natural hosts of *S. pyogenes* and infection or colonization of other animals is rare and is typically a result of close contact with infected humans. The nasopharynx is the commonest site of carriage, and aerosolized nasopharyngeal secretions are the primary means by which group A streptococci (GAS) are spread among humans.[42] The carriage rates of GAS vary with geographic location, season of the year, and age. In children, the rates of pharyngeal colonization vary from 10% to 20%, being most common in Winter and Spring.[43] In adults the carriage rates are much lower. Skin carriage is usually infrequent except for patients who have skin diseases, such as eczema, psoriasis, and wounds or pyoderma.[44] Thus, direct contact with contaminated skin or mucus membranes is of secondary importance, and contact with contaminated surfaces or fomites or via insects are potential sources, but of minor importance.[42] Food-borne outbreaks of GAS pharyngitis from salads, eggs and cheese prepared by infected or colonized food handlers have been reported.[44] Occasionally mini-outbreaks of GAS wound infections in hospitalized patients have been associated with vaginal or ano-rectal colonization in health care personnel.[43,45,46]

Pharyngitis, or sore throat, is a common condition in adults and even more frequent in children. It was estimated that 18 million patients sought care for sore throats in the United States in 1996, making it the sixth leading cause of visits to physicians.[47] Moreover, four to six times more individuals may have, but not seek care for their sore throats. The majority of acute pharyngitis (about two-thirds) are caused by common respiratory viruses (rhinovirus, coronavirus, adenovirus, etc.), and only about 5–10% in adults and 15–30% in children are caused by GAS.[44,48,49] Other bacteria causing pharyngitis include group G and C ß-hemolytic streptococci, diphtheria, *Arcanobacteria hemolyticum, Neisseria gonorrhea, Mycoplasma pneumoniae,* and *Chlamydia pneumoniae.*[44] However, GAS is the most important bacterial cause of sore throat and probably *Fusobacterium necrophorum* (which probably can cause Lemierre syndrome) in adolescents and young adults.[50]

The spectrum of disease of GAS pharyngitis varies from mild sore throat with resolution of symptoms in 3–5 days, to severe tonsillitis with fever and lymphadenitis. Complications such as peritonsillar abscess, scarlet fever, bacteremia with toxic shock syndrome, glomerulonephritis, and rheumatic fever are uncommon complications, and rheumatic fever is now a rare complication in developed countries. The clinical diagnosis of GAS pharyngitis is considered inaccurate without a throat culture; the ability to predict presence of GAS by physicians is limited, with estimated sensitivity of 55–74% and specificity of 58–76%.[44] The Centor criteria proposed to improve the clinical diagnosis of GAS pharyngitis (moderate to severe cases attending an ER) include tonsillar exudates, tender anterior cervical lymphadenopathy, absence of cough, and history of fever.[51] However, if three or four Centor criteria are met, the positive predictive value is still only 40–60%, and in the absence of these criteria, the negative predictive value is only 80%.[44] Because of the low predictive values of clinical criteria, expert panels recommend throat culture and treatment only of confirmed cases of GAS pharyngitis.[52–56] For patients with typical viral syndromes such as rhinorrhea, cough, myalgias, and sore throat (with or without fever), a throat swab is usually not necessary. Cost-effective analysis of the diagnosis and management of pharyngitis in adults have found that throat culture is the least expensive and most effective strategy when the prevalence of GAS pharyngitis is less than 20%.[57]

4.3.3 Medico-legal Discussion

Based on our current knowledge, it was most likely that the source of the GAS in Case three was from the throat (pharyngitis), as the patient had no skin lesions. Although vaginal colonization may result in endogenous disease with GAS, it usually occurs after delivery, intrauterine device insertion, or gynecologic procedure. Furthermore, there was no cluster of cases of postoperative or post-delivery infection in the institution to suggest a health care carrier as a source of the infection.

What was the duty of the FP when he was notified that the patient complained of sore throat just before surgery? An expert witness for the plaintiff in a similar medical practice criticized the care of the FP as being substandard and negligent. He opined that the physician should have arranged to clinically assess the patient or notified the gynecologist of the patient's symptoms of pharyngitis.

Was either the gynecologist or anesthesiologist at fault? Both these physicians denied knowledge of the patient's sore throat before surgery and did not report any abnormal clinical findings. Hence, it is unlikely that they could be held responsible or be considered medically negligent by the court. The anesthesiologist testified that if he were aware that the patient had pharyngitis, he would have postponed the surgery. However, there are no well-documented guidelines for surgeons or anesthesiologists on this issue.

It had been widely believed that general anesthesia should be avoided in patients with upper respiratory tract infections (URTI), although clinical studies found mixed results. An editorial comment in 1991 stated that the current evidence supported the clinical impression of higher risk for pulmonary complications during anesthesia in subjects with respiratory tract infections.[58] The issue was again raised in 2001 after a study in children concluded that recent and active URTI (within 4 weeks) were at increased risk for adverse respiratory events, but most of the children underwent elective procedures safely without increased morbidity.[59] A later study by the same group of investigators in 713 children undergoing cardiac surgery, found the presence of URTI was predictive of postoperative infection and multiple complications.[60] However, there is no similar data in adults and specifically no recent data of the risk of GAS pharyngitis peri-operatively. Guidelines for prevention of surgical site infection noted increased risk of wound infection with any coexistent infections at a remote site,[61] but did not specifically address the issue of pharyngitis. In a previous study of 2,349 patients with clean surgical wounds, wound infection rate in 208 patients with documented remote infections was 14.8% compared with 6.9% in the 2,141 patients without remote infections ($p > 0.001$).[62] However, in this study pharyngitis was not listed as one of the remote infections.

Thus, anesthesiologists and surgeons may face a dilemma with patients presenting with URTI before surgery, because of the fear of complications and litigation. As stated by editorial views, this decision is left to the physician's best clinical judgment about an individual patient undergoing a specific procedure for a specific duration of time.[63] It is evident the surgical team should weigh the risks versus the benefits of proceeding with surgery. Obviously, this would apply mainly to elective procedures rather than urgent or semi-urgent conditions.

Good judgment, common sense, and proper informed consent with adequate discussion involving the patient and family should take precedence in making the decision to proceed with a specific case. In most cases of elective surgery, the final decision to proceed should be left to the patient (or guardian) once the risks have been explained and well documented.

The emotional and economic impact of delaying elective surgical procedures should be considered in the decision making. Overall, the risk of surgical site infection increases by twofold (in clean surgery) in those with remote infection, compared to controls without infection. Pre-operative treatment of remote infection can reduce the risk of subsequent wound infection by 8.5% in treated versus 25% in those not treated.[62] Peri-operative antibiotics for surgical prophylaxis were not helpful in reducing the risk of wound infection for those with remote site infection. Based on limited experience, the authors conclude that remote site infection should be treated for at least 24 h before surgery.[62]

It may be prudent to automatically delay elective procedures for some conditions in patients with remote site infections. These may include surgical procedures with insertion of prosthesis, as the risk of infection can be catastrophic; or in non-essential cosmetic surgery or minor ailments, or where the benefits of the procedure do not warrant even a small increased risk of infection.

4.4 Summary and Final Comments

Complications of pregnancy and delivery are common medical litigation issues, both for infectious and non-infectious adverse events and outcome. Physicians who perform deliveries (obstetricians or family physicians) have one of the highest medical protective fees in Canada. Physicians have to be extra careful when dealing with the pregnant patient. Even though the majority of pregnancies have no significant medical complications during gestation or after delivery, numerous medical adverse events potentiality can occur.

There are several reasons why medical malpractice litigations rulings often go against the healthcare professionals in favor of the plaintiffs and frequently result in large financial compensations. Members of the jury and judges will often be sympathetic to pregnant women who suffer from an adverse event (human nature) and there may be a tendency to rule in their favor. Moreover, the effects of these adverse outcomes may affect not only the mother but also the offspring. The way to limit being sued for medical malpractice when managing or attending pregnant women involves the same principles of good medical practice that should always be followed. Pay attention to detail, (history, examination, and test results), keep an open mind, always look for the worst complication, but do not overlook simple "minor issues" which may become major issues, and always attend and treat promptly. Whenever there is a potential for a catastrophic outcome, (despite the rarity of that event) have a high index of suspicion and do not hesitate to refer or transfer patients quickly to an acute care hospital or tertiary care center, or request a consultation with a specialist. Too often however, basic principles of treatment are neglected until it is too late – such as not opening and exploring a local infected wound. To a certain extent, we as physicians are too over dependent on modern technology (CT, MRI) and specialist opinion before implementing the basic tenets of proven therapy, which have been established for more than a century.

Whether we are practicing medicine in a large urban center or remote medical-outpost, many simple routine screenings can be performed to limit various infectious complications of pregnancy. Tests such as routine midstream urine (MSU) culture in the second trimester (can repeat in third trimester in those with a history of urinary tract infection), routine vaginal culture in late third trimester for group B *Streptococcus* colonization, or assessment for vaginitis, endocervitis and sexually transmitted disease (including syphilis and human immunodeficiency virus [HIV]) can be very informative.

High-risk patients (especially planned or emergency C-section) need to be dealt with cautiously and expectantly. Obstetricians and family physicians who perform deliveries, as well as anesthesiologists, should be familiar with and follow the guidelines for antibiotic prophylaxis to prevent metritis, wound infection, chorioamnionitis and neonatal sepsis. If for some reason indicated antimicrobial prophylaxis were overlooked, the patient should be informed of the oversight and of the possible complications and their early manifestations, and instructed to seek prompt medical attention at the first signs of such complications. The attending medical team and family physician should also be alerted to these potential complications, and maintain a high index of suspicion, whenever the patient presents with a febrile illness within 2 weeks of delivery.

Physicians should be aware that even a mild sore throat could pose a significant risk for patients about to undergo surgery. The risks of the possible complications versus the benefit of the procedure should be discussed with the patient. The final decision should always be made by the patient without coercion. Although most cases develop no significant complications, some can result in severe or catastrophic outcome, which cannot be predicted.

References

1. Kass EH, (1960), Bacteriuria and pyelonephritis of pregnancy. Arch Intern Med. 105: 194–198.
2. Kass EH, (1960), The role of asymptomatic bacteriuria in the pathogenesis of pyelonephritis. In: Quinn EL, Kass EH (eds); Biology of pyelonephritis. Little, Brown, Boston, p399–412.
3. Elder HA, Kass EH, (1965), Renal function in bacteriuria of pregnancy: its relationship to prematurity, acute pyelonephritis and excessive weight gain. In: Kass EH (ed) Progress in Pyelonephritis, FA Davis, Philadelphia, p81–86.
4. Elder HA, Santamarina BA, Smith S, Kass EH, (1971), The natural history of asymptomatic bacteriuria during pregnancy: the effect of tetracycline on the clinical course and the outcome of pregnancy. Am. J. Obstet. Gynecol. 111: 441–462.
5. Sweet RL, Gibbs (ed), (1985), Urinary tract infection. In: Infectious diseases of the female genital tract. Williams & Wilkins, Baltimore, p293–313.
6. Bryant RE, Windom RE Vineyard JP Jr, Sanford JP, (1964), Asymptomatic bacteriuria in pregnancy and its association with prematurity. J. Lab. Clin. med. 63: 224–231.
7. Kincaid-Smith P, Bullen M, (1965), Bacteriuria in pregnancy. Lancet. 1: 395–399.
8. Little PJ, (1966), The incidence of urinary infection in 5000 pregnant women. Lancet. 2: 925–928.
9. Kass EH, (1970), Pregnancy, pyelonephritis and prematurity. Clin. Obstet. Gynecol. 13: 239–254.
10. Cunningham FG, Morris GB, Mickal A, (1973), Acute pyelonephritis of pregnancy: A clinical review. Obstet. Gynecol. 42: 112–117.
11. Cunningham FG, Leveno KJ, Bloom SL, Hauth JC, Gilstrap III LC, Weinstein KD (eds), (2005), Renal and urinary tract disorders. In: Williams Obstetrics, 22nd Edition, MacGraw-Hill, New York, p1093–1110.
12. Hill JB, Sheffield JS, McIntire DD, Wendel Jr GD, (2005), Acute pyelonephritis in pregnancy. Obstet Gynecol. 105: 18–23.
13. Scott CL, Chavez GF, Atrash HK, Taylor DJ, Shah RS, Rowley D, (1997), Hospitalizations for severe complications of pregnancy, 1987–1992. Obstet. Gynecol. 90: 225–229.
14. Gazmararian JA, Petersen R, Jamieson DJ, Schild L, Adams MM, Deshpande AD, Franks AL, (2002), Hospitalizations during pregnancy among managed care enrollees. Obstet. Gynecol. 100: 94–100.
15. Mabie WC, Barton JR, Sibai B, (1997), Septic shock in pregnancy. Obstet. Gynecol. 90: 553–561.
16. Zeeman GG, Wendel GD Jr, Cunningham FG, (2003), A blueprint for obstetric critical care. Am. J. Obstet. Gynecol. 188: 532–536.
17. Harris RE, (1979), The significance of eradication of bacteriuria during pregnancy. Obstet. Gynecol. 53: 71–73.
18. Gilstrap III LC, Cunningham FG, Whalley PJ, (1981), Acute pyelonephritis in pregnancy: an anterospective study. J. AM. Coll. Obstet. Gynecol. 57: 409–413.
19. Freeman CD, Nicolau DP, Belliveau PP, Nightingale CH, (1997), Once-daily dosing of aminoglycosides: review and recommendations for clinical practice. J. Antimicrob. Chemother. 39: 677–686.
20. Cunningham EF, Levens KJ, Bloom SL, Hauth JC, Gilstrap LC III, Wenstrom KD (eds), (2005), Parturition In: Williams Obstetrics, 22nd Edition, McGraw-Hill, New York, p131–186.
21. Cunningham EF, Levens KJ, Bloom SL, Hauth JC, Gilstrap LC III, Wenstrom KD (eds), (2005), Puerperal Infection. In: Williams Obstetrics, 22nd edition, McGraw-Hill, New York, p711–724.
22. Gonçalves LF, Chaiworapongsa T, Romero R, (2002), Intrauterine infection and prematurity. Mental Retard Develop. Disabilities Res. Rev. 8: 3–13.
23. Owen J, Andrews WW, (1994), Wound complications after cesarean sections. Clin. Obstet. Gynecol. 37: 842–855.
24. Gallup DG, Freedman MA, Meguiar RV, Freedman SN, Nolan TE, (2002), Necrotizing fasciitis in gynecologic and obstetric patients: a surgical emergency. Am. J. Obstet. Gynecol. 187: 305–311.

25. Goepfert AR, Guinn DA, Andrews WW, Hauth JC, (1997), Necrotizing fasciitis after cesarean delivery. Obstet. Gynecol. 89: 409–412.
26. Schorge JO, Granter SR, Lerner LH, Feldman S, (1998), Postpartum and vulvar necrotizing fasciitis. Early clinical diagnosis and histopathologic correlation. J. Reprod. Med. 43: 586–590.
27. Stephenson H, Dotters DJ, Katz V, Droegemueller W, (1992), Necrotizing fasciitis of the vulva. Am. J. Obstet Gynecol. 166: 1324–1327.
28. Wall DB, Klein SR, Black S, de Virgilio C, (2000), A simple model to help distinguish necrotizing fasciitis from nonnecrotizing soft tissue infection. J. Am. Coll. Surg. 191: 227–231.
29. Fisher JR, Conway MJ, Takeshita RT, Sandoval MR, (1979), Necrotizing fasciitis. Importance of roentgenographic studies for soft-tissue gas. JAMA. 241: 803–806.
30. Rogers JM, Gibson JV, Farrar WE, Schabel SI, (1984), Usefulness of computerized tomography in evaluating necrotizing fasciitis. South. Med. J. 77: 782–783.
31. Schmid MR, Kossmann T, Duewell S, (1998), Differentiation of necrotizing fasciitis and cellulitis using MR imaging. Am. J. Roentgenol. 170: 615–620.
32. Majeski J, Majeski E, (1997), Necrotizing fasciitis: improved survival with early recognition by tissue biopsy and aggressive surgical treatment. South. Med. J. 90: 1065–1068.
33. Wilson B, (1952), Necrotizing fasciitis. Am. Surg. 18: 416–431.
34. Thompson CD, Brekken AL, Kutteh WH, (1993), Necrotizing fasciitis: a review of management guidelines in a large obstetrics and gynecology teaching hospital. Infect. Dis. Obstet. Gynecol. 1: 16–22.
35. Rea WJ, Wyrick WJ Jr, (1970), Necrotizing fasciitis. Ann. Surg. 172: 957–964.
36. Sudarsky LA, Laschinger JC, Coppa GF, Spencer FC, (1987), Improved results from a standardized approach in treating patients with necrotizing fasciitis. Ann. Surg. 206: 661–665.
37. Treatment Guidelines from The Medical Letter, (2006), Antimicrobial prophylaxis for surgery. Med. Letter; 4: 83–88.
38. American College of Obstetricians and Gynecologist, (2003), Prophylactic antibiotics in labor and delivery. AGOG practice bulletin; no47. National Guideline Clearing house. www.guideline.gov
39. Smaill F, Hofmeyr GJ, (2000), Antibiotic prophylaxis for cesarean section. Cochrane Database Syst. Rev. CD000933
40. Chelmow D, Ruehli MS, Huang E, (2001), Prophylactic use of antibiotics for nonlaboring patients undergoing cesarean delivery with intact membranes: a meta-analysis. Am. J. Obstet. Gynecol. 184: 656–661.
41. French L, (2003), Prevention and treatment of postpartum endometritis. Curr. Womens Health Rep. 3: 274–279.
42. Vosti KL (1982). Streptococcal diseases. In: Hoeprich PD (ed.) Infectious Diseases, 2nd Edition, Harper & Row, Hagerstown, MD: p 235–246.
43. Bisno AL, Stevens DC (2005). *Streptococcus pyogenes*. IN: Mandell GL, Bennett JE, Dolin R (eds). Principles and Practice of Infectious Diseases, 6th Edition, Churchill Livingstone, Philadelphia PA, p2362–2399.
44. Assanasen S, Bearman GML (2007). Group A Streptococcal diseases. IN: Wallace RB, Kohatsu N, Last JM (eds) Wallace/Maxcy-Rossenau-Last; Public Health & Preventive Medicine, 15th Edition, McGraw Hill, New York, p237–245.
45. McIntyre DM (1968). An epidemic of *Streptococcus pyogenes* puerperal and postoperative sepsis with an unusual carrier site – the anus. *Am J Obstet Gynecol* 103: 308–310
46. Grysha PF, O'Dea AE (1990). Postoperative streptococcal wound infection. *JAMA* 213: 1189–1191.
47. Schappert SM (1998). Ambulatory Care Visits To Physician Offices, Hospital Outpatient Departments, And Emergency Departments: United States, 1996. National Center for Health Statistics, Hyattsville, MD.
48. Bourbeau PP (2003). Role of the microbiology laboratory in diagnosis and management of pharyngitis. *J Clin Microbiol* 41: 3467–3472.

49. Huovinen P, Lahtonen R, Ziegler T, Meurman O, Hakkarainen K, Miettinen A, Artila P, Eskola J, Saikku P (1989). Pharyngitis in adults: the presence and coexistence of viruses and bacterial organisms. *Ann Intern Med* 110: 612–616.
50. Centor, RM (2009). Expand the pharyngitis paradigm for adolescents and young adults. *Ann Intern Med* 151: 812–815.
51. Centor RM, Witherspoon JM, Dalton HP, Brody CE, Link K (1981). The diagnosis of strep throat in adults in the emergency room. *Med Decis Making* 1: 239–246.
52. Bisno AL, Peter G, Kaplan EL (2002). Diagnosing of strep throat in adults: are clinical criteria really good enough? *Clin Infect Dis* 35: 126–129.
53. Bisno AL (2003). Diagnosing strep throat in the adult: do clinical criteria really suffice? *Ann Intern Med* 39: 150–151.
54. McIsaac WJ, Kellner JD, Aufricht P, Vanjaka A, Low DE (2004). Empirical validation of guidelines for the management of pharyngitis in children and adults. *JAMA* 291: 1587–1595.
55. Bisno AL, Gerber MA, Gwaltney JM Jr, Kaplan EL, Schwartz RH (2002). Practice guidelines for the management of group A streptococcal pharyngitis. Infectious Diseases Society of America. *Clin Infect Dis* 35: 113–125.
56. Dajani A, Taubert K, Ferrieri P, Peter G, Shulman ST. Committee on Rheumatic Fever, Endocarditis, & Kawasaki Disease of the Council on Cardiovascular Disease in the Young & American Heart Association (1995). Treatment of acute streptococcal pharyngitis and prevention of rheumatic fever: a statement for health professionals. *Pediatrics* 96: 758–764.
57. Neuner JM, Hamel MB, Phillips R, Bona K, Aronson MD (2003). Diagnosis and management of adults with pharyngitis: a cost-effectiveness analysis. Ann Intern Med 139: 113–122.
58. Jacoby DB, Hirshman CA (1991). General anesthesia in patients with viral respiratory infections: an unsound sleep? *J Anesthesiol* 74: 969–972.
59. Tait AR, Malvuja S, Voepel-Lewis T, Munro HM, Stewart M, Pandit UA (2001). Risk factors for perioperative adverse events in respiratory tract infections. *Anesthesiology* 95: 291–306.
60. Malvuja S, Voepel-Lewis T, Stewart M, Pandit U, Riegger LQ, Tait AR (2003). Risk factors for adverse postoperative outcomes in children presenting for cardiac surgery with upper respiratory tract infections. *Anesthesiology* 98: 628–632.
61. Mangram AJ, Horan TC, Pearson ML, Silver LC, Jarvis WR. The Hospital Infection Control Practices Advisory Committee (1999). Guidelines for prevention of surgical site infection. *Infect Control Hosp Epidemiol* 20: 247–270.
62. Valentine RJ, Weigelt JA, Dryer R, Rodgers C (1986). Effect of remote infections on clean wound infection rates. *Am J Infect Control* 14: 64–67.
63. Cotë C (2001). The upper respiratory tract infection (URI) dilemma: fear of a complication and litigation? Anesthesiology 95: 283–285.

Part III
Assortment of Medical Liabilities

Chapter 5
General Practitioner Liabilities

5.1 Case 1: Young Man with Anemia and Low Grade Fever

A 25-year-old male presented to his family physician (FP) with a history of general malaise and fatigue of 6 weeks in duration. He had felt intermittently chilly on occasions and documented a low-grade fever of 38°C once or twice. The patient's past medical history was not significant, and the physician's note provided no details of the physical examination. Routine complete blood count and biochemical tests (glucose, creatinine, liver enzymes, and thyroid stimulating hormone [TSH]) were performed, and were normal except for a mild normocytic anemia of 105 g/L. The young man was therefore referred to a hematologist without any further investigations.

About 2 months after the onset of illness, the patient was assessed by the consultant (hematologist) in his office. A detailed history and physical examination were performed without any significant localizing signs noted. A non-diagnostic bone marrow aspiration was performed, and on the follow up visit, the patient complained of a focal area of tenderness on the sole of his foot. Examination by the hematologist noted a small patch of non-blanching, flat erythematos, and tender area on the sole of his foot. No further investigations or diagnosis was contemplated, but the patient was advised to follow up with his FP. The diagnosis remained mild anemia of unknown cause. Over the next few months, the patient had follow up visit with his FP with no resolution of his symptoms, nor further evolution of signs to make a firm diagnosis. However, there was no record of any detailed examination or investigations besides repeat blood counts showing persistent anemia. About 5 months after onset of illness, the young man checked into a community hospital emergency department. The presenting complaints were chills, generalized weakness, joint pains (but no swelling), shortness of breath on exertion, poor appetite, and some weight loss of 10 lb. The physical examination by the emergency physician noted an unwell, pale looking young male, with a temperature of 38°C, pulse 100/min, blood pressure 110/70 and grade II-III apical systolic murmur. Blood tests and chest-radiograph were performed.

I.W. Fong, *Medico-Legal Issues in Infectious Diseases*, Emerging Infectious Diseases of the 21st Century, DOI 10.1007/978-1-4419-8053-3_5,

While waiting in the emergency department for results of the tests, the young man suddenly collapsed and had a cardiac arrest. Attempts at resuscitation were unsuccessful and the patient died. An autopsy was subsequently performed by the local coroner.

5.1.1 Medical Issues

Symptoms of fatigue and tiredness are very common in the population and are reported in up to 40% of individuals in any given survey. Most of these symptoms are transitory and may be functional or job-stress related without any serious underlying illness. An increasing, but small group of patients (greater in females), have more debilitating fatigue, malaise, history of low-grade fever, and inability to work for several months to years. The great majority with normal examination and numerous blood tests and other investigations, are lumped into the group of patients with Chronic Fatigue Syndrome, depression or somatization disorders. These patients, however, do not have anemia, documented fever or any positive physical finding except tender pressure points consisted with fibromyalgia (in those with generalized aches and pain). Nearly all of these subjects will have a normal erythrocyte sedimentation rate (ESR).

The subject in this report has objective abnormality with a documented anemia, documented fever (as recorded by the patient), and a localized non-blanching, non-palpable, tender lesion of his foot (the only physical sign observed before the emergency visit). The investigations by the hematologist excluded iron deficiency and vitamin deficiencies (vitamin B_{12} or folic acid), as well as acute leukemias. However, several chronic inflammatory and infectious conditions can present with a normocytic anemia, malaise, and low-grade fever. The anemia in these diseases can also mimic iron deficiency with mild microcytosis, low serum iron and iron saturations, but the iron binding capacity is also low, and the serum ferritin is usually high. Moreover, the bone marrow aspirate shows normal or increased iron stores.

The differential diagnosis in these situations is very similar to that of fever of unknown origin, and can be classified as infectious, neoplastic, collagen vascular and other inflammatory disorders.

A few non-rheumatological inflammatory diseases can present with mild anemia, low-grade fever, and malaise. However, other manifestations are usually present such as diarrhea in inflammatory bowel diseases, liver enzyme abnormality in chronic autoimmune hepatitis or biliary cirrhosis, and pulmonary manifestations, as in sarcoidosis. On rare occasion, sarcoidosis can present with fever of unknown origin without pulmonary abnormality, but increased alkaline phosphatase is usually detected due to granulomatous hepatitis.

Collagen vascular disorders may present with non-specific malaise, low grade or high grade documented fever, and anemia. Thus, it is usual practice to investigate for these disorders with a panel of appropriate blood tests, despite the absence of signs of arthritis, or systemic manifestations of skin, lung, sinus, kidney, or nervous system. Although the rheumatoid factor and antinuclear antibodies can be nonspecific, high titers would be suggestive of collagen vascular disorders. Juvenile rheumatoid arthritis, although more commonly seen in women, can present similarly to this case before eventually manifesting signs of inflammatory arthritis or transitory macular-erythematous rash. This disease cannot be confirmed by any serological test, and is a disease of exclusion with an appropriate clinical complex.

The major neoplastic disease that would present in this fashion in a young man without objective signs or familial risk factors would be occult lymphoma; without obvious peripheral enlarged notes. Most of these patients will have well documented, intermittent, or daily fever. However, despite not fulfilling the defined criteria of "fever of unknown origin," imaging studies such as computerized tomography (CT) of the chest and abdomen would have been indicated when infections were excluded. In these cases, enlarged mediastinal or retroperitoneal lymph nodes (with or without splenomegaly) would provide the clue suggestive of lymphoma that eventually should lead to biopsy to confirm the diagnosis.

What infectious diseases should the FP consider for exclusion at the initial presentation? Although the FP may not be knowledgeable about some of the unusual infections, they should be familiar with common and important disease that can present in this manner. After negative investigations without any leads to the diagnosis, then a referral to an internist would be appropriate. There are several fairly common infections that should be considered for investigations, despite absence of persistent documented fever. A detailed history on travel within the past year, animal contacts, consumption of exotic and unpasteurized dairy products, hobbies such as hunting and taxidermy, abuse of recreational drugs, medications, contacts with ill subjects (such as tuberculosis), social history (homeless or use of shelters), and sexual habits should be obtained. A detailed complete physical examination should be well documented in the patients' chart or file. The majority of FP may find these recommendations too time consuming (because of a busy practice) and often perform abbreviated history and examinations. In this situation, the basic tests that should be performed (besides complete blood count and routine biochemical tests) include: blood cultures (two sets separated by hours), urinalysis, monospot test, cytomegalovirus (CMV) IgG and IgM, HIV serology, toxoplasma IgM, chest radiograph, and mantoux skin test. It is often useful for screening purposes to perform an ESR and C-reactive protein (CRP), as these tests are very sensitive (but non-specific) in detecting inflammatory and serious infections. Although the four main conditions causing the mononuclear syndrome (infectious mononucleosis, acute CMV, acute HIV, and toxoplasmosis) usually show lymphocytosis and many atypical lymphocytes on the differential white blood count, these changes are not always present– especially in older subjects. In general, the causes of the mononuclear syndrome produce symptoms that are self-limited and usually resolve by 4–6 weeks, except on some occasions can produce a prolonged post-infectious fatigue or asthenia.

5.1.2 Autopsy Report

The autopsy performed by the coroner found a mildly enlarged heart with evidence of mitral valve prolapse with some destruction, inflammatory reaction, and vegetation indicative of sub-acute bacterial endocarditis. The lungs were only mildly congested (chest radiograph in the emergency department revealed no pulmonary edema). There were no abscesses extending down to the conduction system, but there were focal areas of inflammation scattered throughout the myocardium. The brain revealed early hypoxic changes (secondary to the cardiac arrest) and one or two minor infarcts in non-critical areas. Blood cultures taken in the emergency department and cultures of the mitral valve vegetation grew Streptococcus viridans complex. The conclusion of the coroner was that the young man died from unrecognized sub-acute chronic bacterial endocarditis (SBE), but the immediate cause of his cardiac arrest was unclear. It was postulated that he might have suffered from transitory arrythmia as a result of focal bacterial myocarditis.

The complete blood count before death demonstrated a hemoglobin of 90 g/L, WBC of 12,000/dL, and platelets of 580,000/dL.

5.1.3 Medico-legal Issues

The parents' of the young man (plaintiffs) launched litigation against the family physician and the hematologist, claiming negligence in not making the diagnosis, not performing adequate detailed examination of the heart, and not performing blood cultures that would have led to the diagnosis.

The lawyer representing the FP, in defense, stated the patient had no history of heart disease for SBE to be suspected, and that this is in fact, a rare condition with an unusual presentation. Thus, the FP met the expected standard of care. Furthermore, his blood test results indicated a primarily hematological disorder and the deceased was appropriately referred to a hematologist. The defense of the hematologist stated that the physician performed an adequately detailed history and physical examination and did not detect any signs to suggest bacterial endocarditis (including auscultation of the heart). Furthermore, he was primarily referred to investigate and make an appropriate diagnosis and recommend treatment for the mild anemia. Moreover, the plaintiff (hematologist) stated that SBE is primarily a disease seen and managed by cardiologist or infectious disease specialist, and thus out of the realm of his specialty.

5.1.4 Comments on Medico-legal Issues

I will first deal with the family physician's defense and accepted standard of care. Although, bacterial endocarditis is relatively rare in the community, all physicians during their medical training should be familiar with the clinical manifestations or presentation and method of diagnosis. The fact that the FP may not have seen or cared for a patient with SBE is not an acceptable medical excuse or legal defense. Furthermore, the clinical presentation of the young man is quite typical for SBE, despite the absence of known underlying heart disease. Recent prospective observational studies have found that 47% of patients with bacterial endocarditis have no known previous heart disease.[1] In addition, in developed countries, the majority of patients who develop community-acquired SBE (besides intravenous drug abuser [IVDA]) have subclinical mitral valve prolapse with regurgitation (especially in young adults), or degenerative valvular heart disease in older persons (aortic valve sclerosis or calcified mitral valve annulus).[2]

A major criticism by the plaintiffs' lawyer of the FP's care was that his chart had no documentation of a detailed history or physical examination. It was not noted that a cardiac examination was performed, thus the FP could not have detected any underlying heart disease. There was no evidence that an auscultation of the heart was ever performed by the FP. The transcripts from the examination of the FP revealed that he had no recollection of the initial assessment of the patient, but he may have examined the patient's heart without making any notations. This is a common excuse or theme that I have seen in many medical malpractice suits, where physicians claim in defense that they routinely perform detailed history and physical examinations without documentation of their findings. Most of the instances involve busy medical practices, such as general practitioners office or emergency department. This is a deplorable practice if it were true, and would be a poor legal defense in most cases. In fact, it only suggests an excuse for not performing an adequate clinical assessment, and puts into question the credibility of the physician's testimony. This issue of poor documentation of clinical assessments only provides fodder for litigation lawyers and cannot be overemphasized in its importance.

In his defense, the FP also stated that the blood tests did not reveal evidence to suggest an infectious disease, such as the presence of leucocytosis. As a result, he did not pursue investigations to exclude SBE (i.e. blood cultures). It has been documented in textbooks of medicine for several decades that significant leucocytosis is uncommon in SBE (20–30%), more common with acute endocarditis with *Staphylococcus aureus*; whilst anemia is very common (70–90%), and is higher in the sub-acute form.[3] As previously mentioned, elevated ESR (>90%) and CRP (>90%) are extremely sensitive, but non-specific tests for bacterial endocarditis.[3]

The claims of negligence filed against the hematologist stated that he should have been aware that a chronic (sub-acute) infection or inflammatory condition can produce anemia (anemia of chronic disease), and therefore, he should have performed the appropriate investigations for these conditions. The lawyer for the plaintiffs further charge that a hematologist is first trained in internal medicine before becoming a specialist, and that as a consultant, he should be aware of the differential diagnoses of anemia and ways to distinguish the various conditions. It was further stated, that if the hematologist had performed the appropriate investigations, or referred the patient to an internist or another specialist, the diagnosis could have been made at least 2 months before the young man's death. Thus, appropriate treatment at that time or shortly thereafter would have prevented the adverse outcome.

As a consultant, a specialist is expected to be familiar not only with diseases within his or her own specialty but also with other conditions that can mimic these diseases. It is not sufficient for a consultant to state that the patient does not have a certain condition (which was the reason for the referral) without making suggestions or investigations for other mimics outside his or her specialty. Thus, the courts can hold a physician responsible for missing the diagnosis of disease outside his specialty, if these conditions were expected to be diagnosable by his peers in the same specialty.

5.1.5 Medical Aspects

Anemia of chronic disease is a hypo-proliferative anemia that develops in response to systemic illness or inflammation. It is the second most prevalent cause of anemia after iron deficiency, and is the most common among patients with chronic illness.[4] Various clinical conditions can lead to anemia of chronic disease, including infections, autoimmune or rheumatological disease, and cancer. In a recent review,[4] it was stated that anemia of chronic disease is under-recognized and under-treated by physicians. Patients with anemia of chronic disease typically present with a mild (>100 g/L) or moderate (85–100 g/L) reductions of hemoglobin concentration, but occasionally can be more severe.

The pathogenesis of anemia of chronic disease is more complex than was initially thought. Also termed "anemia of inflammation," it is related to acute or chronic immune activation by infectious agents, auto-immunity, or neoplastic cells. These inciting antigens operate through a common pathway by activation of T cells (CD_{3+}) and monocytes, producing pro-inflammatory cytokines (tumor necrosis factor α [TNF-α], interleukin (1L)-1, 1L-6, interferon [IFN-]-γ), and counter-regulatory cytokine 1L-10.[4] The cytokines act at these four different steps to reduce erythrocyte or hemoglobin production: (1) dysregulation of iron homeostasis, (2) impaired proliferation of erythroid progenitor cells, (3) decrease erythropoietin response, and (4) decreased erythrocyte survival by augmentation of erythrophagocytosis by reticuloendothelial cells.[4,5] Hepcidin, an acute phase reactant produced by the liver in response to 1L-6 and endotoxin, appears to play a key role in iron dysregulation in chronic disease. Hepcidin inhibits iron absorption from the duodenum and by binding to ferroportin, leads to internalization and sequestration of iron within the macrophages, and limits iron availability to erythoid precursors.[4] This sequestration of iron within macrophages limits the availability of free iron to the reticuloendothelial system, and may be an adaptive physiologic response to restrict essential nutrients required for the growth of many microorganism and malignant cells.[5,6]

TNF-α and IFN-γ also inhibit the production of erythropoietin by the kidney and erythropoietin-stimulated hematopoietic proliferation is in turn reduced. There is also direct inhibition of proliferation of erythroid precursors by TNF-α, 1L-1 and IFN-γ.[5,6] The suppressed response of erythroid pregenitor cells to erythropoietin appear to be directly related to the severity of the chronic disease and the amount of circulating cytokines, and so much higher levels of erythropoietin are required to restore erythropoiesis.[7] The pro-inflammatory cytokines diminish the response of erythroid precursor cells to erythropoietin also by down-regulation of erythropoietin receptors.

Thus, the combination of limited availability of free iron to erythroid progenitor cells, decreased biological activity, and concentration of erythropoietin, and erythrocytes damage by cytokines, and free radicals with increased erythrophagocytosis resulting in shortened erythrocyte half-life, all combine to produce anemia.[5,6]

Acute infections and inflammatory condition can also produce anemia by a similar mechanism as chronic disease.[8] However, the prevalence of anemia is less and the reduction in hemoglobin is usually lower. Thus, duration and chronicity of illness seem to play some role.

In the present case under discussion, the anemia was associated with ongoing SBE, and the localized tender erythematous lesion on the sole of his foot was probably an area of cutaneous infarction from minor emboli from the cardiac vegetations. Thus, this sign was also a clue to the diagnosis, but similar lesions can also be present in collagen vascular disorders with vasculitis. Hence, the hematologist should have done investigations to exclude these conditions or refer the patient to an appropriate specialist.

5.2 Case 2: Middle-Aged Male with Bilateral Painful Feet

A 50-year-old businessman presented to his FP with gradual onset of bilateral heel pain, worst on the left side. The patient was previously well with no significant past medical illness. There was no history of previous injuries, arthritic conditions, nor fever or chills. The pain had been developing over the past 2–4 weeks, and was worst in the morning after getting out of bed, and also later in the day. The patient was constantly on his feet and did a great deal of walking around. The physical examination revealed an obese male with no evidence of swelling or redness of his feet. There was localized point of tenderness on the medial side of the inferior calcaneus, most marked on the left foot. Plain radiograph of the feet was performed which showed a calcaneus spur on the left heel. A diagnosis of plantar fasciitis was made.

The physician started treatment with a local steroid injection (mixed with lidocaine) directly in the area of focal tenderness of the heels. The patient was seen a month later for follow-up and reported marked improvement, but in the past week the pain was recurrent in the left heel. Over the subsequent 12 months, the middle-aged man received ten local steroid injections in the left heel, each time with temporary improvement but then progressive worsening. He was subsequently referred to an orthopedic surgeon for consultation. Physical examination by the consultant revealed mild swelling of the left heel with marked tenderness of the medial and plantar surface on deep palpation. By this time, the patient had difficulty bearing weight on his left foot for more than 20 min. A repeat radiograph of the heel showed osteopenia, some soft tissue swelling with areas of sclerosis surrounded by lucency, and periosteal reaction of the entire posterior-medial calcaneus. A diagnosis of chronic osteomyelitis was made, requiring surgical debridement, intravenous antibiotics for 2 weeks, followed by oral anti-staphylococcal antibiotics for a total of 3 months. The patient was kept off his feet for about 2 weeks, and then wore a walking cast for another 4 weeks. Eventually the infection was cured.

5.2.1 Medico-legal Issues

The plaintiff's (patient) lawyer filed litigation for medical malpractice against the FP. The accusation of medical malpractice was based on four issues: (1) the physician was negligent in not informing the plaintiff of potential side effects of the steroid injections, thus a valid informed consent (even verbally) was not obtained to give the injections, (2) the FP was negligent in not recommending standard conservative therapy before instituting steroid injections, (3) the steroid injections were the direct cause of the infection because the physician did not use proper aseptic technique, and (4) the FP was negligent in not referring the patient to a specialist earlier when his symptoms persisted or recurred.

The plaintiff sued the FP for compensation of pain and suffering, loss of income from his inability to properly conduct his business.

5.2.2 Medical Issues

Heel pain or calcaneodynia is a common clinical complaint by patients, usually first seen and managed by general practitioners. Thus, it is expected that FPs should be quite familiar with the differential diagnoses and instituting the appropriate therapy. Although there are several different causes of heel pain, the most common conditions are plantar fasciitis, archilles tendinosis, nerve entrapment, and referred pain from arthritis, gout etc.[9]

Plantar fasciitis or plantar heel pain syndrome occurs in about 10% of runners and a similar proportion of the general population at sometime in their life.[10] The peak incidence is between 40 and 60 years of age, but younger persons who are runners, ballet dancers, and aerobic exercise dancers often present at an earlier age. Risk factors for this common condition include obesity, excessive pronation of the foot (pes planus), high arched foot (pes cavus), prolonged standing and walking on hard surfaces, running on hard surfaces, and faulty shoes.[11]

Pain originates at the plantar fascia attachment to the medial tuberosity of the calcaneus. Plantar fasciitis is believed to be secondary to repetitive micro-trauma of the tissue (at the attachment). Pathology will usually demonstrate degeneration of fibrous tissue and chronic inflammation, with or without fibroblast proliferation.[11] In most cases, the diagnosis can be made by the history and physical examination. The pain or discomfort in the heel usually starts off gradually, with onset first thing in the morning on getting out of bed and standing/walking, or after rising from the sitting position. The pain lessens with weight bearing during the day, but becomes worse with continued activity. The pain is also worse on walking barefoot or going upstairs. Examination of the foot is usually normal except for a localized tenderness at the inferior, medial calcaneus over the tuberosity. Plain radiographs are useful mainly to exclude a stress fracture, and the presence of a calcaneal spur is non-specific and not useful for diagnosis. Technetium bone scan typically shows a focal area of increased uptake over the medial tuberosity of the calcaneus, and ultrasound can demonstrate a thickening of the fascia and edema at the attachment. MRI is very sensitive and specific, but is rarely required except in atypical cases.[11,12]

The treatment of plantar fasciitis should be conservative, symptom based, and aimed for source control, as symptoms will resolve in over 80% within 12 months with this approach.[11] Activities that cause or aggravate symptoms should be reduced or avoided. Weight loss for obese patients should be helpful but has not been systematically studied. Symptomatic therapies often used include local ice for acute pain followed by local heating, massage and stretching of the plantar fascia and calf muscles. Non-steroidal anti-inflammatory agents (NSAIDS) are used in short courses to control pain and discomfort. Various maneuvers have been used such as foot strapping, night foot splints to maintain neutral position of the foot, orthotics to provide arch support, and ultrasound therapy, but their value have not been established.[11,13] Use of steroid injection is often of temporary benefit, and casting and surgery (plantar fasciitomy) is a last resort for those failing conservative therapy.[14] However, most interventions used to manage plantar fasciitis have not been studied adequately, and studies do not support the effectiveness of any one treatment. Conservative therapy such as shoe inserts and exercise plus NSAIDS should be (and avoid aggravating habits) the initial therapeutic approach.[11,14]

Corticosteroid injections may provide short term benefits, but do not improve long term outcomes and can be associated with rupture of the plantar fascia[15] and secondary infection. In a recent Cochrane review of interventions for treating plantar heel pain or plantar fasciitis, 19 randomized trials involving 1,626 subjects were analyzed.[16] The conclusion was that present treatments had marginal benefit over no treatment or control therapies such as stretching exercises. Steroid injections seem to be useful in the short term, but only to a small degree. There was limited evidence that heel pads and stretching exercises were associated with better outcomes than custom-made orthoses for people who stand for more than 8 h per day.[16] The complications of steroid injections into soft tissue, bursa, or joints are very low, but best documented in the rheumatology literature, as it is most commonly used in arthritis. It is estimated that iatrogenic infection is (surprisingly rare) <1/10,000[17] injections, but is more common with severe rheumatoid arthritis, in 1/2,000–1/10,000.[18] Other complications include atrophy of soft tissue, local nerve damage, tendon (or fascia) rupture, avascular necrosis of bone, and systemic steroid absorption with rare adrenal insufficiency after withdrawal from prolong repeated injections.[17] A few cases of fatal *Staphylococcus aureus* sepsis have also been described after intramuscular or intra-articular steroid injections.[19]

In a microbiological study assessing two different methods of skin preparation (alcohol swipe for a few seconds vs. chlorhexidine in alcohol cleaning for 1 min), the recovery rate of bacteria from the injecting needles were compared.[20] The rate of bacterial recovery was 5/35 (14%) from proper aseptic technique and 8/29 (27%) with the alcohol swipe (almost double the rate).[20] The investigators reported no statistical difference, but this is likely secondary to low sample size or inappropriate statistical methods (Chi-square analysis). If the two methods were compared by Fisher's exact test (which is more appropriate for the sample size), then the results would be significantly different ($p = <0.01$). The current recommendations for skin preparation for arthrocentesis or steroid injections by a rheumatology textbook is an aseptic technique using two swipes with iodine followed by cleaning with alcohol.[17]

5.2.3 Comments on Medico-legal Issues

It is a wise principle for physicians to always discuss the benefits and risk of any interventions with their patient. This is most important when embarking on invasive therapy, potentially harmful treatment, or where the management is not of established value, but maybe useful. Based on the examination for discovery of the plaintiff, there was no information or discussion by the defendant on the merits and risks of the steroid injections. Moreover, multiple repeated steroid injections into the patient's foot would result in greater accumulative risk of complications without any long-term value on the underlying condition. Although the complications of a single steroid injection are very low, multiple injections would result in added risk each time.

The other accusation against the physician was his failure to properly "sterilize" the area before injecting the steroid. Thus, his failure to take proper aseptic precautions resulted in introduction of bacteria into the plaintiff's foot and led to chronic osteomyelitis of the calcaneus. Statements from the examination of discovery from both the defendant and plaintiff indicated that the physician only used alcohol swipe for skin preparation before the steroid injections. Although the data from the rheumatologic literature indicate that infections are rare, even with alcohol swipe before arthrocentesis or intra-articular injections, most of the data is collected from knee joints. It is quite likely that bacterial colonization of the foot, especially the plantar surface, would be greater in concentration and more diverse. Hence, an alcohol swipe would be very ineffective in decolonizing the sole of the foot. It is most surprising that corticosteroid injections into joints and soft tissue are not more frequently complicated by local infections. Based on the microbiological data of bacterial recovery from sterile needles after skin preparation, there appears to be a 14–27% risk of introducing bacteria.[20] Although it is possible that a very low concentration of skin bacteria could be cleared by the innate immune mechanism, the corticosteroid itself, in high local concentration would counteract this effect and compromise bacterial clearance. Since most bacteria such as coagulase-negative *Staphylococci* and diphtheroids account for the most of the normal skin flora, they are not virulent or pathogenic except in the presence of a foreign body or prosthesis. Another possible explanation for the low incidence of infection after steroid injection is the presence of lidocaine (xylocaine), which is often mixed into the solution to provide local anesthesia. Multiple local anesthetics at concentrations typically used in the clinical setting (lidocaine 1–3%) inhibit the growth of numerous bacteria and fungi under various condition.[21]

In summary, although the FP made a reasonable and valid diagnosis, the subsequent management of the patient's condition fell below the standard of care. The physician failed to discuss the potential complications of steroid injections, he did not initiate appropriate conservative standard treatment first, he should have referred the patient to a specialist after the first or second steroid injection, and he definitely should not have pursued repeated corticosteroid injections. It is also very likely that if the physician had followed standard treatment recommendations for plantar fasciitis, that the patient's symptoms would have resolved and he would not have developed chronic osteomyelitis of the heel.

5.3 Case 3: A Young Healthy Male with Pain in His Upper Thigh

While traveling in Europe, a 30-year-old male slipped but did not fall, and over the next 24 h noticed a nagging pain in his right upper thigh. This was associated with some chills, fever, and sweating which lasted for a few days. The subject sought attention after the sixth day of onset of his symptoms at an emergency department in a hospital in Paris. The attending emergency physician noted that the subject was a healthy, athletic young man (jogging at least once a week for several kilometers), with no significant past illness, nor on any medications. The recorded vital signs, including temperature were normal, and the only physical abnormality was the presence of mild swelling and tenderness on palpation of the right upper anterior thigh. A complete blood count revealed a hemoglobin level of 128 g/L, a leucocyte count of 14,500 cells/μL, with 85% neutrophils, platelet count of 430,000/μL ESR of 44/h and normal coagulation screen.

An ultrasound of the right thigh revealed an echogenic complex cystic mass of about 5 cm by 6 cm in diameter. The Parisian physician advised the patient to have surgical or percutaneous drainage of the fluid filled mass on that day. However, the patient deferred the drainage, as he decided that he would return home and have definitive treatment in Canada.

The patient was subsequently evaluated at an emergency department of an urban tertiary care university teaching hospital in his resident city. A note of the Parisian hospital's investigations and findings with recommendations was provided to the emergency physician. This assessment occurred about 8 days after onset of his symptoms, and there was increasing pain and fever the day before. The physical examination findings were temperature of 37.2°C, pulse of 107/min and the right upper thigh was tense and indurated over the lateral region. A repeat ultrasonography of the patient's thigh showed no deep vein thrombosis, but two well-defined echogenic collections within the muscle most likely interpreted as hematomas, but infection could not be excluded. However, the emergency physician came to the conclusion that the young man had a spontaneous hematoma that required no drainage, as the patient was afebrile and not toxic. He thus advised the patient to use non-prescription analgesics (naproxen 500 mg twice daily) and Cephalexin 500 mg every 6 h for 2 weeks for possible cellulitis and to follow up with his FP.

The young man was seen by his FP a week later with persistent pain but no recurrent fever. No further investigation was performed and no change in physical findings was noted. The physician prescribed naproxen 500 mg every 8 h and a muscle relaxant cyclobenaprine (flexeril) 5–10 mg every 8 h, with a return appointment in 1 month. Over the next 2 months, the patient was seen for reassessment on three separate visits, a visit to the same emergency department a month later, and to his FP, and each time increasing pain and discomfort was noted. The thigh and gluteus muscles were described as hardened, swollen, and tender. Repeat ultrasound showed a large fluid collection (from the buttock to the knee (19 × 6 × 3 cm), interpreted as consistent with liquefaction of hematoma. Treatment consisted of increasing analgesics of greater potency from naproxen to acetaminophen 600 mg/codeine 60 mg every 6 h and compression bandages. Finally, after about 3 months from the onset of his initial symptoms, the patient could not bear weight on the right lower limb, complained of malaise, fatigue, and weight loss of about 10 lb. He was then referred to an orthopedic specialist.

5.3.1 Comments

When assessing a patient with localized muscle pain, swelling, and tenderness with or without fever, several conditions should be considered in the differential diagnosis. A history of trauma and vigorous exercise may suggest contusion or rupture of muscle fibers, or hematoma, but occasionally hematoma may occur spontaneous after minor trauma or muscle exertion in subjects with bleeding tendencies (i.e. hemophiliac, severe thrombocytopenia and chronic or acute anti-coagulation). Infectious diseases will commonly present with intermittent or persistent fever, but may on occasion give a history consistent with transient, short-lived fever. Although underlying renal failure, hypothyroidism, and absence of recurrent fever is more commonly found in the elderly, sometimes it is present in healthy young adults. Initial differential diagnosis should therefore consist of necrotizing fasciitis and pyomyositis, the former condition usually being more rapidly progressive with severe pain and toxicity. Other differential diagnoses might include compartment syndrome, muscle infarction (primarily in diabetics with renal impairment), and benign and malignant muscle tumors (sarcoma). The latter condition usually presents with no or minimal pain, but with the presence of a mass in the muscle. Cellulitis and thrombophlebitis should be easily recognized and excluded.

On presentation to the hospital in Paris, the results of the ultrasound narrowed the diagnosis to two main entities, muscle abscess (pyomyositis) or spontaneous hematoma. Rarely do sarcomas present with a complex mass with a central area of necrosis and liquefaction. The presence of leucocytosis (absence of significant traumas) and normal coagulation parameters would be against hematoma. Thus, the Parisian recommendation was the correct approach; surgical or percutaneous drainage and antibiotics for staphylococcal coverage.

5.3.2 Clinical Course

The patient was seen by an orthopedic surgeon who found the young man to be chronically ill looking, limping, with moderate swelling and increased warmth of the right upper thigh and hip. There was pain on palpation and severe pain and limitation of movements of the hip. Complete blood count showed a hemoglobin count of 100 g/L, a leucocyte count of 15,700 cells/μL, platelet count of 650,000/μL, and ESR 110/h. A CT scan of the hip and thigh demonstrated a fluid filled, heterogeneous, complex mass 8 × 10 cm, tracking and communicating with the hip and with destruction of the head and proximal femur, indicative of chronic osteomyelitis. Surgical drainage and debridement of the excision of the proximal femur was performed. Cultures grew S. aureus fully susceptible to most antibiotics, and he was treated with 4 weeks of intravenous cloxacillin, followed by 2 months of oral cloxacillin. The young man subsequently required a total hip arthoplasty (prosthetic hip) for ambulation and functioning. He spent a total of 4 weeks in hospital and another month in rehabilitation.

5.3.3 Medico-legal Issues

Litigation was launched by the young man against the emergency department and hospital of the tertiary care, university teaching center, and the FP for negligence. Compensation was sought for prolonged pain and suffering, time lost from work, permanent impairment from inability to play sports and participate in his leisure activity of jogging. The charges were that the physicians involved were negligent in not recognizing the seriousness of his illness, the need for hospitalization and surgical treatment of the muscle abscess. It was obvious from the letter by the Parisian's physician that he had a muscle abscess which required drainage and antibiotics. Failure of his FP and emergency physicians to refer him to an orthopedic or general surgeon much earlier led directly to extension of the infection and resulted in destruction of his hip joint.

The emergency physicians' defense was that the ultrasound report (performed locally) indicated that the heterogeneous complex mass was consistent with a hematoma, and that their treatment of the patient was therefore appropriate for that condition. In the statement from the examination of discovery of the defendant (FP), his defense said that he was following the directions and diagnosis provided by the emergency physicians from a tertiary care center. Therefore, he felt that the onus was on the hospital physicians to admit and arrange for appropriate consultations and treatment.

5.3.4 Medical Aspect

The young man in case 3 represents a typical case of pyomyositis, or spontaneous development of muscle abscess. This is an uncommon condition in North America or Western Europe, but it is probably under reported. It is more prevalent in tropical and subtropical countries, hence also called "tropical myositis." This is really a misnomer as it is not confined to the tropics, nor related to travel to warmer climate. The reason for the greater incidence in tropical countries is unknown but it tends to occur in younger ages (children, adolescents, young adults) in the tropics, whereas in North America it mainly occurs in adults and older persons (predominantly males).[22] Postulation on the greater prevalence and younger age groups in tropical countries include high incidence of malnutrition and parasitic diseases, but there is no direct link with parasitic infections. Most patients in warmer climates are healthy and very active.

In North America, a high proportion of patients (50–60%) have underlying conditions such as diabetes, HIV infection, intravenous drug abuse (IVDA), leukemias or other immunosuppression.[22] About 20–50% have a history of preceding blunt trauma or vigorous exercise, but up to 40% of patients are healthy and active. The pathogenesis has not been established, but it is postulated that injury to the muscle (most commonly large muscle groups of the lower limbs) by vigorous exercise or trauma allow bacteria (from transient bacteremia) to become established. In animal models, the muscle is resistant to infection by bacteremia unless there is preceding injury.[22] Pyomyositis in tropical countries is nearly always caused by *S. aureus* (95%), whereas in temperate regions, *S. aureus* accounts for 66–70%, with the remainder caused by streptococci, coliforms and anaerobes.[22,23] However, even in temperate climates, pyomyositis in healthy adults is mostly due to *S. aureus* or *Streptocccus*. Whereas, bacteremia at presentation is rare in tropical pyomyositis (5%) and in North America it is present in 31%.[23] In a recent prospective study from Taiwan (area with temperate climate), 35 cases of pyomyositis were enrolled over a 16 year period.[24] Patients with underlying disease (66%) were older (mean age 47.8 year), with higher prevalence of bacteremia (52.2%), and gram-negative infections (30.4%), compared to the healthy patients (34%) who were younger (mean age 27 year), with lower prevalence of bacteremia (8.3%), and no gram-negative bacilli infection and lower mortality.[24] Hence, older adults with chronic underlying disease with pyomyositis should be treated empirically with broad-spectrum antibiotics, but healthy younger adults can be treated with anti-staphylococcal penicillin or first generation cephalosporin (except where MRSA is suspected).

The clinical progression of pyomyositis appears to advance in three overlapping, but somewhat distinct stages.[23,25] The first invasive stage has a sub-acute onset with variable fever, local swelling with or without erythema, mild pain and tenderness, and with overlying skin and soft tissue often indurated or of woody consistency. This stage is frequently misdiagnosed or overlooked, and there is pus to drain. During the second, suppurative stage (10–21 days after onset), the patient is usually febrile with distinct muscle swelling (often bulging), tenderness, and increased warmth, but skin erythema is often absent, and localized pus can be aspirated or drained. In the third stage, the systemic manifestations become more florid with high spiking fever, sepsis syndrome, or development of metastatic infection. The area of the overlying muscle has increased bulge or marked swelling, is extremely tender with increased erythema and warmth, and flocculation may be detectable. The current case 3 was recognized in Paris during the second stage (abridged first stage), and remained in this stage for a few months (probably modified by the oral antibiotics), with consequent extension of the infection to the hip.

Ultrasonography is very sensitive and useful for guiding aspiration or percutaneous drainage of muscle abscess (second and third stage),[26] but CT or MRI are more sensitive in the first invasive stage, and MRI should be used in selected cases especially to differentiate from diabetic muscle infarction or other inflammatory myositis.[27] Leucocytosis, elevated CK, and acute phase reactants (ESR and CRP) are frequently seen in all stages of pyomyositis, but leucocytosis may be more variable in the initial invasive stage. In the early stage of pyomyositis, ultrasound usually reveals diffuse muscle swelling with edema and diffuse hyperemia (manifested by hyperechogenicity, with or without localized hypoechogenicity [early necorsis]).[28] At this stage, pyomyositis will usually respond to antibiotic treatment alone. During stages 2 and 3, a muscle abscess is easily outlined by ultrasound as a round or tubular shape, with variable echogenicity from hypoechoic to isoechoic or hyperechoic.[28] Internal debris is a common feature and septae are more common with chronic abscesses. Color doppler imaging usually reveals variable hyperemia of the abscess wall and the immediate surrounding tissues. The threshold for aspiration should be low, as sometimes muscle abscesses may appear quite solid with no discernible fluid, yet still yield pus on aspiration.[28]

5.3.5 Comments on Medico-legal Aspect

Based on the clinical presentation and ultrasound findings, leucocytosis pyomyositis (muscle abscess) should have been strongly suspected from the first emergency visit in the patient's resident city. If a simple aspirate had been performed, the diagnosis would have been easily confirmed, and definitive drainage and appropriate antibiotic course would have prevented the adverse outcome. In this case (like so many others), three physicians involved in the patients assessment and care missed the obvious diagnosis. Physician should learn from this case that imaging, even with modern technology cannot discern the pathology of many lesions, and can only narrow the differential diagnosis and (as commonly mentioned in many radiological reports) the physician should interpret the report with the clinical context of the case. In this case, infection or muscle abscess should have been the primary diagnosis to exclude. The fact that the emergency physicians missed the correct diagnosis on two separate occasions is no excuse for the FP not to consider the proper differential diagnosis and refer the patient to a specialist (orthopedic surgeon) much earlier. The medical report from the Parisian physician (who made the correct diagnosis and offered definitive treatment), was ignored and represented a strong supportive evidence for the plaintiff's lawyer.

References

1. Hoen B, Alla F, Selton-Suty C, Béguinot I, Bouvet A, Briançon S, Casalta JP, Danchin N, Delahaye F, Etienne J, Le Moing V, Leport C, Mainardi JL, Ruimy R, Vandenesch F; Association pour l'Etude et la Prévention de l'Endocardite Infectieuse (AEPEI) Study Group, (2002), Changing profile of infective endocarditis: results of a 1-year survey in France. JAMA. 288: 75–81.
2. Fong IW, (2008), New Perspectives of Infections in cardiovascular disease. Curr. Cardiol. Rev. 5: 87–104.
3. Karchmer AW, (2008), Infective endocarditis. In: Fauci AS, Braunwald E, Kasper DL, Hauser SC, Longo DL, Jameson JL, Loscalzo J (eds); Harrison's Principles of Internal Medicine, 17th Edition, McGraw-Hill Companies, Inc., USA; p789–798.
4. Weiss G, Goodnough LT, (2005), Anemia of chronic disease. N. Engl. J. Med. 352: 1011–1023.
5. Zarychanski R, Houston DS, (2008), Anemia of chronic disease: a harmful disorder or an adaptive, beneficial response? CMAJ. 179: 333–337.
6. Weinberg ED, (1992), Iron depletion: a defense against intracellular infection and neoplasia. Life Sci. 50: 1289–1297.
7. Means RT Jr, Krantz SB, (1991), Inhibition of human erythroid colony-forming units by gamma interferon can be corrected by recombinant human erythropoietin. Blood; 78: 2564–2567.
8. van Iperen CE, van de Wiel A, Marx JJ, (2001), Acute event-related anaemia. Br J Haematol. 115: 739–743.
9. Price MD, Chiodo CP, (2009), Foot and ankle pain. In: Finestein GS, Budd RC, Harris ED Jr, McInnes IB, Ruddy S, Sergent JS; Kelley's Textbook of Rheumatology, 8th Edition, Saunders/Elsevier, Philadelphia, p643–650.
10. D'Maio M, Paine R, Mangine RF, Drez D, (1993), Plantar fasciitis. Sports. Med. Rehabil. Series 16: 137–142.
11. Longford CA, Gilliland BC, (2009), Periarticular disorders of the extremities. In; Fauci AS, Braunwald E, Kasper DL, Hauser SC, Longo DL, Jameson JL, Loscalzo J (eds); Harrison's Principles of Internal Medicine, 17th Edition, McGraw-Hill Companies, Inc, USA, p2184–2186.

12. DiMarcangelo MT, Yu TC, (1997), Diagnostic imaging of heel pain and plantar fasciitis. Clin Podiatr Med Surg. 14: 281–301.
13. Wolgin M, Cook C, Graham C, Mauldin D, (1994), Conservative treatment of plantar heel pain: long-term follow-up. Foot Ankle Int. 15: 97–102.
14. Cole C, Seto C, Gazewood J, (2005), Plantar fasciitis: evidence-based review of diagnosis and therapy. Am Fam Physician. 72: 2237–2242.
15. Acevedo JI, Beskin JL, (1998), Complications of plantar fascia rupture associated with corticosteroid injection. Foot Ankle Int. 19: 91–97.
16. Crawford F, Thomson CE, (2003), Interventions for treating plantar heel pain. Cochrane Database Sys Rev. CD000416
17. Wise C, (2009), Arthrocentesis and injections of joints and soft tissue. In: Feinstein GS, Budd RC, Harris ED, McInnes IB, Ruddy S, Sergent JS; Kelly's Textbook of Rheumatology, 8th Edition, Saunders/Elsevier, Philadelphia, p721–739.
18. Ostensson A, Geborek P, (1991), Septic arthritis as a non-surgical complication in rheumatoid arthritis: relation to disease severity and therapy. Br. J. Rheumatol. 30: 35–38.
19. Kortelainen ML, Särkioja T, (1990), Fatal complications of intramuscular and intra-articular injections. Z Rechtsmed. 103: 547–554.
20. Cawley PJ, Morris IM, (1992), A study to compare the efficacy of two methods of skin preparation prior to joint injection. Br J Rheumatol. 31: 847–848.
21. Johnson SM, Saint John BE, Dine AP, (2008), Local anesthetics as antimicrobial agents: a review. Surg Infect. 9: 205–213.
22. Pasternack MS, Swartz MN, (2005), Myositis In: Mandell GL, Bennett JE, Dolin R (eds); Principles and practice of Infectious Diseases, 6th Edition, Elsevier, Churchill-Livingstone, Philadelphia, p1194–1204.
23. Christin L, Sarosi GA, (1992), Pyomyositis in North America: case reports and review. Clin. Infect. Dis. 15: 668–677.
24. Chiu Sk, Lin JC, Wang NC, Peng MY, Chang FY, (2008), Impact of underlying diseases on the clinical characteristics and outcome of primary myositis. J. Microbiol. Immun. Infect. 41: 286–293.
25. Filler S, Dolgin SE, Saphir RL, Shlasko E, Midulla P, (2003), A case confirming the progressive stages of pyomyositis. J. Pediatr. Surg. 38: 1551–1553.
26. Robben SGF, (2004), Ultrasonography of musculosketal infections in children. Eur. Radiol. 14 (Suppl 4): 65–77.
27. Scott DL, Kingsley GH, (2004), Use of imaging to assess patients with muscle disease. Curr. Opin. Rheumatol. 16: 678–683.
28. Chau CLF, Griffith JF, (2005), Musculosketeal infections: ultrasound appearance. Clin. Radiol. 60: 149–159.

Chapter 6
Looking Beyond the Obvious

6.1 Case 1: Middle-Aged Man Presenting with Palpable Purpuric Rash on His Legs

A 41-year-old male, previously well and employed as a university lecturer, was referred by his family physician to a rheumatologist at an urban university teaching hospital medical clinic. The patient had a history of feeling unwell (general malaise), occasional symptoms of chills and night sweats (but no documented fever), joint pain of the large joints on his lower limbs, and a recent rash on his lower legs, all appearing gradually over the preceding 2–3 weeks. The significant findings noted by the rheumatologist were a short midsystolic murmur at the cardiac apes (grade II/VI), no actively swollen joints, and the presence of scattered discrete palpable purpuric (non-blanching) papules of the legs. A tentative diagnosis of vasculitis was made, and blood tests, urinalysis, chest radiograph, and skin biopsy were performed. Complete blood count revealed mild anemia (hemoglobin 110 g/L), leucocyte count of 7,500 cells/μL, platelet count of 130,000/μL and an erythrocyte sedimentation rate (ESR) of 98/h and normal creatinine, urinalysis, and liver enzymes. The chest radiograph was normal and hepatitis B antigen and hepatitis C antibody were negative. The anti-nuclear antibody (ANA) was 1:30 with a speckled pattern, but the double stranded DNA antibody was negative. The rheumatoid factor was 1:120 and the anti-neutrophil cytoplasmic autoantibodies (C and P ANCA) were negative. Skin biopsy was reported as showing typical changes of leukocytoclastic angiitis or necrotizing vasculitis. The patient had no symptoms of dry eyes, xerostomia, or Raynaud's phenomenon, and cryoglobulins and anti-phospholipid antibodies were negative. In view of the patient's constitutional symptoms, he was placed on Prednisone 60 mg daily for 2 weeks, and then tapered by 5 mg per week.

6.1.1 Comments

Palpable purpura is the classic cutaneous manifestation of vasculitis, and it usually represents affliction of the capillaries and venules of the skin. It is most commonly

I.W. Fong, *Medico-Legal Issues in Infectious Diseases*, Emerging Infectious Diseases of the 21st Century, DOI 10.1007/978-1-4419-8053-3_6,

seen with large blood vessel vasculitis, but it can occur occasionally with large blood vessel arteritis (i.e. Wegener's granulomatosis, Churg-Strauss syndrome, polyarteritis nodosa, and Behcet's disease). The diagnosis of cutaneous necrotizing vasculitis can be a manifestation of several different diseases, all which require different forms of treatment. Hence, it is important to investigate and elucidate the underlying disease. The differential diagnosis of conditions presenting with cutaneous necrotizing vasculitis can be classified as follows[1,2]: (1) isolated primary skin disorder (idiopathic), (2) allergic reaction to drugs or chemicals (diagnostic agents), (3) systemic auto-immune disorders (i.e. systemic lupus erythematosus (SLE), rheumatoid arthritis, dermatomyositis, Sjögren's syndrome, cryoglobulinemia, antiphospholipid auto-antibody syndrome), (4) various viral and bacterial infections, (5) paraneoplastic vasculitis, secondary to hematological, lymphoid, or solid organ malignancies.

Based on the clinical history, manifestations, and investigations, several disorders could be excluded, such as drug-induced vasculitis (he was on no medications) and clinically overt collagen vascular disorders (SLE, rheumatoid arthritis, dermatomyositis, Sjögren's syndrome etc.). There were no findings to suggest neoplasms, but no aggressive investigations were performed to exclude occult malignancy (such as computerized tomography [CT] of the chest and abdomen, upper and lower intestinal endoscopies). The only investigations performed for an infectious etiology were the hepatitis B and C serologies, but no blood cultures or other investigations were performed to exclude other possible infections that could produce vasculitis. Rickettsial agents could be excluded by the epidemiology, because although they can produce cutaneous vasculitis, they are not endemic in Canada, and the subject had no travel to endemic areas in North America.

Once various underlying diseases have been excluded, the management approach is usually a conservative, symptom based therapy (if there were no evidence of systemic involvement) such as a non-steroidal, anti-inflammatory agents (NSAID).[2] Failure to respond to NSAIDs is an indication for institution of corticosteroids, or evidence of systemic vasculitis.[1,2] It is generally recommended to perform a Mantoux skin test to exclude latent tuberculosis before starting systemic corticosteroid for more than a couple of weeks. Tuberculosis itself can produce an infectious vasculitis, but not usually palpable purpura. However, erythema induratum, which is associated with tuberculosis (*Mycobacterium tuberculosis)* may be detected by polymerase chain reaction [PCR] is considered a form of nodular vasculitis.[2]

6.1.2 Clinical Course of Case 1

After 4 weeks of corticosteroid therapy, the patient was admitted via the emergency department to another urban university teaching hospital. Apparently, there was initial improvement in the rash of his legs, but in the last 2 weeks, he had experienced increased malaise, fatigue, and drenching night sweats with chills. On the morning of admission, the patient suddenly collapsed with inability to communicate with his wife, and he was brought to the hospital by an ambulance. Examination by the physician revealed a temperature of 38.2°C, pulse of 120/min, blood pressure of 110/60 mmHg. Positive physical findings revealed pallor of the conjunctivae, with two small petechiae, mild clubbing of the fingers, a grade III/VI pansystolic apical murmur, and dense right hemiparesis. The patient was unable to speak (aphasic), and was drowsy, but had movements on the left side of the body. No palpable purpura was present.

Blood tests revealed a hemoglobin of 9.5 g/μL (95 g/L), leucocyte count of 12,300 cells/μL, platelet count of 270,000/μL, and creatinine of 1.5 mg/dL (1.35 μmol/L). A CT scan of the brain revealed a large sub-cortical infarction on the left side with mild hemorrhage. Blood cultures (three sets) all grew a fully susceptible Streptococcus viridians complex and an echocardiogram revealed evidence of a myxomatous mitral valve with moderate to severe regurgitation, and a 1 cm mobile mass of the leaflet consistent with vegetation. The patient was started on intravenous penicillin (18 million units/day) and became afebrile in a few days. However, after about 5 days, his neurological status deteriorated and a repeat CT scan showed multiple areas of cerebral infarction and a very large left cerebral hematoma. Within a few hours, the patient died and no autopsy was performed.

6.1.3 Medico-legal Issues

The wife of the deceased man sought legal counsel to consider litigation against the rheumatologist for negligence in mismanagement and failure to make the proper diagnosis. The attorney for the plaintiff requested an independent medical review of the case to determine whether there were sufficient grounds for medical malpractice litigation.

Opinions were requested by the lawyer on the following issues:

1. What was the cause of death?
2. Was there a relationship between the underlying condition causing death and the cutaneous necrotizing vasculitis?
3. Were the clinical assessment and investigations by the rheumatologist appropriate and meet the standard of care for a specialist?
4. Could the proper diagnosis have been made earlier, and if so, would appropriate treatment have made a difference in the outcome?
5. Was the rheumatologist negligent in his medical assessment and management?
6. Should the rheumatologist have referred the patient to another specialist (i.e. infectious disease specialist or a cardiologist)?

On review of the case, it was evident that the deceased male suffered from sub-acute bacterial endocarditis (SBE), with multiple cerebral emboli that led to brain infarction and hemorrhage, and were the immediate cause of death.

Likely through brain ischemia, swelling and tentorial herniation of the brainstem were the mechanisms producing death. The obvious predisposing factor for the SBE was the mitral prolapsed valve with regurgitation, which was seeded by transient bacteremia from oral *Streptococci* with normal daily activities (chewing, teeth brushing or flossing). The subject and his wife were unaware of any underlying heart disease, and there was no recent history of oral-dental procedures or regular routine dental hygienic assessments.

Since there was no evidence of any other concomitant disease that could account for the initial presentation with palpable purpura, SBE would have to be considered the direct cause of the necrotizing cutaneous vasculitis. This is a known complication of bacterial endocarditis, but it rarely presents in this manner (as a primary cutaneous vasculitis).

6.1.4 Medical Aspects

Palpable purpura (reddish-purple, non-blanching papule) is the hallmark of cutaneous leukocytoclastic vasculitis. These lesions are distinguishable from erythema nodosum, which are erythematous subcutaneous nodules, and erythema induratum, which can form nodules but are usually larger, and present with indurated skin and subcutaneous plaques bilaterally. Erythema induratum is a form of reactive inflammatory lobular panniculitis associated with focal nodular vasculitis, that later may present with granulomatous reaction.[3] It is strongly associated with tuberculosis (which may not be active), and can be idiopathic and occasionally associated with hepatitis C, propylthiouracil and rheumatoid arthritis.[3]

Vasculitides are a heterogenous group of relatively rare disorders, many of which are of unknown cause (idiopathic), but are classified according to clinical complex as distinct entities (described under auto-immune or collagen vascular disorders). Microbial organisms of nearly all classes are capable of causing vascular inflammation in vessels of any size. Ancient diseases such as syphilis and tuberculosis have long been recognized to produce aortitis and arteritis.

Infectious agents can produce vascular disease by three main mechanisms: (1) direct invasion of the blood vessel from adjacent soft tissue or organ infection (or blood stream), (2) invasion and occlusion of the vasa vasorum (small nutrient vessels of the larger arteries), i.e. syphilitic aortitis, and (3) immune complex microbial vasculitis. There is also evidence that microbial agents can damage vascular endothelium or precipitate thrombosis by stimulating the coagulation cascade and may initiate or accelerate atherosclerosis.[4] Some microbes such as viruses[5,6] and rickettsiae[7] initiate damage to the endothelium directly or via immune complexes. In other organisms such as syphilitic arteritis[8] and ecthyma gangrenosum (secondary to pseudomonas bacteremia in neutropenia), the agents cause greater damage on the tunica media and tunica adventitia of the vessel wall.

Leukocytoclastic vasculitis (LCV) is the most common vasculitic manifection of acute or chronic infections.[9] LCV is an immunologic vasculitic response to antigens in an immune complex (containing immunoglobulins and complement components), which can be found in sera and in vasculitic deposits.[10,11] There are several viruses associated with vasculitis which are believed to be causally linked. Hepatitis B virus (chronic infection) can cause small vessel vasculitis (with diffuse purpura) and more commonly, medium-size vessel vasculitis (polyarteritis nodosa [PAN]).[12] Chronic hepatitis C virus has been associated with mixed cryoglobulinemia (composed of IgG and IgM) that are precipitated by the cold. Cryoglobulin deposits consisting of cryoglobulins binding to portions of the hepatitis C virion, an activated complement leading to cryoglobulinemia vasculitis of small and medium-sized vessels.[13] Hepatitis C is mainly associated with type II (90% caused by hepatitis C) and type III cryoglobulinemia (which can be related to other infections, SLE and Sjögren's syndrome).[13] Patients with cryoglobulinemia vasculitis often present with confluent purpura, arthralgias, myalgias, and malaise. Parvovirus B-19 chronic infection can produce vasculitis resembling polyarteritis nodosa in manifestation and histopathology.[12]

Human immunodeficiency virus (HIV) has been associated with various types of vasculitis in adults and children at various stages of the disease. The majority of reports have been case reports or small case series, and no large prospective comparative studies have been performed with the general population to determine the relative increase or prevalence of this complication. The pathogenesis of vasculitis with HIV infection can be related to specific infectious vasculitis (i.e. cytomegalovirus [CMV] and *M. tuberculosis* are the most common); drug-induced and idiopathic (classic inflammatory vasculitis), or possibly the HIV itself;[14] and isolated intracranial vasculitis (related to *Varicella zoster* virus or idiopathic).[14] The types of vasculitis described in HIV subjects are diverse and include: hypersensitivity vasculitis, polyarteritis nodosa, Henoch Schönlein purpura, Behcet's disease, Wegener's granulomatosis and pulmonary microscopic angiitis (in patients with high CD4 cell count and during immune reconstitution), rapidly progressive necrotizing vasculitis of the aorta and large arteries with aneurysm, giant cell arteritis with aortic root dilatation, and Kawasaki syndrome. The largest retrospective case series was an evaluation of 148 selected HIV-infected subjects who underwent muscle or peripheral nerve biopsies because of neuromuscular symptoms (N = 63). Eighty-five underwent skin biopsies because of cutaneous lesions. Thirty-four (23%) patients had inflammatory vascular disease, necrotizing arteritis (N = 3), mononuclear inflammatory vascular disease (N = 17), and other small vessel inflammatory changes. However, only 11 patients could be classified as having a distinct category of vasculitis (polyarteritis nodosa, Henoch=Schönlein purpura, drug-induced hypersensitivity vasculitis) and 23 were classified as "other" vasculitis, type unspecified).[15] The etiology factors associated with the vasculitis were drugs (N = 6), hepatitis (N = 1), cryoglobulinemia (N = 2), CMV (N = 1), human T lymphotropic virus type-I (N = 2), and HIV antigen (N = 2).[15] Immune deposits were found in small vessel walls of five skin biopsy samples and the muscle of five patients with necrotizing arteritis.

Less commonly associated with small vessel vasculitis (mainly isolated case reports) are hepatitis A virus, Epstein-Barr virus, herpes simplex virus, rubella virus, Hantavirus, and even influenza virus vaccination,[9] but whether some of these episodes represent coincidental temporal relationship or cause and effect is unclear.

LCV and Henoch-Schönlein purpura have been attributed to many common pathogenic bacteria with gram-positive cocci predominating.[15] Most of the cases of vasculitis in these circumstances have been related to sub-acute bacterial endocarditis (SBE) or acute endocarditis.[16] However, chronic meningococcemia and gonococcemia can present with cutaneous and visceral vasculitis without endocarditis.[17–19]

Rickettsiae typically involves the endothelium of smaller arteries, veins, and capillaries that supply the skin, and may also affect the CNS, skeletal muscle, myocardium, lung, and kidney.[9,20] Cutaneous lesions with macules, petechiae, purpura and sometimes necrosis of the skin, is a signature feature of many rickettsial infections, especially Rocky Mountain spotted fever and typhus. However, a non-spotted fever rickettsial infection can also present with visceral vasculitis.[21]

M. tuberculosis can rarely be associated with large and small vessel vasculitides, but can involve major visceral and peripheral arteries.[9] Aortitis appears to be the most common vascular involvement, either by direct extension or hematogenous seeding to the vasa vasorum. Tuberculous aortitis can involve the thoracic or abdominal aorta with saccular aneurysm or pseudo-aneurysm, and histopathology characteristically shows a granulomatous panvasculitis.[9] The cutaneous manifestation of tuberculosis consists mainly of erythema nodosum and erythema induration (nodular and lobular panniculitis with nodular vasculitis, respectively). Rarely, military tuberculosis can present with cutaneous vasculitis[22] and even Henoch-Schönlein purpura.[23,24] *Mycobacterium leprae* (lepromatous and multibacillary stage) can develop small vessel vasculitis of the skin, either as a reaction to treatment in the form of erythema nodosum leprosum, or *de novo* as Lucio's reaction with focal patchy vasculitis resulting in skin necrosis and ulceration.[25]

Vascular complications of SBE are the most frequently recognized clinical infectious vasculitides. The pathogenesis varies from emboli to immune complex LCV, and direct invasion of the blood vessel wall or mycotic aneurysm from vasa vasorum occlusion. The skin manifestations of SBE are variable and include petechiae (the most frequent, occurring in 20–40%), subungual splinter hemorrhages (often traumatic and non-specific), Osler's nodes which occurs in $<$10–15% (perivasculitis or necrotizing vasculitis), Janeway lesions in $<$10–15% (vasculitis with microabscesses), focal dermal infarcts with skin necrosis (from septic emboli or immune complex vasculitis), and raised or flat purpuric lesions or papules (with leukocytoclastic vasculitis).[26]

Another extremely rare infectious cause of cutaneous vasculitis is parasites (*Strongyloides stercoralis,* and *microfilaria*e *Acanthamoeba*).[9] Although invasive fungi can cause larger vessel vasculitides (aspergillosis, cryptococcosis, mucormycosis) from direct invasion or vasa vasorum occlusion (mycotic aneurysm), skin manifestations are distinctly rare (ecthyma gangrenosum in a neutropenic host, although associated with *Pseudomonas aeruginosa* septicemia can be secondary to systemic candidiasis). In these lesions, the organisms invade the medial adventitial vessel layers in the skin producing hemorrhagic necrosis.[9]

6.1.5 Comments on the Medico-legal Aspect

Based on the sequence of events and our understanding of the pathogenesis of endocarditis, it is highly likely that SBE was the cause of the patient's rash (palpable purpura) and arthralgia, on presentation to the rheumatologist. Thus, it is relevant to address the next issue. Should the rheumatologist have considered the diagnosis and performed the appropriate investigations for SBE? It can be argued that as a specialist in internal medicine, the rheumatologist should have been aware of the infectious causes of cutaneous LCV. The expectations and the standards set for a rheumatologist in a university teaching hospital may also be higher than that for a community-based rheumatologist.

Any physician investigating a patient with cutaneous or visceral vasculitis should perform a batch of tests to exclude infectious causes (as they are often curable and missing the diagnosis can be lethal). The tests should include two to three sets of blood cultures, (separated by hours) and serologies for the following viruses: hepatitis B and C (A, if there was evidence of acute hepatitis), HIV, CMV, Epstein-Barr virus, and parvovirus B-19. A syphilis serology, Mantoux skin test, sputa for *M. tuberculosis,* smear and culture for patients with cough and abnormal chest radiograph, open lymph node biopsy of enlarged nodes for histopathology, and mycobacterial and fungal smears and cultures should all be considered as well. Routine echocardiogram is not necessary unless there is evidence of significant heart valve disease, regurgitant cardiac murmur, or positive blood culture for typical microorganisms of SBE. Investigations for other infections are not necessary unless there is clinical or epidemiological evidence to suggest their possibility.

It is likely that if these guidelines were followed, that the diagnosis of SBE could have been confirmed a month earlier in the present case (before starting corticosteroids). Appropriate treatment with intravenous antibiotics (penicillin) would likely have prevented the adverse outcome, as the risk of systemic emboli dramatically decreases after 2 weeks of antibiotic therapy.

Hence, the plaintiff and legal counselor may have sufficient grounds to pursue medico-legal litigation against the rheumatologist for failing to diagnose SBE as the underlying cause of the LCV and for instituting inappropriate treatment (steroids) that could have aggravated the infectious disease. Furthermore, his negligence to make the correct diagnosis was more than likely to have contributed to the adverse outcome, which could have been prevented by earlier diagnosis and treatment. In situations such as this case, it would be best for the defendant's attorney to seek an outcome settlement, although there may still be a chance that the court would rule in their favor.

6.2 Case 2: Acute Loin Pain

A 26-year-old male presented to his family physician (FP) with recent onset of loin pain, anorexia, and nausea. The abdominal examination was noted to be normal. A routine blood count, urinalysis, urine culture and abdominal radiograph were performed. One week later, the patient was reassessed in the FP's office, and he was noted to have persistence of the same symptoms, plus development of a bruise on his left hand. There was no detailed re-examination or detailed history provided in the physician's office file, but the investigations revealed a mild anemia (hemoglobin 11.6 g/dL) and elevation of the leucocyte count (15,100 cells/μL) with predominant neutrophils. The urinalysis and abdominal radiograph were noted to be normal. Although there was no review of the young man's past history, in the medical file it was noted that 6 years ago he had a fall with apparent loss of consciousness (no recall of the event). Investigations at that time included an echocardiogram and electrocardiogram which were noted to be normal by the FP. Otherwise, there was no significant past illness. No diagnosis, management plan, nor further investigations were proposed at this visit.

The patient was again seen by the FP the following week. His abdominal pain, nausea, and anorexia persisted with mild diarrhea that was noted. Stool specimens for culture and examination for parasites were negative for any pathogens, and serum vitamin B12 and red blood cell folate were reported as normal. Four days later, he returned to the FP office with symptoms of worsening abdominal pain and tenderness and night sweats. A chest radiograph was performed, which was normal, and so the physician prescribed ciprofloxacin 1 g daily for 3 days but did not request any blood cultures. There was no noted improvement in the patient's symptoms a few days later, and again at 1 month later. Repeat examination of the chest and abdomen were noted to be normal, and the FP prescribed HpPAC (lansoprazole/clarithromycin/amoxicillin) for 7 days, presumably for Helicobacter pylori infection (empiric treatment).

About 2 months after the initial visit to the FP, the patient admitted to a suburban hospital emergency department, with complaints of slurred speech and left-sided weakness. An internist was then consulted and a detailed history revealed that the young man had been feeling unwell for 3 months, with weakness, malaise, intermittent chills and fever, and 30 lb weight loss. He also reported having chest pain of several days duration, shortness of breath on exertion, and episodes of shortness of breath at night consistent with paroxysmal nocturnal dyspnea.

Examination in the emergency department revealed a pale, ill-looking young man with a temperature of 38°C, pulse of 120/min, and blood pressure of 90/60 mmHg. There was obvious marked clubbing of the digits of the hands and feet, elevated jugular venous pressure, some basal crackles in the lungs, palpable systolic thrill over the precordium, and a grade 4/6 pansystolic apical murmur, mild diffuse abdominal tenderness and left-sided hemiparesis.

Investigations then showed an anemia of 9.7 g/dL, leucocyte count of 20,000 cells/μL, and a platelet count of 590,000/uL. A chest radiograph revealed cardiomegaly with mild pulmonary edema and an electrocardiogram demonstrated S-T elevations indicative of inferior myocardial ischemia. Blood cultures were obtained, which subsequently grew Streptococcus viridans, and the patient was then transferred to an urban tertiary care center (critical care unit). Cardiac catheterization revealed an embolus to the right coronary artery and echocardiogram showed large vegetation on the mitral valve with severe regurgitation and flail anterior valve leaflet. Emergency mitral valve replacement was performed a few days later and he received 6 weeks of intravenous antibiotics. Three months later, he required reconstructive surgery for bilateral popliteal mycotic aneurysms, and closure of a perivalvular leak from around the prosthetic mitral valve.

6.2.1 Medico-legal Issues

Legal counsel was sought by the patient to pursue malpractice litigation against the FP. The lawyer of the plaintiff requested an expert opinion on the case. Specific questions that were posed by the attorney were:

1. Did the FP meet the standard of care?
2. Could the diagnosis of SBE have been made earlier?
3. Would earlier treatment have prevented the complications of stroke, heart attack, and mycotic aneurysms?
4. If the diagnosis and treatment were implemented 1–2 months before, would the need for heart valve surgery have been averted?

To assess the first issue of competence and the standard of care of the FP, it is important to analyze the information available to the physician at each visit, such as the completeness of the history and physical examination, the appropriateness and thoroughness of his investigations, and the formulation of a differential diagnosis and plan of action. His failure to recognize the nature and seriousness of the patient's condition and the reason for failing to refer the patient to an internist or other specialist are also to be considered. All the above factors have to be taken into consideration in the context of the expectations from a FP comparative to the expected standard of his peers.

At the first visit, soon after onset of the young man's symptoms, there was insufficient information to make a clinical diagnosis. Since the patient's symptoms were mild and non-specific, a localized examination of the abdomen and the initial investigations were reasonable for a GP's practice. At the second visit a week later, when the patient's symptoms were no better, new signs had appeared (bruise on the left hand without trauma), and with the presence of mild anemia and leukocytosis, it should have alerted the physician to a possible serious infection or inflammation. At this visit, the physician should have reviewed his previous notes on file, performed a detailed history, functional inquiry and complete physical examination. At the examination of discovery of the FP and the plaintiff, it was evident that no detailed history or complete physical examination (including the heart) was ever performed at any of the visits. Moreover, the previous notation by the FP (stating that 6 years prior the echocardiogram was normal) was incorrect. The official echocardiogram report (which was available in the physician's office file) indicated that there was evidence of mitral valve prolapse with mild regurgitation, thus providing evidence of underlying cardiac valvular disease that predisposes to the risk of bacterial endocarditis. Even in the absence of this evidence, if the FP had performed a complete examination, he would more likely than have detected a significant regurgitant murmur that should have highlighted his suspicion for possible SBE. Thus, failure of the FP to obtain a detailed history and perform a complete examination at the second and subsequent visits represented substandard care even for a FP.

There was sufficient clinical information available to the FP that indicated a possible serious infection (history of chills and night sweats, raised leucocyte blood count). Thus, the minimum laboratory investigations that should have been performed by the second visit were blood cultures (at least two sets, taken hours apart), and an abdominal ultrasound. If the FP were cognizant of mitral regurgitation, then an echocardiogram or referral to a specialist (internist, cardiologist, infectious disease) or to a hospital emergency department would have been warranted. Failure of the FP to perform simple investigations (blood cultures) and to consult a specialist resulted in misdiagnosis, delayed treatment and subsequent harm to the patient. These factors contributed to the adverse outcome, which could have been avoided by earlier diagnosis and appropriate management. This dereliction of action represents medical negligence and substandard care.

Thus, failure to diagnose and consult an internist (or other specialist) was directly responsible for injury to the patient. It was highly likely that earlier treatment (1–2 months before) would have prevented the complications of stroke, heart failure, mycotic aneurysms and need for cardiac surgery.

6.3 Case 3: Sudden Loss of Vision

While watching a movie at the cinema, a previously healthy, 23-year-old male suffered from sudden transient loss of vision (lasting 5–8 min) in the right eye. This was associated with a sensation of chills. That night he attended the emergency department of an urban tertiary care hospital. The patient was noted to have a heart murmur, but no other illness, and there was persistent blurred vision in the right eye. A cursory eye examination was performed by the emergency physician, and he reassured the patient that the visual impairment would resolve and was probably due to stress. He was advised to follow up with his FP sometime in the near future.

Over the next 2 days, the young man experienced progressive worsening of the blurred vision with some discomfort and redness of the eye. He also experienced intermittent episodes of feverish sensation and chills with malaise. On the 3 day, he sought medical attention at another emergency department of a tertiary care university teaching hospital. A past history of heart murmur was obtained, but the nature of the cardiac abnormality was unknown. Physical examination at that time revealed a temperature of 38.8°C, pulse of 105/min, and blood pressure of 128/50 mmHg. There was evidence of 1–2 focal conjunctival bilateral petechiae, injection and suffusion of the right conjunctiva with cloudiness of the vitreous. He had markedly decreased vision and could barely count fingers at a close range (within a foot). The only other significant abnormality included a grade 3/6 systolic ejection murmur along the left sternal border and radiating to the neck with grade 2/6 early blowing diastolic murmur at the left sternal border. Blood tests revealed a hemoglobin of 12.5 g/dL, leucocyte count of 16,300 cells/μL, platelet count of 530,000/μL. Serum creatinine, electrolytes and glucose were all normal. A chest radiograph was normal and three sets of blood cultures were obtained. An ophthalmology consultant diagnosed acute bacterial endophthalmitis and vitreous aspirate was performed. All the blood cultures and vitreous aspirate grew Staphylococcus aureus (methicillin sensitive). Antibiotics were started on admission consisting of vancomycin/amikacin which was later changed to cloxacillin intravenously. A subsequent cardiology consultation resulted in an echocardiogram which showed bicuspid aortic valve with mild regurgitation and a 1 cm diameter mobile mass, consistent with a vegetation. The patient received intravenous cloxacillin 8 g/day for 6 weeks and survived, but remained mostly blind in the right eye. He was able to see only light and shadows with perception of movement.

6.3.1 Medico-legal Issues

The young man retained counsel who requested an independent medical assessment in order to determine legitimacy of malpractice litigation against the initial hospital and emergency department attending physician. Furthermore, it was also questioned whether earlier recognition of the ophthalmologic condition 2–3 days earlier would have resulted in better outcome with respect to the plaintiff's vision.

Based on the medical records of the initial emergency department visit, the symptoms and clinical findings were nonspecific and insufficient to make a diagnosis. This visit occurred just after midnight and the clinical eye examination by the attending physician was not indicative of an ophthalmologic emergency to request an immediate consultation. However, there were several issues raised by the plaintiff's lawyer that are worth discussing.

1. Should the physician have performed a more detailed examination?
2. Were the provisional diagnosis and assessment reasonable and at an acceptable standard of care?
3. In view of the sudden onset of loss of vision, should the physician not have considered a vascular event?
4. In view of the patient's age and previous well being, should the physician have considered a differential diagnosis of embolus from the heart?

6.3.2 Medical Issues

The acute transient visual loss in the patient's right eye was most consistent with a vascular event, and with subsequent development of endogenous bacterial endophthalmitis from a septic embolus. However, transient visual loss is a relatively common complaint seen by ER (emergency room) physicians. This can represent a serious disease or a benign condition and a meticulous history and examination is needed to determine the etiology. Benign or non-organic (functional) visual loss varies from mild blurring of vision to complete loss of light perception, and pose a difficult challenge. The two main causes of functional visual loss are malingering and conversion reaction (hysterical blindness). In malingering, there is an external secondary gain (often associated with a recent injury) and the visual complaint is out of proportion to the underlying injury. Patients with malingering try to circumvent diagnostic ophthalmic tests and have variable findings on visual acuity assessment.[27] In a conversion reaction, there is a flat, relaxed affect, despite severe visual loss. The patient is usually very cooperative and his or her behavior conflicts with the seriousness of the complaint. In functional visual loss, the ocular examination and papillary reaction are critical in the assessment. Normal papillary reaction and ocular examination with markedly poor vision in one eye suggest a non-organic visual loss. Further testing by an ophthalmologist (e.g. 4 diopter prism) can confirm the diagnosis.

Organic transient monocular visual loss can be the first manifestation of a serious underlying illness (as in this case). The age of the patient is very important in assessment of the differential diagnosis. The history is also important in defining the type of visual loss (monocular or bilateral), length of the episode, frequency of the episode, associated symptoms and underlying risk factors. Monocular transient visual loss can occur in amaurosis fugax, papilledema due to increased intracranial pressure, and conditions associated with prolonged visual loss such as retinal migraine, severe hypertension and blood dyscrasias.[28] By definition, the young man had an episode of amaurosis fugax (transient loss of vision for <10 min). This is a rare condition in young adults and is most commonly seen in those >50 years old from carotid artery or aortic arch emboli. In young adults (as in this case) the conditions that should be screened for include cardiac disease (as in SBE), arrhythmias (such as atrial fibrillation or mural thrombus), history of migraine, hyperviscosity syndrome, hypercoagulable disorders, and increased antiphospholipid cardiolipin antibody.[28]

The fact that the patient had a history of heart murmur should have alerted the initial ER physician to the possibility of cardiac valvular disease with emboli (such as SBE), despite absence of previous fever, rather than attribute the visual loss to stress. Based on his actions, the initial ER physician evidently considered functional visual loss as the primary diagnosis without a thorough history and physical examination. With the presence of a significant cardiac murmur, blood cultures and echocardiogram should have been performed at the initial presentation. Earlier diagnosis and treatment for bacterial endocarditis may have prevented progression to a flagrant bacterial endophthalmitis and loss of vision. It should be recognized however, that acute *S. aureus* endophthalmitis, even with appropriate treatment infrequently results in full recovery of vision. Thus, it may be difficult to determine whether or not earlier treatment by 1–2 days would have resulted in much better visual outcome.

It is evident from the case however, that physicians should not consider transient visual loss as being functional, without first excluding organic causes. This cavalier approach to a patient's worrisome complaint is a set up for malpractice litigation.

6.4 Case 4: Drug-Related Skin Rash

In the fall of 2007, a 54-year-old male attended the emergency department of a small town community hospital, with recent onset of a total body rash. The patient first noticed fever and left shoulder pain for 2 days. He initially tried an old supply of flurbiprofen (NSAID) he had used intermittently for many years, with no relief. He had a supply of acetaminophen 300 mg/codeine 30 mg from 2006 (prescribed after ganglionectomy), which he also tried. He had used both medications before without any drug rash or allergic reaction. The ER physician noted a pulse of 76/min, blood pressure of 115/75, temperature of 35.8°C and a diffuse confluent erythematous rash on his body, face and arms. A diagnosis of allergic drug reaction was made. No investigation was performed and he was prescribed prednisone 50 mg daily for 3 days, and diphenhydramine 50 mg every 6 h for 7 days, and then discharged.

The patient returned to the same emergency department the following morning with worsening symptoms of fever, chills, nausea, vomiting, cough and shortness of breath, muscle and joint pains. A past history revealed a history of allergy to eggs, horse and environmental pollen. There was also a history of gout and alcohol consumption of ≥6 beers/day. Examination on this visit revealed a pulse of 115/min, temperature of 38.8°C, respiratory rate of 32/min, blood pressure of 100 mmHg systolic, and a diffuse red body rash with stress tenderness of the wrist, knees, and ankles. The internist was consulted and blood tests and a chest radiograph was obtained. Soon after admission to the emergency department, intravenous saline was started and intravenous methylprednisolone 80 mg and diphenhydramine 50 mg were administered. Complete blood count showed a hemoglobin of 15.2 g/dL, leucocyte count of 6,600 cells/μL with a shift to the left (bands 900) and platelet count of 71,000 μL. Serum creatinine was 1.44 mg/dL, creatinine kinase (CK), 1,347 U/L (normal 55–170 μL), and oxygen saturation of 100% on inhaled oxygen. The chest radiograph was reported as normal. He was transferred to the intensive care unit and given epinephrine in the form of EpiPen intramuscular injection.

Overnight, the patient became confused and agitated with a temperature of 38.6°C, heart rate of 145/min, blood pressure of 130/94, and a diagnosis of delirium tremens was entertained and he was treated with intravenous diazepam 30 mg. Blood cultures were obtained. Twenty-four hours after admission, he was still febrile and sedated and sepsis syndrome was considered. The patient was placed on ertapenem and later vancomycin was added. The condition of the patient deteriorated later that day with worsening shock, mottled periphery, decreased level of consciousness and hyperventilation. He was transferred to a tertiary care center but died less than 48 h after the transfer.

Blood cultures subsequently all grew S. aureus (MSSA), and autopsy revealed multiple skin blisters with desquamation (epidermal-dermal separation and micro-abscess on microscopy), pulmonary edema, mitral valve prolapse with vegetation, multifocal myocardial microabscesses, meningeal inflammation with neutrophils and microabscesses of the brain, and multiple renal microabscesses. Conclusion by the pathologist was the patient died from S. aureus endocarditis with meningitis, myocarditis and presenting as staphylococcal toxic shock syndrome.

6.4.1 Medico-legal Issues

The spouse of the patient (plaintiff) consulted a lawyer to initiate medical malpractice litigations. The claims filed against the ER physician and internist of the local community hospital were: (1) misdiagnosis and mistreatment of his illness from the outset, thus negligence in not considering and diagnosing a severe infection in a timely manner; (2) her husband had been using the analgesics intermittently for years without any adverse reaction, therefore it was negligent to consider drug-rash (allergic reaction) as the primary diagnosis; (3) the initial ER physician was negligent in not considering infection as a differential diagnosis, as the fever started before the use of the analgesics; (4) the same ER physician was negligent in not performing any routine blood investigations or blood cultures; (5) both the other ER physician and internist at the second emergency visit should have considered and treated the patient immediately for an infection, as he was not improving after treatment for an allergic drug reaction; (6) if the plaintiff's husband were investigated and treated appropriately at the first emergency visit, he likely would have survived, and (7) even on the morning of admission, before his mental status and vital signs deteriorated, initiation of appropriate antibiotics more likely than not, would have resulted in a better outcome.

6.4.2 Medical Issues

The two main aspects of this case that resulted in misdiagnosis and the adverse outcome were assuming that the patient had a cutaneous drug eruption, and not considering toxin-induced rash of staphylococcal or streptococcal infection.

Cutaneous drug reactions are fairly common and occur in about 2.2% of hospitalized patients. New drugs started within 6 weeks are mostly responsible, except for drug-induced lupus or drug-induced cutaneous pseudolymphoma, or drugs used intermittently (as in this case). There are three main morphological types of cutaneous drug reactions.

The first type is exanthematous eruption with erythematous morbilliform or maculopapular rash is the most common manifestation in about 95% of skin reactions. Typically, the exanthema or erythema starts on the trunk and spreads peripherally in a symmetrical fashion. Usually, this rash occurs within a week of drug initiation and resolves in 7–14 days. The color changes from erythema to bright red, to brownish-red, and followed by scaling or desquamation. A severe form of this manifestation is the hypersensitivity syndrome reaction (HSR). HSR consists of an exanthematous eruption, fever and internal organ dysfunction (liver, kidney and central nervous system). Although the present case could have represented an HSR, this occurs most frequently in first exposure to the drug with initial symptoms starting 1–6 weeks after the exposure. Fever and malaise are often the presenting symptoms, and the skin eruption can progress into other forms (Stevens-Johnson Syndrome [SJS]), toxic epidermal necrolysis [TEN], pustular eruption). Thus, the fact that the patient had been on the analgesics intermittently for years and the fever started before the medications were against a diagnosis of drug-related HSR.

Even when HSR appears to be the most likely diagnosis (especially with common implicated drugs), the differentials that should be considered for exclusion include viral exanthema, bacterial infection (especially streptococcal and staphylococcal infections, rarely rickettsial infection), and collage vascular disorders. The most common drugs associated with HSR are: (1) anticonvulsants (phenytoin, phenobarbital, carbamazepine), (2) sulfonamides, (3) dapsone, (4) allopurinol (strongly associated with Han Chinese or HLA-B5801 allele), (5) minocycline, and (6) lamotrigine.[29]

The second type is urticarial reactions with pruritic red wheals of various sizes lasting <24 h (but new lesions can appear), is indicative of IgE mediated immediate reaction (i.e. penicillin and other antibiotics). Angioedema is a severe manifestation with deep dermal and subcutaneous tissue swelling, frequently unilateral, non-pruritic and lasting for 1–2 h, but can persist for 2–5 days. Besides urticaria, cutaneous flushing, pruritus, nausea/vomiting, abdominal pain, nasal congestion, laryngeal edema, bronchospasm and hypotension (anaphylaxis) can occur. Serum sickness-like reaction is a rare form with fever, rash (urticaria), and arthralgias occurring 1–3 weeks after initiation of the drugs, with lymphadenopathy and eosinophilia possibly being present. Cefaclor is associated with increased risk of serum sickness-like reaction (up to 0.2%), due to the metabolite binding to tissue proteins and eliciting an inflammatory response. Other at risk drugs include cefprozil, bupropion, minocycline, rituximab and infliximab.[29] The patient in this case did not fit into this category of manifestation.

The third type is pustular eruptions most commonly present with acneiform lesions that are usually monomorphous and may appear on the arms and legs. This reaction is most commonly associated with iodides, bromides, steroids, isoniazid, phenytoin and lithium.[29]

Acute generalized exanthematous pustulosis (AGEP) is a rare form of drug reaction after 1–3 weeks of initiation and accompanied with acute febrile reaction often with leukocytosis. Desquamation usually occurs about 2 weeks later. AGEP is most commonly associated with β-lactams, macrolides and calcium channel blockers.[29] The differential diagnosis of AGEP should include pustular psoriasis, sub-corneal pustular dermatosis, pustular vasculitis and TEN (desquamative phase).

6.5 Misdiagnosis of Infective Endocarditis

Despite the fact that infective endocarditis (IE) was first described in the mid-sixteenth century and the protean manifestations of this disease have been well documented since the early 1900s,[30] the diagnosis still eludes physicians in remote and urban health care centers. The four illustrative cases in this chapter were misdiagnosed, or the diagnosis overlooked as physicians failed to look beyond the obvious presentations. In other words, they did not generate an adequate differential diagnosis that could explain the patient's initial symptoms. In all four cases, failure to consider the diagnosis of IE was at least partially (or wholly) due to incomplete history and physical examination, thus attributable to human error. It could be argued by the defendants' lawyer that IE is a relatively rare condition and these cases represent unusual or atypical manifestations, thus failure to recognize the diagnosis should not be considered negligence. To a large degree, this may be a valid argument, except there were evidences in each of these cases to indicate an infectious or cardiac abnormality that could explain the patients' symptoms. Therefore, although the misdiagnosis of a specific condition such as IE may be understandable, not performing blood cultures or referring to a suitable consultant, which could have resulted in the proper diagnosis, were the main acts of negligence.

Despite improvements in antibiotic and surgical therapy of IE over the decades, this disease still carries a high morbidity and mortality. This may partly be attributable to delayed diagnosis. Theoretically, all physicians should be familiar with the manifestations and methods of diagnosing IE, as they are present in all textbooks of medicine and taught in medical school and postgraduate training. One of the reasons physicians may be misdiagnosing IE is that the classic signs that are considered diagnostic of IE are in fact infrequent (such as splinter hemorrhages, Osler's nodes, Janeway lesions, etc). In a recent large prospective cohort of 2,781 adults with definitive IE, most patients (77.0%) present early (<30 days) without the classic hallmarks of IE.[31] Fever was the most common finding (96%), with new murmur in only 48%, degenerative valve disease being the most common predisposing factor, and significant valvular regurgitation found in 63.8% of the patients.[31] Surprisingly, in this large study, an elevation of the ESR was only found in 61%, and elevated CRP in 62%. However, it is unclear what proportion of patients had both normal ESR and CRP, and whether or not the normal first phase reactants were confined to patients with severe heart failure (32%).

Physicians should be cognizant of the fact that IE can present in a variety of different ways depending on the organ(s) involvement and that nearly any organ or viscera in the body can be affected. The symptoms or manifestations of IE are the results of one or more of the following processes: (1) invasiveness of the microorganism (valvular destruction resulting in heart failure, abscesses causing conduction disturbance), (2) immune response to invading microbes (fever, malaise, anorexia, weight loss, secondary anemia, etc), (3) embolic phenomena (strokes, abdominal pain, cutaneous lesions, hematuria, visual disturbance, etc.), (4) immune complex disease (purpura, vascular disease, kidney disease [glomerulonephritis], rheumatologic symptoms, etc), (5) secondary seeding by invasive microbes (especially *S. aureus)*, such as distant foci of infection (septic arthritis, osteomyelitis, bacterial meningitis, renal carbuncle, bacterial endophthalmitis, etc), (6) toxin induced disease (rarest form) from *S. aureus* such as toxic shock syndrome, etc.

A cardinal feature of many cases of misdiagnosed IE is the bad practice of many physicians of prescribing empiric antibiotics for fever in patients, without a clear idea of the cause or diagnosis and failure to obtain blood cultures beforehand. The majority of culture-negative IE (11.1%) with negative blood cultures was related to previous antibiotics in the preceding 7 days (61%).[31]

References

1. Mandell BF, Hoffman GS (2003). Systemic necrotizing arteritis. In: (eds) Fitzpatrick's Dermatology in General Medicine, 6th Edition (Freedberg IM, Eisen AZ, Wolff K, Austen KF, Goldsmith LA, Katz SI), McGraw-Hill, New York, p. 1718–1727.
2. Soter, NA (2003). Cutaneous necrotizing venulitis. In: (eds) Freedberg IM, Eisen AZ, Wolff K, Austen KF, Goldsmith LA, Katz SI: Fitzpatrick's Dermatology in General Medicine, 6th Edition, McGraw-Hill, New York, p. 1727–1735.
3. Camilleri MJ, Su DW (2003). Panniculitis. In: (eds) Freedberg IM, Eisen AZ, Wolff K, Austen KF, Goldsmith LA, Katz SI: Fitzpatrick's Dermatology in General Medicine, 6th Edition, McGraw-Hill, New York, p. 1047–1063.
4. Fong, IW, (2009). New perspectives of infections in cardiovascular diseases. *Curr Rev Cardiology*.
5. Grefte A, van der Giessen M, van Son W, The TH (1993). Circulating cytomegalovirus (CMV)-infected endothelial cells in patients with an active CMV infection. *J Infect Dis* 167:270–277.
6. Henson JB, Crawford TB (1974). The pathogenesis of virus-induced arterial disease – Aleutian disease and equine viral arteritis. *Adv Cardiol* 13:183–191.
7. Walker DH, Mattern WD (1980). Rickettsial-vasculitis. *Am Heart J* 100:896–906.
8. Jackman JD, Radolf JD (1989). Cardiovascular syphilis. *Am J Med* 87:425–433.
9. Somer T, Finegold SM (1995). Vasculitides associated with infections, immunization and antimicrobial drugs. *Clin Infect Dis* 20:1010–1036.
10. Reed WP, Williams RC (1977). Immune complexes in infectious diseases. *Adv Intern Med* 22:49–69.
11. Kammer GM, Soter NA, Schur PH (1980). Circulating immune complexes in patients with necrotizing vasculitis. *Clin Immunol Immunopathol* 15:658–672.
12. Sargeant JS (2009). Polyarteritis and related disorders. In: (eds) Firestein GS, Budd RL, Harris ED Jr, McFinnes IB, Ruddy S, Sargeant JS. Kelly's Textbook of Rheumatology, 8th Edition. Saunders/Elsevier, Philadelphia, p. 1453–1460.
13. Stone JH (2009). Immune complex mediated small vessel vasculitis. In: Firestein GS, Budd RL, Harris ED Jr, McFinnes IB, Ruddy S, Sargeant JS. (eds) Kelly's Textbook of Rheumatology, 8th Edition. Saunders/Elsevier, Philadelphia, p. 1465–1473.

14. Reveille JD, Maganti RM (2009). Rheumatic manifestations of human immunodeficiency virus infection. In: (eds) Firestein GS, Budd RL, Harris ED Jr, McFinnes IB, Ruddy S, Sargeant JS. Kelly's Textbook of Rheumatology, 8th Edition. Saunders/Elsevier, Philadelphia, p. 1453–1460.
15. Gherardi R, Belec L, Mhiri C, Gray F, Lescs MC, Sobel A, Guillevin L, Wechsler J (1993). The spectrum of vasculitis in human immunodeficiency virus-infected patients. A clinicopathological evaluation. *Arthritis Rheuma* 36:1164–1174.
16. Rubenfeld S, Min K-W (1977). Leukocytoclastic angiitis in subacute bacterial endocarditis. *Arch Dermatol* 113:1073–1074.
17. Sotto MN, Langer B, Hoshino-Shimizu S, de Brito T (1976). The pathogenesis of cutaneous lesions in acute meningococcemia in humans: light, immunofluorescent, and electron microscopic studies of skin biopsy specimens. *J Infect Dis* 113:506–510.
18. Keeping PWN, Aston NO, Croft DN (1980). Steroid responsive vasculitis and arthritis in meningococcal septicaemia. *Postgrad Med J* 56:736–738.
19. Ostlere LS, Harris D, Johnson M, Rustin MHA (1993). Gastrointestinal and cutaneous vasculitis associated with gonococcal infection in an HIV-seronegative patient. *J Am Acad Dermatol* 29:276–278.
20. Walker DH, Mattern WD (1980). Rickettsial vasculitis. *Am Heart J* 13:183–191.
21. Linnemann CC Jr, Pretzman CI, Peterson ED (1989). Acute febrile cerebrovasculitis. A non-spotted fever group rickettsial disease. *Arch Intern Med* 149:1682–1684.
22. Fisher JR (1977). Miliary tuberculosis with unusual cutaneous manifestations. *JAMA* 238:241–242.
23. Chan CHS, Chong YW, Sun AJM, Hoheisel GB (1990). Cutaneous vasculitis associated with tuberculosis and its treatment. *Tubercle* 71:297–300.
24. PachecoA (1991). Henoch-Schonlein vasculitis and tuberculosis. *Chest* 100:293–294.
25. Quismorio FP Jr, Rea T, Chandor S, Levan N, Frion GJ (1978). Lucio's phenomenon: an immune complex deposition syndrome in lepromatous leprosy. *Clin Immunol Immunopathol* 9:184–193.
26. Franks AG Jr, Werth VP (2003). Cutaneous aspects of cardiopulmonary disease. In: (eds) Freedberg IM, Eisen AZ, Wolff K, Austen KF, Goldsmith LA, Katz SI: Fitzpatrick's Dermatology in General Medicine, 6th Edition, McGraw-Hill, New York, p. 1625–1624.
27. Lawton AW (2004). Retrochiasmal pathways, higher cortical function, and non-organic visual loss. In: (eds) Yanoff M, Daker JS. Ophthalmology, 2nd Edition, Mosby, St. Louis, p. 1295–1303.
28. Hedges TR Jr (2004). Vascular disorders. In (eds): Yanoff M, Daker JS. Ophthalmology, 2nd Edition, Mosby, St. Louis, p. 1401–1409.
29. Shear NH, Knowles SR, Shapiro L (2008). Cutaneous reactions to drugs. In (eds): Wolff K, Goldsmith LA, Katz SI, Gilchrist BA, Paller AS, Lefell DJ Fitzpatrick's Dermatology in General Medicine, 7th Edition, McGraw-Hill, New York, p. 355–362.
30. Osler W (1909). Chronic infective endocarditis. *Q J Med* 2:219–230.
31. Murdoch DR, Corey GR, Hoen B, Miro JM, Fowler VG Jr, Bayer AS, Karchmer AW, Olaison L, Pappas PA, Moreillion P, Chambers ST, Chu VH, Falco V, Holland DJ, Jones P, Klein JL, Raymond NJ, Read KM, Tripodi MF, Utili R, Wang A, Woods CW, Cabell CH, for the International Collaboration on Endocarditis – Prospective Cohort Study (ICE-PCS) Investigators, (2009), Clinical presentation, etiology and outcome of infective endocarditis in the 21st century. *Arch Intern Med* 169:463–473.

Chapter 7
Failure to Counsel Can Lead to Litigation

7.1 Introduction

Most senior physicians still in practice can recall their earlier experiences when patients were rarely informed about their disease, much less their medications. Most patients then (and still common in developing nations) were unaware of their diagnosis, names or types of medications and potential side effects. We have come a long way since then, but some patients still remain in the dark about their illnesses or medications. In the distant past, when physicians were held in high esteem and given unquestionable authority, treatment and investigations were ordered empirically without explanation or the patients' input. Nowadays, the expectations are different and physicians are held accountable for making unilateral decisions. At present, in most developed countries, the approach to management of a patient's condition is expected to be a joint venture between the physician and patient or family. This chapter will give case scenarios where the health care system (particularly physicians) have failed the patients by omission in providing adequate information and counselling.

7.2 Case 1: Combination Treatment of Infective Endocarditis

A previously well 63-year-old male was admitted to a suburban hospital with history of fever and weight loss of several weeks duration. Several blood cultures grew a penicillin sensitive streptococcus. The patient had evidence of mitral valve regurgitation and vegetation, thus a diagnosis of sub-acute bacterial endocarditis was made. The patient was obese and weighed 105 kg with a height of 175 cm (5 ft 9 in.). Estimated lean body (or ideal body weight) was 72 kg, and the baseline serum creatinine was 1.0 mg/dL (95 μmol/L). Treatment was initiated with intravenous penicillin (18 million units/day) and gentamicin 80 mg, every 8 h after a loading dose of 100 mg. After the third day, a gentamicin blood level was reported with a peak post-dose level of 5.7 μg/mL and trough pre-dose level of 3.2 μg/mL. A dose adjustment was made and 160 mg every 24 h of gentamicin was instituted.

I.W. Fong, *Medico-Legal Issues in Infectious Diseases*, Emerging Infectious Diseases of the 21st Century, DOI 10.1007/978-1-4419-8053-3_7,

After the fourth day, the patient was transferred to a tertiary care university teaching hospital for consideration of mitral valve replacement. By then, the patient had developed evidence of congestive heart failure requiring daily dosing with furosemide. About 9 days after the onset of treatment, the serum creatinine rose to 1.6 mg/dL (144 μmol/L) and the gentamicin level showed peak of 6.0 μg/mL and trough 0.5 ug/mL. The dose of gentamicin was changed again to 80 mg every 12 h. Repeat gentamicin levels 2 days later showed a peak of 4.6 μg/mL and trough of 1.2 μg/mL. Around this time (11 days on therapy), the patient developed symptoms of dizziness, especially on standing, but there was no evidence of postural hypotension. The gentamicin was continued for 16 days and was discontinued when the dizziness, vertigo, and nausea became intolerable. The patient had successful treatment of the endocarditis and heart failure with surgical replacement of the mitral valve and 4 weeks course of antibiotics. However, he was left with permanent dizziness and unsteady gait that was diagnosed by an ear, nose and throat (ENT) specialist as permanent vestibular damage caused by gentamicin toxicity.

7.2.1 Medico-legal Issues

The patient and his wife (plaintiffs) launched litigation against the physicians involved in his care at both the suburban hospital and the tertiary care center. The charges against the physicians were mismanagement and negligence in giving treatment (gentamicin) that resulted in permanent disability. This has resulted in impairment of his function both socially and at work. Furthermore, none of the physicians ever mentioned possible adverse effects of the medications or gave him a choice on treatment. The plaintiff further stated that if he knew of the possible side effects (such as permanent vestibular damage), he would never have agreed to the gentamicin treatment.

In their defense, the lawyers for the defendants stated that the underlying disease (SBE) is a life-threatening serious medical condition, and the combination of antibiotics used was recommended by treatment guidelines. Moreover, the caring physicians were prudent and were monitoring his kidney function and gentamicin blood levels. At no time were the gentamicin blood levels high enough to cause toxicity, and the defendants argued that reasonable care was taken and they were not negligent in their management. None of the health care personnel (including physicians) denied the claim that they failed to inform or counsel the patient on the potential gentamicin toxicity, or offer alternative therapy.

7.2.2 Medical Issues

There are two main medical issues in the case that should be addressed, besides discussion on gentamicin toxicity and monitoring serum concentrations, which will be dealt with later in this chapter. First, is the issue of treatment of choice in penicillin susceptible streptococcal IE and secondly, the value of adding gentamicin to the treatment. Penicillin administered intravenously for 4–6 weeks is considered the therapy of first choice for streptococcal IE, with cure rates over 90–95%. However, non-randomized observational studies have found that combination of an aminoglycoside with penicillin can be used in sensitive strains of streptococcal IE for 2 weeks of parenteral therapy with similar cure rates.[1,2] Hence, previous and current guidelines[3] list the combination regimen as an alternative treatment. The only advantage of the combination regimen is to shorten the duration of parenteral therapy for convenience. It does not improve the morbidity and outcome. The main disadvantage of the combination therapy is the result of gentamicin ototoxicity (predominantly vestibular disturbance which can be permanent), and chance of nephrotoxicity. Hence, it is most important to counsel and fully explain to the patients the benefits, risks, and alternate therapy. In the past 20 years that I have counseled patients and offered combination therapy with gentamicin for 2 weeks versus 4 weeks of monotherapy with penicillin for *Streptococcus viridans* IE, and nearly all patients chose the longer course penicillin therapy rather than risk the chance of gentamicin toxicity.

Although the side effects and potential toxicity should be explained to patients for all forms of therapy, it is more pressing and of greater importance when there is equally effective alternate therapy that may be safer. Thus, failure to discuss the potential adverse effects of the medications (with well-known toxicity) and offer alternative standard treatment could be considered negligence by the courts.

7.3 Case 2: Ventricular Shunt Infection

A 62-year-old female with a longstanding indwelling ventricular peritoneal (VP) shunt presented to the emergency department of a tertiary care teaching hospital with fever, headaches, nausea and vomiting. She had a history of a third ventricle colloid cyst resected 9 years before and placement of the VP shunt, with recent surgical repair of a diaphragmatic hernia complicated by post-operative pneumonia just 3 months before. Physical examination and cerebrospinal fluid (CSF) examination were compatible with an infected VP shunt and meningitis. Initial therapy consisted of ceftriaxone, vancomycin and oral rifampicin, but was changed 1–2 days later to ceftazidime 6 g/day, ampicillin 12 g/day, and gentamicin 100 mg every 8 h because the CSF cultures grew Pseudomonas aeruginosa and Enterococcus faecalis. Around the same time, the VP shunt was externalized as a drain. After 48 h, the gentamicin blood levels were noted to be low and the dose was increased to 120 mg every 8 h. Subsequent gentamicin levels 5 days later reported a peak blood level of 9.8 μg/mL and trough of 2.1 μg/mL. The infectious disease consultant changed the gentamicin dose to 160 mg every 12 h and repeat levels taken 4 days later were 8 μg/mL and 1.1 μg/mL, peak and trough respectively. The serum creatinine throughout this time remained normal. The CSF cultures continued to grow the same bacteria after 10 days of therapy, and the old VP shunt was removed and replaced with a new ventriculostomy drain. Two consecutive CSF cultures subsequently failed to grow any microorganisms, and the external drain was replaced by a permanent VP shunt after 2 weeks of therapy. The patient was discharged home after spending 25 days in hospital with the plan to continue intravenous antibiotics at home consisting of piperacillin 18 g/day and gentamicin 160 mg every 12 h for another 4 weeks.

About 12 days on home intravenous therapy, the patient developed dizziness, unsteady gait and difficulty walking. Gentamicin peak level 3 days before was 11.4 μg/mL and the dose was reduced to 120 mg every 12 h. About 5 days after onset of dizziness when her symptoms persisted, the gentamicin was discontinued. An ENT consultation was obtained and testing confirmed she had gentamicin ototoxicity with vestibular damage and high frequency hearing loss. She had received a total of 5 weeks of gentamicin. Three months later, there was some improvement in the dizziness, but she had a wide-based gait and positive Romberg's test with her eyes closed.

7.3.1 Medico-legal Issues

Counsel for the plaintiff filed litigation malpractice suit against the physicians involved in her hospital care (including the infectious disease consultant who recommended the antibiotics) and the family physician (FP) monitoring the outpatient care.

The medico-legal issues raised were:

1. Was the standard of care provided by her attending and consultant physicians adequate?
2. Were the antibiotics and the treatment given appropriate and necessary?
3. Why did the treating physicians fail to inform the patient, family, and home-care nurses about the side effects of gentamicin?
4. Why was the gentamicin continued when symptoms of toxicity developed?

The defendants' defense counsel argued that the physicians were acting in good faith and provided satisfactory standard of care. Moreover, blood levels were being monitored regularly and changes in doses adjusted accordingly to prevent toxicity. Furthermore, since the blood levels were not excessively high, her side effects likely represented an idiosyncratic reaction.

7.3.2 Medical Issues

Before dealing with the issue of gentamicin adverse effect in this case (which represent a toxic reaction and not an idiosyncratic reaction [an abnormal susceptibility or reaction peculiar to the individual]), I will first discuss the appropriateness of the therapy.

VP shunt infection with meningitis is a relatively rare condition and as a result, experience in any one center is limited. Thus, most studies are based on retrospective case series collected over many years. Guidelines and recommendations for treatment are based on clinical experience, empiricism, and expert opinions; rather than controlled, randomized trials.

CSF shunt infection varies from 5% to 41%, but the usual incidence is generally 5–15%.[4] Most of the cases occur within the first month of placement, but late infection can occasionally occur. Recent series have reported a post-operative CSF shunt infection of less than 4%.[5] Increased risk of CSF shunt infection has been associated with premature birth, previous shunt infection (12–20% of new shunt infection), revision of shunt, preoperative skin preparation (including shaving), operative procedure (length of operative time, human traffic in the operating room, intra-operative use of neuroendoscope) and insertion of ventricular-atrial (VA) shunt below T7.[4] Some studies show a greater risk of infection with VP shunts compared to VA shunt.

Most studies on the microbiology of CSF shunt infections report a predominance of coagulase-negative *Staphylococci* (CoNS, in 47–64%), *Staphylococcus aureus* (12–29%), gram-negative bacilli (6–20%), mixed infection (10–15%), anaerobes (6%), and *Corynebacteria* (1–14%).[4] Both *P. aeruginosa* and *Enterococcus* species are relatively rare causes of CSF shunt infection. It is likely in this case that the recent abdominal surgery (repair of diaphragmatic hernia) was the predisposition for CSF shunt infection. Late VP shunt can also occur from blockage or dysfunction with recent revision of the shunt, but rarely from erosion of the catheter tip into the bowel, or from frank peritonitis, secondary to ruptured viscus.

Once a CSF shunt infection is suspected, it is best to obtain CSF from the shunt reservoir for analysis, cell count, glucose, gram stain, and culture. Empiric therapy for those consistent with bacterial meningitis consists of broad-spectrum antibiotics such as vancomycin (for staphylococcal coverage) with ceftazidime, cephapirin or carbapenem (imipenem/meropenem).[4] The antibiotics are then adjusted according to culture and susceptibility. For *P. aeruginosa* meningitis and CSF shunt infection, the best choice of antibiotic (and largest experience) is ceftazidime, but there is no good evidence that adding an aminoglycoside improves the outcome.[6] There is limited data on enterococcal meningitis or CSF shunt infections, but ampicillin 12 g/day for susceptible strains would be recommended; or vancomycin for more resistant strains and penicillin allergies. Although many guidelines and textbooks recommend adding gentamicin for life-threatening enterococcal infections, there is no evidence to support this advice except for enterococcal endocarditis (where it appears that gentamicin for 2 weeks as part of the combination therapy is optimal).[7]

The largest clinical study of enterococcal meningitis collected 39 cases at two hospitals over 25 years, and included 101 cases from previous reports in their review.[8] Among the 140 cases reviewed, 82 cases (59%) were post-operative (or shunt related) and 58 cases (41%) were spontaneous. The outcome or mortality was not improved with combinations of gentamicin (22%) versus monotherapy (16%), with either ampicillin (18% mortality) or vancomycin (14% mortality).[8] In this study, the median duration of therapy was 18 days and the mortality was higher in spontaneous meningitis (33%) versus post-operative meningitis (12%), $p < 0.001$.[8]

An important aspect of the management of CSF shunt infection is removal of the entire shunt and insertion of an external ventricular drain, as this represents a biofilm infection which cannot be cured without removal of the foreign body. Attempts to preserve the shunt and treat with antibiotics alone have resulted in failure, longer hospital stays and greater mortality.[4] Externalization of the infected shunt alone is usually ineffective and results in persistent infection with continued recovery of the organism.

Some antibiotics penetrate the blood-brain barrier poorly and systemic administration results in very low CSF concentration (all aminoglycosides). Therefore, if gentamicin is deemed necessary for management of meningitis or CSF shunt infection, it is best given via the intraventricular drain (1–8 mg/day),[4] and this would avoid the systemic toxicity of the drug. Occasionally, it is necessary to give additional doses of vancomycin via the ventricular drain if there is persistent growth of the bacteria (i.e., CoNS) after removal of the infected shunt. Some studies have found variable penetration of vancomycin in the CSF and the intraventricular dose from 5 to 20 mg/day.[4] Another approach that is being commonly used for biofilm, device-related staphylococcal infections is the addition of oral rifampicin as combination therapy, despite lack of clinical studies to support the value.

What should be the duration of therapy for CSF shunt infection? There is no valid study to assess the duration of antibiotic treatment in meningitis or CSF shunt infection. For a CoNS infection, Turkel and Kaufman[4] recommend at least 7 days of antibiotics, plus negative CSF cultures for 48 h before re-shunting. In gram-negative bacilli infections they also recommend 3 weeks of antibiotics, while others suggest 2 weeks of treatment, then stopping the antibiotics for 3 days and repeating CSF culture. Then, if negative, proceed to re-shunting.[9] Overall, most guidelines for management of CSF shunt infections only recommend antibiotics for 2–3 weeks.

Based on the current scientific medical data, the case under discussion should not have required any antibiotics on discharge. The case should have been managed with prompt complete removal of the VP shunt and insertion of an external ventricular drain. Systemic gentamicin was never indicated, and if it were used, the aminoglycoside should have been administered intraventricularly for a few days. In this case, it would have been reasonable to treat with antibiotics for 2–3 weeks, then after discontinuation to repeat the CSF culture before re-shunting. An alternate approach would be to give 2 weeks of antibiotics, and if three daily CSF cultures were negative, to re-insert a new shunt with continuation of the antibiotics until 1 day after re-shunting (personal experience).

7.4 Case 3: Pelvic Infection

A 42-year-old obese female (with a BMI of 43) was admitted via the emergency department to a community hospital with acute lower abdominal pain and fever. She was febrile, with a fever of 39°C, lower abdominal tenderness and guarding, and cervical tenderness on bimanual palpation. The white blood cell count was markedly elevated (24,000 cells/dL), and a pelvic ultrasound revealed a complex ovarian cystic mass. She was admitted to the gynecology service with a diagnosis of pelvic inflammatory disease (PID). An emergency laparoscopy was performed the same day for drainage of a tubo-ovarian abscess; and treatment was initiated intra-operatively with gentamicin 500 mg every 24 h, clindamycin 900 mg every 8 h, and cefotaxime 2 g every 8 h. Her baseline serum creatinine was normal.

There was no significant past illness such as sexually transmitted disease (STD), no intrauterine device (IUD), and no sexual activity for over 6 weeks, but her menstrual period had occurred a few days before. The patient had a history of penicillin allergy (drug rash), migraine, and ovarian cyst. Results of the drained pus were available 4 days later as growing only group A Streptococcus resistant to clindamycin and the macrolides. Clindamycin was discontinued and metronidazole added instead to the two other antibiotics. Gentamicin concentration was being monitored with acceptable blood levels – trough of 0.3–0.8 μg/mL and peak concentration of 20.5–24.6 μg/mL. Serum creatinine was initially 0.7 mg/dL (61 μmol/L) and gradually increased to 1.1 mg/dL (98 μmol/L) by day 10 and 1.3 mg/dL (118 μg/L) on day 20. The combination with gentamicin was finally discontinued after 20 days of treatment. Three days before stopping antibiotics, the patient experienced nausea, vomiting, and dizziness. She was assessed by an ENT specialist 2 weeks later and was found to have imbalance, unstable gait, and oscillopsia, but normal hearing. A diagnosis of gentamicin vestibular toxicity was made. Six months later, she was found to have some improvement, but she still could not drive and suffered from imbalance at night.

7.4.1 Medico-legal Issues

The patient hired a lawyer to institute medical malpractice litigation against the gynecologist for negligence of care, failure to inform the patient of adverse effects of the gentamicin, and failure to use a safer alternative antibiotic. The patient sought compensation for inability to perform her job (her employment required frequent traveling by car), and significant impairment of her social life.

The physician's defense was that gentamicin is recommended as part of a combination regimen that is standard treatment for severe PID. Moreover, careful monitoring of her renal function and gentamicin levels did not reveal excessive concentrations. Thus, the defendant's lawyer argued that the charges of negligence and medical malpractice against his client should be dismissed.

7.4.2 Medical Issues in PID

Pelvic infections (PID) can be related to STDs, post-delivery or caesarian section, or following other gynecologic surgery (hysterectomy); and sometimes occurring spontaneously without any of the above. Spontaneous PID in healthy young women is most commonly due to *Neisseria gonorrheae* (in the first episode) or *Chlamydia trachomatis*. The latter infection is most commonly sub-acute or sub-clinical and usually is not associated with peritoneal signs such as guarding or very high fever and leukocytosis.

Previous PID with *N. gonorrheae*, IUD, post-delivery or post-hysterectomy are commonly associated with mixed bacterial infection in the setting of acute PID. The microorganisms commonly recovered from pelvic drainage usually include non-group A *Streptococci*, coliforms and anaerobes including *Bacteroides* species. Hence, guidelines developed in the early 1980s recommended clindamycin and gentamicin as a suitable regimen for hospitalized patients with severe PID.[10] It has always been the standard practice and recommendation to alter the initial empiric treatment to a more streamlined and safer regimen once culture and susceptibility of the bacterial isolates were available (usually within 3–4 days). Recently, there has been availability of several different regimens that are just as effective as clindamycin and gentamicin for mixed infection, but with safer profiles (i.e., cefoxitin, ampicillin-sulbactam, piperacilin-tazobactam, ertapenem, moxifloxacin or levoquin + metronidazole). Thus, in most cases, gentamicin would be discontinued after 3–4 days (when used) and substituted with a safer agent (i.e., quinolone). The duration of therapy for even severe PID is usually 10–14 days, with switching from parenteral to oral therapy at least 48 h after defervescence.[10]

Streptococcus pyogenes is an unusual cause of PID, but is occasionally found in puerperal infections or post-operatively; either in patients endogenously colonized with group A *Streptococcus* (vagina or rectum), or introduced nosocomially. In this case, the patient probably had vaginal colonization with group A *Streptococcus* and the organism ascended in the uterine cavity around the time of her menstrual period (when there is less protection by the normal cervical mucus plug). Once the culture report was available, the optimal therapy would have been to use monotherapy with cefazolin (intravenously) until the patient was afebrile for 48 h, followed by oral cephalexin for another 10 days (as the patient had mild penicillin drug reaction in the past).

In summary, the management of the case under discussion may be considered reasonable for the initial few days, but once the culture results were available, the continuation of a combination of antibiotics with gentamicin was inappropriate and unwarranted. Failure to utilize the culture report in the decision on definitive therapy and unnecessarily prolonged use of a toxic agent could be considered adequate criteria for medical malpractice or negligence on the part of the treating physician. Even the initial combination of cefotaxime with gentamicin was unnecessary, and one of the two could have been selected for the gram-negative (aerobic) bacilli coverage.

7.5 Case 4: Diabetic Foot Infection

In the summer of 2005, a 34-year-old diabetic male was admitted to a small community hospital with evidence of foot infection. Radiography of his foot revealed evidence of localized septic arthritis and osteomyelitis of the fourth metatarsal-phalangeal joint. Clinically, he was not in a "toxic" condition and his vital signs were normal except for a temperature of 38°C. His left foot revealed a superficial necrotic ulcer on the sole, some swelling, and redness of the foot. A swab from the foot ulcer and blood cultures was obtained in the emergency department before starting intravenous clindamycin and oral ciprofloxacin. The patient was admitted under the care of an orthopedic physician and an internist was consulted.

Review of past history revealed that the patient had suffered from a nail puncture injury 4 months before and was admitted to a hospital with an abscess and cellulitis of the same foot. Previous surgical drainage (2½ months before the latest admission) and culture of the pus grew S. aureus, group B Streptococcus and mixed gram-negative anaerobes. He was treated at that time with oral ciprofloxacin and intravenous cefzalin, and then discharged on oral cephalexin. A rash developed soon after discharge and the cephalexin was changed to ciprofloxacin 500 mg twice a day for 10 days.

On the final admission, no surgical procedure was performed (besides superficial debridement of the ulcer), and the patient was discharged after 4 days on gentamicin 500 mg once daily, ceftazidime 2 g every 8 h, and metronidazole 500 mg every 8 h. Culture report of the wound swab revealed S. aureus and group B Streptococcus only and the blood culture was negative. This report was available on the patient's chart on the day of discharge. The baseline serum creatinine was 0.8 mg/dL (76 μmol/L), and the gentamicin blood levels showed a peak concentration of 14 μg/mL and trough of 0.6 μg/mL. The medication was changed 1 day after discharge to gentamicin 300 mg once daily and cloxacillin 1 g every 6 h intravenously. A month later (on home intravenous therapy), the patient returned to the emergency department with symptoms of vertigo (aggravated by head movements), nausea, vomiting, hearing loss and tinnitus lasting 4 days. He was treated with dimenhydrinate 50 mg every 6 h, and referred to the orthopedic clinic for follow up.

Nine days after, while still on the same antibiotic, the patient returned to the emergency department with persistent and worsening vertigo, nausea and unsteady gait. Serum creatinine then had risen to 1.5 mg/dL (138 μmol/L), but the gentamicin trough level was 0.6 μg/mL. Gentamicin toxicity was then suspected, the antibiotic discontinued, and he was referred to an ENT specialist. Bilateral ototoxicity and vestibular toxicity attributable to gentamicin was confirmed. Two years later, the patient was noted to have persistent bilateral vestibular disturbance causing nausea, vertigo and unsteady gait.

7.5.1 Medico-legal Issues

The patient subsequently sued the orthopedic surgeon, internist and the emergency attending physician for medical malpractice. Claims were for financial compensation for loss of income from unemployment, impaired future prospects of employment, and suffering that affected his lifestyle.

Specific charges claim that both the orthopedic surgeon and the internist were negligent and provided substandard care for prescribing long-term gentamicin and not informing the patient of the potential side effects. Moreover, the plaintiff's lawyer had obtained expert opinion that gentamicin was not needed in the first place. The emergency physician was negligent in not recognizing that the patient had gentamicin toxicity when he attended the ER with vertigo, thus allowing continued use of the drug for another 9–10 days after presenting with ototoxicity. This resulted in greater and permanent vestibular damage. The ER physician care was substandard, as he should be aware of the symptoms of gentamicin toxicity, and should have discontinued the drug by contacting the FP as soon as the symptoms of ototoxicity occurred.

7.5.2 Medical Aspect of Diabetic Foot Infection

Foot infections in diabetic patients are common, usually resulting from a break in the integument (skin ulceration), secondary to neuropathy and vasculopathy. Acute superficial ulceration or paronychia (or no obvious skin defects) are usually caused by single organism infection, most commonly *S. aureus* or *Streptococcus* species. Deep penetrating, chronic, ulcers (particularly on the sole of the foot) are usually complicated with mixed infection (with three to four organisms) such as *S. aureus, Streptococci,* coliforms and anaerobes.[11] These chronic ulcers or sinuses are frequently complicated by septic arthritis or osteomyelitis of the foot.

Although cultures of the foot ulcer can be misleading in determining the etiologic organisms of the infection, it can be helpful if interpreted with proviso. For instance, mixed growth of coliforms on ulcers may be over-represented and are often superficial colonizers, but their absence on the surface is against deeper involvement. On the other hand, superficial swabs will fail to grow most anaerobes, but their absence does not exclude their role in infection. When *S. aureus* or *Streptococci* are recovered as the only organisms from open wounds, ulcers, or sinuses, they are usually the etiology of the associated cellulitis or deep soft tissue infection.[12]

In the present case, cultures of the patient's foot only grew *S. aureus* and group B *Streptococcus,* but no coliforms. Thus, antimicrobial coverage for coliforms was not necessary once the culture report was available. However, mixed infection with anaerobes would not be excluded. In fact, the previous cultures from a deep abscess in the preceding hospital admission had grown the same gram-positive bacteria, plus anaerobes. Thus, a suitable agent while in the hospital would have been intravenous cloxacillin plus oral metronidazole, and continued oral therapy upon discharge with amoxicillin/clavulanic acid for 2–3 months for osteomyelitis/septic arthritis.

In this case, there was no medical indication for continued gentamicin therapy. Gentamicin should be limited to short-term therapy (unless strongly indicated with no alternative), especially in diabetics and the elderly (who are more prone to renal impairment and ototoxicity), for only 3–5 days, as safer alternatives are commonly available.

7.6 Gentamicin Toxicity and Litigation

Aminoglycosides, such as gentamicin, are potentially toxic to the kidney and the auditory and vestibular apparatus. Studies in experimental animals and humans demonstrate that aminoglycoside-induced damage to the kidney is primarily a defect of the proximal tubular cells.[13] Accumulation of the aminoglycoside takes place in the renal cortex, and the most important correlations with nephrotoxicity are duration, total dose, and state of hydration.[13] A threshold of duration for nephrotoxicity has not been clearly defined, but development of nephrotoxicity before 5 days of therapy is unusual.[14] Although previous studies with multiple daily dosing of gentamicin had suggested that a pre-dose level of ≥2.0 μg/mL and a post-dose level >10 μg/mL were associated with increased nephrotoxicity, it is believed that the rise in serum levels in patients experiencing toxicity is probably due to renal impairment, rather than the toxicity being its cause.[14]

The aminoglycosides are ototoxic drugs with a narrow margin of safety and therapeutic index, and this is a major limiting factor in the clinical use of these antibiotics. This is of special clinical significance because their effects result in permanent loss of inner ear function. The ototoxic effects of the aminoglycoside usually become apparent after repeat doses of the drug and onset may be delayed even after discontinuation. The ototoxic effects of the aminoglycoside (based on animal experiments) is related to high perilymph and endolymph concentrations and persistence of these antibiotics in the inner ear fluid (half-life of 15–25 h).[15] The aminoglycosides cause the death of hair cells in the organ of Corti and chronic administration may lead to extensive damage and hearing loss (initially high frequency sound). The initial impairment may improve, but there is a point at which the hair cell damage is so severe it is irreparable.

Clinical studies have demonstrated that vestibular toxicity is more frequent than cochlear toxicity with gentamicin.[16] Endolymphs containing high concentrations of aminoglycoside bathe the vestibular neuroepithelium tissues of the inner ear to maintain equilibrium and reflex control of the eyes. The reason why a particular aminoglycoside may predominantly affect the vestibular or cochlear apparatus is not clear. The initial manifestation of toxicity is probably an alteration in the ionic balance of the endolymph, followed by interaction between the aminoglycoside and the cell membrane lipids which may alter permeability, and subsequent respiratory functions.[16] The patient initially may present with fullness in the ear, tinnitus, dizziness, nausea, and vomiting. The subsequent high frequency hearing loss is followed by hearing loss of conversational tones and impaired equilibrium.

Ototoxicity due to aminoglycosides is closely associated with renal impairment, probably from failure to properly reduce the dose. When the aminoglycoside levels are maintained within a predetermined range, ototoxicity and nephrotoxicity are independent events.[16] Although initial studies suggested that peak concentration of gentamicin >13 μg/mL correlated with ototoxicity, trough concentrations >2 μg/mL are more closely related to toxicity than peak serum concentrations. However, in a review on the subject of vestibular toxicity of gentamicin, it was concluded that cumulative dose is the greatest risk factor and the toxic effect was unrelated to serum concentrations.[17] Currently, ototoxicity and nephrotoxicity of gentamicin are considered to be best correlated with duration, total accumulated dose, age, and renal function.[18] Accurate rates of nephrotoxicity and ototoxicity to gentamicin have been difficult to establish, but mild renal impairment has been reported in 8–26%; ototoxicity with standard duration of therapy (about 7 days) is usually low at 2–5%, but can be as high as 25%, with longer duration of therapy.

There is recent evidence that some people have an inherited predisposition that renders them highly susceptible to the ototoxic effect of the aminionglycosides.[19] Even a single dose in predisposed individuals can result in permanent hearing loss.[20] In countries where aminoglycosides are used widely because of their low cost, familial cases of aminoglycoside hearing loss occurs in about 25% of the ototoxic cases and are associated with much shorter courses.[19] The most common predisposing mutation is the m.1515A > G polymorphism, a mitochondrial DNA mutation. In China, where aminoglycosides are commonly used, this mutation accounts for at least 22–59% of aminoglycoside ototoxicity.[19] This mutation of the mitochondrial DNA makes the human mitochondrial ribosome similar to the bacterial one, facilitating aminoglycoside binding to the hair cells of the inner ear (half-life of several months).[19] In some populations, the m.1555A > G mutation seems to be a common cause of deafness by itself. In Spain, 27% of families with at least two deaf individuals were positive for this mutation, and everyone with the mutation who was exposed to aminoglycosides became deaf.[21] Hence, aminoglycosides should be avoided in patients with familial history of deafness, unless rapid testing can be performed to exclude the 1555A > G mutation. The reports so far have mainly found enhanced cochlear damage (deafness) and not vestibular disturbance with this mutation.

Gentamicin is often used to treat patients with IE such as *S. aureus*, *Streptococcus* species, and *Enterococcus* species. In a recent prospective cohort study of 373 patients with IE, 289 (77%) received gentamicin (mean duration of 14 days).[22] The nephrotoxic effect of gentamicin was directly related to the treatment duration, with a decrease of endogenous creatinine clearance of 0.5% per day of treatment. In another recent prospective study, even adding initial low-dose gentamicin for a few days in *S. aureus* bacteremia and endocarditis is nephrotoxic. A total of 22% of patients who received low dose gentamicin versus 8% of patients not receiving gentamicin, experienced decreased creatinine clearance (mild renal impairment).[23] It should be noted however, that evidence from randomized trials does not support the use of aminoglycosides in staphylococcal or streptococcal endocarditis, and the evidence is limited in enterococcal IE (based on retrospective and prospective cohort non-randomized studies).[24]

7.6.1 Once-Daily Dosing of Aminoglycosides

For more than a decade, once-daily dosing of aminoglycosides has replaced multiple daily doses in North America and Western Europe. Numerous in vitro and animal studies have supported using once-daily aminoglycoside dosing except for synergistic effect in enterococcal IE.[25] The use of a high single dose of aminoglycoside over an extended interval appears to optimize bacterial killing (concentration dependent killing) with an extended period of post-antibiotic effect (0.5–8 h) of bacterial inhibition. Animal testing has largely supported the concept that once-daily dosing is more effective and appears less toxic than more frequent dosing. Clinical studies show at least equal effectiveness and no greater toxicity when compared with traditional dosing.[25] Theoretically, less frequent dosing of aminoglycoside is desirable to allow efflux of the aminoglycoside molecules form the renal proximal tubules and cochlear or vestibular cells. The clinical benefit of a high, once-daily dose of aminoglycoside seems to be greatest for severe gram-negative bacillary infections and to achieve a high peak concentration of eight to ten times the usual MIC (minimal inhibitory concentration) of gentamicin or tobramycin for susceptible strains (0.5–4 μg/mL). For gentamicin or tobramycin, 5–7 mg/kg is given in a single daily dose in these circumstances. Monitoring peak levels of the aminoglycoside is not considered necessary and pre-dose concentration of <1 μg/mL has been recommended for once-daily dosing of gentamicin or tobramycin.[26]

7.6.2 *Medico-legal Aspects of Gentamicin Toxicity*

In this chapter, four cases of gentamicin toxicity were described that resulted in medical malpractice litigation under different conditions. A striking feature of these cases is twofold: (1) gentamicin was not necessary, especially for prolonged management of their infections; (2) the patients were never counseled on the potential side effects or offered alternative therapy.

In a recent report from the Canadian Medical Protective Association (CMPA) during a 5 year period from 2002 to 2006, 423 legal actions were initiated against physicians.[27] Of these, 116 (27%) involved antibiotic administration and aminoglycosides were cited in 16 (14%) of the antibiotic cases.

A review of the CMPA experience from 1984 to 2006 identified 65 cases of medico-legal problems resulting from alleged aminoglycoside toxicity, with 62 resulting in legal actions.[27] A variety of disciplines were involved in these cases (family medicine 18%, internal medicine 15%, general surgery 12%, other surgical subspecialties 17%, medical subspecialties and others 29%, and obstetrics and gynecology 5%). The average time between onset of therapy and first appearance of toxicity symptoms was 24 days (range 3–43 days). A remarkable feature (similar to the cases described) was that in 24 cases (43%) the drug was continued for more than 72 h after onset of symptoms or signs of toxicity. Vestibular disturbance accounted for 86% of the toxicity, with nephrotoxicity in 7% and cochlear toxicity in 7%. Gentamicin was the antibiotic used in 53 cases (95%) and the average duration of therapy was 25 days (range 4–56 days). Although frequent monitoring of aminoglycoside levels and creatinine were only performed in 26 cases (46%), the drug levels were usually within acceptable therapeutic ranges.[27] Six of the cases involved topical aminoglycosides and all were in the presence of non-intact tympanic membrane with average duration of 86 days. Oral neomycin was used for 240 days in a patient with hepatic and encephalopathy which resulted in cochlear toxicity. Of major importance in this report from CMPA, is that consent (or failure to discuss risk-benefit ratio) was identified as an issue in 92% of the legal actions. Furthermore, an organism was cultured in only 33 cases (50%) and an alternative antibiotic could have been be used (based on susceptibility data) in 87% of those with an isolate.[27]

What are the take home messages for physicians? Physicians should always discuss the risk versus benefit of any therapy with their patients. It is very uncommon for there to be one primary agent for treatment. The healthcare team has to be cognizant of the known adverse effects of the medications and to discontinue them at the first sign or symptom of toxicity (if possible). Aminoglycosides should be avoided for prolonged use (hardly any indication for prolonged use) in the elderly and in patients with underlying renal and auditory impairment or dizziness. If aminoglycosides are initiated, their use should be curtailed to the first 3–4 days, and then switched to safer alternatives once culture and susceptibility results are available. There is very little evidence to support combined use of aminoglycoside for synergistic effect and this trend should be discouraged. Standard care of patients receiving aminoglycosides include monitoring serum creatinine and blood levels one to two times per week, but keep in mind that toxicity often occurs within acceptable drug levels.

Topical aminoglycosides should be avoided in the presence of a non-intact tympanic membrane (and their value in these circumstances is very questionable). In recent years, oral neomycin for hepatic encephalopathy has largely been replaced by lactulose, which is safer.

References

1. Wilson WR, Thompson RL, Wilkowske CJ, Washington JA 2nd, Giuliani ER, Geraci, JE (1980). Short-term therapy for streptococcal infective endocarditis: combined intramuscular administration of penicillin and streptomycin. *JAMA* 245:360–363.
2. Sexton DJ, Tenebaum MJ, Wilson WR, Steckelberg JM, Tice AL, Gilbert D, Dismukes W, Drew RH, Durack DT, and the Endocarditis Treatment Consortium Group (1998). Ceftriaxone once daily for four weeks compared with ceftriaxone plus gentamicin once daily for two weeks for treatment of endocarditis due to penicillin-susceptible streptococci. *Clin Infect Dis* 27:1470–1474.
3. Baddour LM, Wilson WR, Bayer AS, Fowler VG Jr., Bolger AF, Levison ME, Ferrieri R, Gerber MA, Tanl LY, Gewitz, MH, Tong DC, Steckelberg JM, Baltimore RS, Shulman ST, Burns JC, Falace DA, Newburger JW, Pallasch J, Takahshi M, Tambert KA, (2005). Infective endocarditis: diagnosis, antimicrobial therapy and management of complications: a statement for healthcare professionals from the Committee on Rheumatic Fever, Endocarditis and Kawasaki Disease, Council on Cardiovascular Disease in the Young, and the Councils on Clinical Cardiovascular Surgery and Anesthesia, American Heart Association: Endorsed by the Infectious Disease Society of America. *Circulation* III:e394–e432, DOI:10.1161/ *Circulation AHA* 105.165564.

4. Tunkel AR, Kaufman BA (2005). Cerebrospinal shunt infections. In: (eds.) Mandell GL, Bennett JE, Dolin R. Principles and Practice of Infectious Diseases, 6th Edition. Elsevier/ Churchill/Livingstone, Philadelphia, p. 1126–1132.
5. Morrisette I, Gourdeau M, Francoeur J (1993). CSF shunt infections: a fifteen-year experience with emphasis on management and outcome. *Can. J Neurol Sci* 20:118–122.
6. Fong IW, Tomkins KB (1985). Review of *Pseudomonas aeruginosa* meningitis with special emphasis on treatment with ceftazidime. *Rev Infect Dis* 7:604–612.
7. Fong IW (ed) (2009). Combination antimicrobial therapies. In: Emerging issues and controversies in infectious disease. Springer, New York, p. 337–377.
8. Pintado V, Cabellos C, Moreno S, Meseguer MA, Ayats J, Viladrich PF (2003). Enterococcal meningitis: a clinical study of 39 cases and review of the literature. *Medicine*; 82:346–364.
9. Chapman PH, Borges LF (1985). Shunt infections: prevention and treatment. *Clin Neurosurg* 32:652–664.
10. Sweet RL, Gibbs RS (eds) (1985). Pelvic inflammatory disease. In: Infectious diseases of the female genital tract; Williams & Wilkins, p. 53–77.
11. Lipsky BA, Berendt AR, Deary HG, Embil JM, Joseph WS, Karchmer AW, LeFrock JL, Lew DP, Mader JT, Norden C., Tan JS (2004). Diagnosis and treatment of diabetic foot infections. *Clin Infect Dis* 39:885–910.
12. Fong IW (1996). Management of diabetic foot infection: a position paper. *Can J Infect Dis* 7:361–365.
13. Whelton A (1982). Renal tubular transport and intrarenal aminoglycoside distribution. In: (eds.) Whelan A, Neu HC. The aminoglycoside: microbial, classical use, and toxicology. Marcel Dekker, Inc., New York, p. 191–221.
14. Smith TR, Smith CR (1982). Risk factors for aminoglycoside nephrotoxicity. In: (eds.) Whelan A, Neu HC. The aminoglycoside: microbial, classical use, and toxicology. Marcel Dekker, Inc., New York, p. 401–415.
15. Brummett RE, Fox KE (1982). Studies of aminoglycoside ototoxicity in animal models. In: (eds.) Whelan A, Neu HC. The aminoglycoside: microbial, classical use, and toxicology. Marcel Dekker, Inc., New York, p. 453–486.
16. Blendush CL (1982). Ototoxicity: clinical considerations and comparative information. In: (eds.) Whelan A, Neu HC. The aminoglycoside: microbial, classical use, and toxicology. Marcel Dekker, Inc, New York, p. 453–486.
17. Guiliunas P, Dominguez-Moreno M, Lazarus JM, Lowrie EG, Gottlieb MN, Merrill JP (1978). Vestibular toxicity of gentamicin. *Arch Intern Med* 138:1621–1624.
18. Chambers HF (2005). Aminoglycosides. In: (eds) Brunton LL, Lozo JS, Parker KL Goodman and Gilman: The pharmacological basis of therapeutics, 11th Edition, McGraw-Hill, New York, p. 1155–1171.
19. Bitner-Glindzicz M, Rachman S (2007). Ototoxicity caused by aminoglycosides is severe and permanent in genetically susceptible people. *BMJ* 335:784–785.
20. Usami S, Abe S, Shinkawa H, Kimberling WJ (1998). Sensorineural hearing loss caused by mitochondrial DNA mutations: special reference to A1555G mutation. *J Commun Disord* 31:423–434.
21. Estivill X, Goven N, Barcelo F, Badenas C, Romero ER, Moral L, Scozzri R, D'Urbano L, Zevianni M, Torroni A (1998). Familial progressive sensorineural deafness is mainly due to mt DNA A1555G mutation and is enhanced by treatment of aminoglycosides. *Am J Hum Genet* 62:27–33.
22. Buchholtz K, Larsen CT, Hassager C, Bruun NE (2009). Severity of gentamicin's nephrotoxic effect on patients with infective endocarditis: a prospective observational cohort study of 373 patients. *Clin Infect Dis* 48:65–71.
23. Cosgrove SE, Vigliani GA, Campion M, Fowler VG Jr, Abrutyn E, Corey R, Levine DP, Rupp M, Chambers HF. Karchmer AW, Boucher HW (2009). Initial low-dose gentamicin for *Staphylococcus aureus* bacteremia and endocarditis is nephrotoxic. *Clin Infect Dis* 48:713–721.

24. Leibovici L, Vidal L, Paul M, (2009). Aminoglycoside drugs in clinical practice: an evidence-based approach. *J Antimicrob Chemother* 63:246–251.
25. Freeman CD, Nicolau DP, Belliveau PP, Nightingale CH, (1997). Once daily dosing of aminoglycosides: review and recommendations for clinical practice. *J Antimicrob Chemother* 39:677–686.
26. Reeves DS, MacGowan AP, (1993). Once-daily aminoglycoside dosing. *Lancet* 341:895–896.
27. Berringer R, (2008). Aminoglycoside therapy balancing risk versus benefit. Risk identification for all physicians. *Can Med Protect Assoc* R10814E: www.cmpa-acpm.ca.edu.

Chapter 8
Litigations for Unexpected Adverse Events

8.1 Case 1: Drug-Induced Hepatitis

A 53-year-old Iranian female who immigrated to Canada about 3.5 years before was referred to an internist for a positive Mantoux skin test (11 mm in diameter). The subject was previously well with no symptoms indicative or suggestive of active tuberculosis. A routine tuberculosis skin test was performed because the patient had applied to be a volunteer at a local hospital. She had no significant past illness or known allergies, and she was never diagnosed with nor had known contact with anyone with active tuberculosis. The subject never ingested alcohol and was not known to have hepatitis or be a carrier of any hepatitis virus. Baseline investigations performed by the internist included routine complete blood count, routine biochemical tests (liver enzymes, creatinine, and glucose), serum ferritin, and thyroid-stimulating hormone – all of which were normal. A chest radiograph was reported to be normal.

The patient was prescribed isoniazid 300 mg once daily and pyridoxine 25 mg once daily to be taken for 9 months as treatment for latent tuberculosis. At the follow-up appointment 3 months later, her only symptom was that of knee pain, which was treated as osteoarthritis with diclofenac, a non-steroidal anti-inflammatory drug (NSAID). Five weeks later, she returned to see the internist with a history of increased dark colored urine and yellowish skin discoloration for a few weeks. Blood tests were ordered and patient was referred to a gastroenterologist. She was assessed by the gastroenterologist a week later, who noted symptoms of dark urine and yellowish skin discoloration for a month. The specialist noted the patient to be mildly icteric with a bilirubin of 51 μmol/L (normal $<$20 μmol/L), the alkaline phosphatase (ALP) was 352 μ/L (normal 25–96 μ/L), and the serum glutamate-oxaloacetate transaminase (SGOT) was 1,102 μ/L (normal 4–28 μ/L). The isoniazid was then discontinued and further investigations were performed. Serologies for the hepatitis viruses (A, B, and C) revealed no acute infection but immunity to hepatitis A and B, and a liver ultrasound was normal.

The patient's symptoms over the following week worsened with jaundice, anorexia, malaise, and distention of the abdomen. She was then admitted to a hospital emergency department. Repeat blood tests revealed normal blood count,

I.W. Fong, *Medico-Legal Issues in Infectious Diseases*, Emerging Infectious Diseases of the 21st Century, DOI 10.1007/978-1-4419-8053-3_8,

creatinine, glucose and electrolytes; but the bilirubin had risen to 219 μmol/L, the SGOT was 978 μ/L, the serum alanine aminotransferase (ALT) was 641 μ/L (normal 10–45 μ/L), ALP 300 μ/L, and the prothrombin time 2.2 s. A repeat ultrasonography of the abdomen revealed large ascites and a liver of 13 cm in length with normal contour. Over the next 2 weeks, she became drowsy and encephalopathic, and was transferred to a tertiary care hospital where a liver transplantation was successfully performed (live donor from the patient's daughter). Pathology of the liver showed a markedly shrunken liver with signs of fulminant hepatitis, with negative stains for hepatitis B antigens.

8.1.1 Medico-legal Issues

A lawsuit was subsequently launched by the patient (plaintiff) against the physician who prescribed the isoniazid. The statement of claim alleged the following: (1) isoniazid was directly responsible for the plaintiff's fulminant hepatitis which resulted in the need for a liver transplant, (2) informed consent was never obtained to prescribe the drug, as the plaintiff was never counseled on the adverse effects, nor given a choice of treatment, (3) use of the isoniazid was never indicated, as the patient had no symptoms or signs of active disease, (4) the physician should have realized that the positive Mantoux test was due to a previous BCG vaccination as a child (the defendant was informed of this fact) and therefore there was no need to treat the plaintiff for latent tuberculosis.

Based on the above facts, the internist was negligent in prescribing isoniazid and he should have monitored her liver enzymes after initiation of treatment (according to the statement of claims). The lawyer for the plaintiff further stipulated that if his client were never treated unnecessarily for latent tuberculosis, she would not have suffered from fulminant hepatitis or required a liver transplant. Hence, the treating physician provided substandard care and compensation was sought for pain and suffering of the plaintiff, as well as for the daughter who underwent partial hepatectomy for liver donation.

8.1.2 Medical Issues

There are several medical issues that need to be addressed here in order to make a valid judgment of the plaintiff's claims. (1) First, were there medical indications for the use of isoniazid? (2) Did the physician obtain adequate consent for treatment? (3) Can we be sure that the fulminant hepatitis was due to isoniazid? (4) And lastly, could the severe adverse event have been avoided with proper monitoring?

The indications for treatment of latent tuberculosis as recommended by the Center for Disease Control and Prevention (CDC) and the American Thoracic Society are shown in Table 8.1.[1]

Table 8.1 Indications of treatment of latent tuberculosis

	Mantoux reaction (size)
Very high risk	
HIV/immunosuppression, anti-TNF drugs	≥5 mm
Close contact of active pulmonary TB	≥5 mm
Fibrotic changes on chest x-ray	≥5 mm
Children <5 years old	0–4 mm
Start INH for close contact	
Repeat Mantoux in 8–12 weeks, if negative can be stopped.	
High risk	
Recently infected (≤2 year)	≥10 mm
IVDA/other drug abuse (i.e. crack cocaine)	≥10 mm
High risk conditions (chronic renal failure, diabetes, silicosis, short gut, intestinal bypass, post-gastrectomy, gastric stapling, malnutrition)	≥10 mm
Immigrants from high endemic areas (<5 year)	≥10 mm (Asia, Africa, Latin America)
Residents of long term care (nursing homes, mental institutions, chronic health care)	≥10 mm
Institutions (homeless shelters, correctional facilities)	≥10 mm
Health care workers	≥10 mm
Low risk	Not needed

The case under discussion does not fall into the high-risk category for treatment of latent tuberculosis, but may be considered as an intermediate risk on cursory assessment. Although employees and staff of healthcare facilities, especially those involved in direct patient contact, should be offered treatment of latent tuberculosis, there is no such stipulation for volunteers in hospitals. Most healthcare facilities screen volunteers for active tuberculosis by Mantoux skin test and chest radiograph for those with positive reaction. Another category under which the subject could be considered is an indication for treatment of latent tuberculosis include persons from highly endemic countries within 5 years of immigration with a positive Mantoux test (≥10 mm), irrespective of previous BCG vaccination. This group of people represent one of the largest segment of newly diagnosed patients with active tuberculosis in North America and Europe.[2,3] In 2006, 57% of all tuberculosis cases in the United States were among foreign-born persons,[4] and in several European countries >50% of tuberculosis case occur among foreign-born people.[5]

There are 22 countries with a high burden of tuberculosis (TB) that account for 80% of the TB cases globally.[6] These countries are located predominantly in Asia (South East Asia and Western Pacific regions) Africa, Brazil (South America), the Russian Federation (Eastern Europe), and Afghanistan (Middle East). The estimated new TB cases (all forms) per 100,000 people per year in Iran is 22, which falls in the low risk category (0–24) as present in North America and Western Europe.[6] The incidence and prevalence of TB in the Middle East varies from country to country, and Iran actually falls into the relatively lower risk group of nations. Thus, persons from Iran would not be offered treatment for latent TB based only on country of origin and immigrating to North America within 5 years. The indication for treatment of latent TB in this case is borderline or very debatable, but most physicians (including internists) may not be aware of this fact.

The treatment of choice for latent TB is now standardized to a 9-month course of isoniazid (INH) 300 mg once daily for adults, with or without pyridoxine (vitamin B6) to prevent peripheral neuritis. This is believed to be about 90% effective in preventing future reactivation of TB; but it does not prevent re-infection (with a new strain), which is a risk mainly in highly endemic countries. The main worrisome adverse effect of INH is clinical hepatitis, which can be fatal or lead to fulminant hepatitis that requires liver transplantation. There are two types of hepatic toxicity seen in INH; a common transient elevation of the transaminases seen in 10–30% of patients that occurs within 4–6 months and is benign and asymptomatic, and clinical hepatitis (symptomatic) which is much less common, age-related, and only occurs in about 1% of treated patients. Clinical hepatitis with INH is rare under 20 years of age and increases to about 2–2.3% above 50 years, and in persons >65 years, the risk increases to about 4.5%.[1] About 50% of INH hepatitis occurs in the first few months of treatment and the remainder occurs later up to 12 months (if still on INH).[7] The prognosis of overt INH hepatitis is usually very good if the drugs are discontinued promptly with the first sign of clinical hepatitis. The overall mortality is about 10% or 4.2 per 100,000 patients treated with INH.[7] Middle-aged black women seem to have the worst prognosis from this complication. In the majority of patients, there is clinical and biochemical resolution of signs and laboratory abnormality within 1–2 months of stopping the drug. Occasionally, patients can present or develop a sub-acute, more protracted course that mimics chronic viral hepatitis and leads to cirrhosis.[7]

The pathogenesis of INH hepatotoxicity was initially considered to be an idiosyncratic reaction, but there is increasing evidence that this is a direct toxic effect of metabolite(s). There appears to be a higher risk and greater severity with higher doses, and higher incidence in slow acetylators.[8,9] Animal experiments show that INH metabolism leads to acetyl hydrazine, which after oxidation forms toxic intermediates. These are thought to produce damaging effects by acetylating or alkylating macromolecules within liver cells, but the exact mechanism of liver cell injury is unknown.[7] In slow acetylators, acetyl hydrazine accumulates and predisposes to hepatotoxicity. Another metabolic pathway involves hydrolysis of INH to hydrazine and isonicotinic acid. Hydrazine is known to be directly hepatotoxic and hydrolysis of INH is increased by alcohol and rifampin.[9] The mechanism of age-related hepatotoxicity is unclear, but could possibly be related to the slowing of acetylation with advancing age.

Most guidelines and recommendations of latent TB strongly discourage treatment with INH in patients with active liver disease. Close clinical and biochemical monitoring for liver toxicity are mainly recommended for subjects with high risk for clinical hepatitis, such as older people (≥65 years), those with history of liver disease, chronic carriers of hepatitis B and C, alcohol abusers, concomitant users of other hepatotoxic drugs, and subjects who suffer from malnutrition or AIDS.

Current textbooks of medicine do not recommend routine biochemical monitoring for healthy adults being treated with INH.[10] In these circumstances, baseline liver tests are performed and patients should be counseled on symptoms of clinical side effects and should be monitored clinically. Some experts and the manufacturer recommend biochemical monitoring for persons >35 years old, pregnant women, (and those within 3 months post-partum), monthly for 3 months, then afterwards at 1–3 month intervals.[1,11]

INH should be discontinued promptly at the first sign of clinical hepatitis. Symptoms of hepatitis may include fatigue, weakness or fever >3 days, malaise, unexplained anorexia, right upper quadrant pain or discomfort, and jaundice. If the ALT is ≥3–5 times the upper limit of normal, the drug should be discontinued, even if the patient is asymptomatic. Restarting INH at a small dose has been recommended by some experts in asymptomatic patients. It is of interest to note that the American Thoracic Society, the British Thoracic Society, and the Task Force of the European Respiratory Society only recommend regular biochemical monitoring of liver function on multidrug treatment for TB in patients with chronic liver disease or increased serum transaminases prior to treatment.[12] In the case of symptoms of hepatotoxicity, the liver function should be examined. This may be based on the fact that there is no good evidence that routine monitoring of liver function will decrease the chance of fulminant hepatitis or fatality, and prompt discontinuation of medications with first onset of symptoms usually results in full recovery in those with clinical hepatitis.

8.1.3 Hepatitis due to NSAID

The defendants' lawyer raised a critical question. Is it absolutely certain that the fulminant hepatitis suffered by the patient was due to isoniazid? With any serious adverse event, to make an assessment requires several steps and investigations to reach a valid conclusion. This involves a process of deduction and exclusion of other etiologies (such as hepatitis virus), other agents, and use of Bayes theorem to assess overall probability (definite, probable, or possible), as well as posterior and prior probability (based on known literature reports). Other considerations include temporal relationship with use of the medication, compatibility of clinical features and laboratory data, histopathology data and previous reports, and reproduction of the event by re-challenge with the putative agent. Although this is the most definitive method of proving cause and effect, it is the least used because of the potential risk of harm to the patient and the ethical and moral issues.

The temporal relationship, clinical features, laboratory data, and histology of the liver are all compatible with INH - induced hepatitis. However, the investigation excluded well-known causes of viral hepatitis. The patient was also receiving diclofenac, which started 5 weeks before the clinical diagnosis of hepatitis and 2–3 weeks before the onset of symptoms. Thus, there is a temporal relationship with diclofenac treatment and the onset of clinical hepatitis. NSAIDs in general are known, but rare causes of drug-induced hepatitis.[7] The incidence of diclofenac-induced clinical hepatitis is about 1–5 per 100,000 users, and the incubation period varies from 3 to 12 weeks (consistent with the present case).[7] Data from the diclofenac monograph (Novartis Pharmaceuticals) indicates that there is a higher incidence of moderate to severe (3–8 times upper limit of normal) and marked (>8 times normal) elevation of transaminases when compared to other NSAIDs. In addition, rare causes of severe hepatic reactions, including liver necrosis, jaundice, and fulminant fatal hepatitis (or requiring liver transplant) have been reported with diclofenac. To date, there is no evidence of enhanced risk of clinical hepatitis in patients receiving both INH and diclofenac or other NSAIDs. Elderly women are more susceptible to NSAIDs-induced hepatitis. Histopathology of the liver usually reveals zone 3 or 5 spotty acute hepatocellular necrosis, but there can be granulomas, cholestasis, hepatic eosinophilia, and even chronic active hepatitis with overuse of NSAIDs.[7] The prognosis is usually very good from withdrawal of NSAIDs. There is no evidence that concurrent treatment with INH and NSAIDs increased the risk or severity of hepatitis.

8.1.4 Summary and Conclusion of Medico-legal Aspects

Treatment for latent TB in the case under discussion was not indicated, but the circumstances could be interpreted as representing a borderline indication to use INH. However, the patient should have been offered the choice of no treatment versus therapy for latent TB. The risk versus benefit should have been discussed and the potential side effects explained to the patient. The patient should have been counseled to discontinue the medication at the first symptoms suggestive of clinical hepatitis.

Monitoring for liver disturbance by biochemical tests is not routinely recommended for patients at low risk for clinical hepatitis, and the physician should not be held responsible for his failure to order these tests. Clinical monitoring however is standard and the physician can be held responsible for either failure to recognize the manifestations of hepatitis, or his failure to promptly withdraw all drugs once these signs appear.

It cannot be concluded that INH was irrefutably culpable for the fulminant hepatitis, but based on the relative risk and incidence, it was more likely the cause than diclofenac. In any case, both drugs should have been discontinued immediately with the first signs of clinical hepatitis.

8.2 Case 2: Severe Drug Rash

For 2 years, a 35-year-old male had suffered from recurrent bouts of nasal congestion, nasal discharge, and post-nasal drip with only partial, temporary relief from decongestants, antihistamines, and topical corticosteroids. His FP referred him to an internist and clinical allergist for further management. His past history was negative for any significant medical illness, but the patient had previous surgery for nasal septal deviation, and had stopped smoking 2 years before.

Examination by the allergist revealed inflamed edematous nasal mucosa with some purulent discharge, and a radiograph of the sinuses demonstrated mucosal thickening of both maxillary antra. Based on these findings, the consultant made a diagnosis of chronic rhino-sinusitis with an allergic and infectious component. The consultant prescribed intranasal corticosteroids and a 2-week course of trimethoprim-sulfamethoxazole (TMP-SMX). The patient reported that he was treated by his FP 2 months before with triple sulfonamide antibiotics (trisulphamine) for 7 days without any side effects. He had no known drug allergies before this visit.

Towards the end of the 2-week course of TMP-SMX, the patient developed malaise, low-grade fever, and a body rash that started on the face and trunk. This rash rapidly progressed over the next 48 h to involve his limbs, mouth, and eyes, with blistering of the skin. He was admitted to the emergency department of a hospital with a diagnosis of sulfonamide-induced toxic epidermal necrolysis (TEN). Further care was performed in the burn unit. As a consequence of this adverse reaction, the patient developed bilateral corneal ulcerations requiring repeated corneal transplants. Despite this, he remained blind in the left eye and had severe visual impairment on the right side.

8.2.1 Medico-legal Issues

Medico-legal actions were launched by the patient's lawyer claiming medical malpractice against the allergist in failing to warn the patient of the potential adverse effects of TMP-SMX. Moreover, the plaintiff claimed that antibiotics were never needed in the first place and if he had known of these potential side effects, he would not have agreed to be treated with the TMP-SMX.

The defense retorted that the adverse reaction suffered by the patient was extremely rare, and that the patient had previously been treated with sulfonamides, without any reaction. They claimed this reaction could not have been predicted and that it was not the standard medical practice for physicians to list all the rare side effects of licensed drugs on the market.

8.2.2 Medical Issues

The first relevant issue in this case is the following question: Should any antibiotic have been prescribed? If antibiotics were indicated, was the choice of the TMP-SMX appropriate? Current consensus is that antibiotics are overused and prescribed unnecessarily for sinus disease.

Sinusitis is commonly due to respiratory viruses and allergic reaction (as in hay fever), and antibiotics are of no value in these situations. The presence of purulent nasal discharge can be seen in the above conditions, but is not diagnostic or indicative of bacterial sinusitis.[13] Radiographs of sinuses showing thickened mucosa or fluid in the chambers are non-specific and not diagnostic of bacterial sinusitis, as these changes can also be seen in viral infection and allergic sinusitis.

The etiology of chronic sinusitis is complex and there is a lack of consensus of the pathogenesis. Multiple factors may predispose to chronic sinusitis and allergy appears to play a prominent role, with or without polyps.[13] Other factors include structural abnormalities (outflow obstruction, retention cysts, etc.) and irritants such as smoking. Chronic sinusitis is usually defined as having symptoms of sinus inflammation lasting longer than 12 weeks, with documented inflammation (by imaging techniques) at least 4 weeks after appropriate therapy with no intervening acute infection.[14] Computerized tomography (CT) is the preferred imaging technique to identify any obstruction and polyps. Although antibiotics are commonly used in chronic sinusitis, their benefits have not been established by randomized trials, and the role of bacterial superinfection has not been well-defined.[13] The best microbiological data from patients with chronic sinusitis have found aerobic (52.2%) and anaerobic pathogens (47.8%) are common in these cases.[15] The most common aerobes were *Streptococcus* species and *Hemophilus influenzae* (non-typable strains), and the most common anaerobes were *Prevotella* species, anaerobic *Streptococci* and *Fusobacterium* species.

Management of chronic sinusitis is challenging and involves combined medical and surgical therapy. For surgical cases where there is good clinical and imaging evidence of chronic bacterial sinusitis, empiric antibiotics should be effective against *Streptococci, H. influenzae* and anaerobes. Amoxicillin-clavulanate would be a suitable choice, and for β-lactam allergic patients, a new fluoroquinolone with anaerobic activity (moxifloxacin) would be an acceptable alternative.[13] Failure to respond usually indicates the need for surgery which can be performed by endoscopy, and in these cases, antibiotic treatment should be guided by sinus culture (by puncture or endoscopy-guided). Although antimicrobials are commonly used for extended periods (3–4 weeks) for acute superinfection or exacerbation, no studies have addressed the issue of duration of therapy.

Although the case under discussion may not meet the diagnostic criteria for chronic bacterial sinusitis, making this diagnosis and instituting antibiotic therapy (although a judgment error) should not be considered gross negligence, or represent substandard care to merit malpractice litigation. The choice of antibiotic (even if the diagnosis of chronic sinusitis were correct), however, would not be a suitable selection. For acute bacterial sinusitis, amoxicillin/ampicillin is considered the drug of choice and TMP-SMX is recommended as an alternative agent for subjects allergic to penicillin.

What counseling should patients receive when prescribing an antibiotic, and specifically TMP-SMX? Most physicians do not spend time to inform their patients about the adverse effects of prescribed medications. On the other hand, most pharmacists do provide written information on new prescriptions. Physicians cannot depend on this fact though, nor rely on this service for defense in a court of law. In most situations, physicians may counsel patients on drugs with known high risk of toxicity or side effects. For frequently prescribed medications (such as most oral antibiotics), counseling often is neglected, or only the common adverse effects are mentioned.

The incidence of uncomplicated skin reaction (allergic skin rash) to TMP-SMX (mainly due to the sulfonamide component) in the general population is about 1–4% of recipients.[16] This consists of mainly toxic erythema, a maculopapular eruption, infrequently urticaria, erythema nodosum, and fixed drug eruption.[16] Severe skin reactions in TMP-SMX recipients are rare and include Steven's-Johnson Syndrome (SJS), toxic epidermal necrolysis (TEN), exfoliative dermatitis, and necrotizing cutaneous vasculitis. Previous estimates of severe skin reaction were 1 in 100,000 recipients.[16] Patients with HIV infection have a much higher incidence of cutaneous reaction to TMP-SMX (especially those with AIDS).

8.2.3 Epidermal Necrolysis

Epidermal necrolysis (EN) is a rare and life-threatening reaction, mainly drug induced, which encompasses SJS and TEN. These two conditions represent severity variants of identical process and differ only in the percentage of body surface involved.[17] The incidence of SJS and TEN are estimated at 1.6 per million person-years and 0.4–1.2 cases per million person-years, respectively.[17] Although EN can occur at any age, it increases in prevalence after the fourth decade, and is more frequent in women. There is some evidence that the risk of EN increases with HIV, collagen vascular disorders, and cancers.

The clinical features of EN are characterized by skin and mucous membrane involvement. Initially, the skin reaction begins with macules (mainly localized to the trunk, face, and proximal limbs), and then progresses to involve the rest of the body and become confluent with flaccid blisters leading to epidermal detachment.[17] Patients may become systematically ill with fever, dehydration, hypovolemia, secondary bacterial infection, esophageal and pulmonary involvement, and complications and death from sepsis.

The pathogenesis of EN is not completely understood, but studies indicate cell mediated cytotoxic reaction against keratinocytes leading to massive apoptosis. Early in the process, there is a predominance of CD_8 killer T lymphocytes in the epidermis and dermis of bullous lesions, and later monocytes develop. Cytotoxic CD_8 T cells express α–β Tcell receptors are able to kill cells through production of perforin and granzyme B. Drugs are the most important causes of EN and TEN and >100 different drugs are implicated. CD_8 oligoclonal expansion corresponds to a drug specific, major histocompatibility complex (MHC) – restricted cytotoxicity against keratinocytes.[17] Pro-inflammatory cytokines IL-6, TNFα, and Fas ligand are also present in skin lesions. Genetic susceptibility appears to be important, and there is strong association with Han Chinese with HLA-B5802 leucocyte antigen and SJS induced by carbamazepine, and HLA-B5801 antigen and SJS induced by allopurinol.[17]

High-risk drugs (about 12) from six different classes, account for 50% of EN reactions. These include allopurinol, sulfonamides, anticonvulsants (carbamazepine, phenobarbital, lamotrigine), nevirapine (non-nucleoside analog), oxicam NSAIDS, and thiacetazone.[18] The incubation period for EN ranges from 4 to 30 days, but most cases occur within 8 weeks of starting the medication. Rare cases can appear within hours of use, or same day if they had prior reaction. Early, non-specific symptoms (fever, headache, rhinitis, myalgias) may precede mucocutaneous lesions by 1–3 days. Some patients may also present with pain on swallowing or stinging of the eyes. About one third of patients begin with non-specific symptoms, another third with primary mucous membrane involvement, and the rest present with an exanthema.[17] Progression from a localized area to full body involvement can vary from hours to days. The classification of EN depends on areas of detachable epidermidis by a positive Nikolsky sign (dislodgement of epidermidis by lateral pressure) and flaccid blisters. The diagnosis of SJS is made when there is less than 10% body surface area (BSA) involvement; SJS/TEN overlaps with 10–30% BSA, and TEN for >30% BSA involvement.[17] In severe cases of EN, the mucous membranes (buccal, ocular, genital) are involved in about 90%, and 85% have conjunctival affliction consisting mainly of hyperemia, erosions, chemosis, photophobia, and excessive lacrimation. Severe form of eye involvement can result in shedding of eyelashes, corneal ulceration (as in Case 2), anterior uveitis, and purulent conjunctivitis.[17]

Extra-cutaneous complications mainly seen in severe TEN may include pulmonary disease (25%) with hypoxia, hemoptysis, bronchial mucosal casts, interstitial changes, and acute respiratory distress syndrome (ARDS), which carries a poor prognosis. The gastrointestinal tract involvement is less common, but can include esophageal necrosis, small bowel disease with malabsorption, and colonic disease (diffuse diarrhea and bleeding). Renal involvement is mainly proteinuria and hematuria, but proximal renal tubular damage can sometimes cause renal failure.

Late ophthalmic complications occur in about 20–75% and consist of abnormal lacrimation with dry eyes, trichiasis (ingrowing eyelashes), entropion (inversion of eyelid), and visual impairment or blindness from scarring of the cornea.

Prognosis of EN varies with the severity of illness and prompt withdrawal of the offending agent. The overall mortality of EN is 20–25%, but for SJS it is lower, at 5–12%, and higher for TEN >30%. Development of a prognostic scoring system (SCORTEN) for TEN,[19] has recently been found useful, but the performance of the score in prediction is best on day 3 of hospitalization.[20] The prognostic factors that are each given one point include the following: age >40 years, heart rate >120/min, cancer, or hematologic malignancy, BSA involved >10%, serum bicarbonate of <20 mM/L, and serum glucose of >14 mM/L. The mortality rate in TEN increases with accumulation of points as follows: 0–1 point has a mortality rate of 3.2%, 2 points has a mortality rate of 12.1%, 3 points has a mortality rate of 35.8%, 4 points result in a mortality rate of 58.3%, and >5 points result in nearly uniform mortality of 90%.[19]

Management of EN or TEN consists of prompt removal of the offending agent and symptomatic therapy. Patients with a SCORTEN of 0–1 can be managed on the regular medical wards, whereas those with >2 points should be transferred to a burn center or intensive care unit (ICU).[17] It is most important to maintain hemodynamic support with adequate fluids and electrolyte balance. Central venous lines should be avoided because the risk of superinfection is high, and so peripheral intravenous access should be used. Moreover, the rash and blistering is greatest proximally. Nutritional support should be maintained orally or by nasogastric tube, and use of prophylactic heparin is warranted, and also an air-fluidized mattress preferable. Unlike severe burns, extensive and aggressive debridement of necrotic epidermidis is not recommended.[17] There is no indication for prophylactic antibiotics, but patients should be monitored diligently for infection and treated promptly when present. There is no standard protocol for skin dressing, and antiseptic is used depending on the individualized center's experience. Eye care should consist of a daily examination, artificial tears, antiseptic and vitamin A drops every 2 h. Regular mouth rinse with antiseptic solution several times a day is recommended.

There is no proven specific therapy for any form of EN. Steroids were initially considered for SJS, but their value is unproven, controversial, and they are not routinely recommended. Intravenous immunoglobulin (IVIG) is also very controversial, and although initial retrospective studies suggested benefit, recent prospective, non-randomized studies have not confirmed any definite value, and some studies showed increased renal failure and mortality with IVIG.[21] In one of the largest studies from a single center, IVIG was assessed in a prospective non-comparative study of 34 patients with EN, and 20 subjects with TEN. There was no evidence of improvement in mortality, progression of detachment, nor re-epidermalization. Most deaths occurred in elderly patients with initially impaired renal function. Thus, IVIG is not recommended for EN unless being assessed in a randomized clinical trial. The death rate with IVIG was 32%, which was higher than the historical death rate in the same center (20%), in historical controls with TEN not treated with IVIG.[22] Thus, IVIG may be harmful in patients with EN.

8.2.4 Discussion of Medico-legal Issues

One of the issues raised by the plaintiff was that he was not counseled on the potential severe side effects of the TMP-SMX, and that if he were aware of the risk, he would not have agreed to take it. Is it the responsibility of the physicians to explain all potential albeit rare adverse effects of any treatment? The courts may take in consideration the standard practice of the physician's peers, or what is considered accepted practice. Most physicians (if they do counsel patients on medications) would mention the most common side effects, but would not usually mention rare adverse effects. For instance, it would be justifiable to mention that a drug rash could be seen with TMP-SMX, if the patient happens to be allergic to the drug (which should be discontinued as soon as this occurs). As physicians, we would not usually mention that there is a rare risk of shedding of the skin, blindness, or death. Similarly, when prescribing penicillin in patients not known allergic to the drug, we generally do not counsel that there is a 1:50,000 to 1:300,000 risk of dying from anaphylaxis (which is treatable). Yet, if we were to order or prescribe chloramphenicol, it is expected that we should counsel the patient that there is a 1:50,000 to 1:300,000 risk of aplastic anemia, which is not treatable except by bone marrow transplantation. Hence, it may be asked; what is the best method of informing patients on medication toxicity? It is acceptable to leave this to pharmacists to provide literature on these drugs as the sole form of counseling. It is the prescriber's responsibility to obtain informed consent before ordering the medications.

It may be the best policy for prescribers to list the most common side effects, then occasional severe adverse reactions, and mention a possibility of other rare unforeseen adverse reaction (without specifying these latter reactions unless requested by the patient). The details of the counseling may vary on several factors, such as the relative safety profile (therapeutic to toxic ratio), enhanced risk factor for side effects (which may depend on underlying comorbidities or genetic predisposition), and the expected duration of treatment; as the longer an individual is exposed to a drug, the greater the potential for some side effects.

The CMPA have provided some guidelines for risk management considerations in prescribing opioids[23] that are useful for all medication orders and may curtail medico-legal cases from drug adverse events. These medico-legal considerations are:

1. Is there an appropriate indication for this drug?
2. Is the starting dose and need for continuation appropriate?
3. Have you considered the need for monitoring that would be reasonable for your patient?
4. Have you considered the potential effect of any concomitant medication that might influence the dosing, monitoring, and side effects?
5. Have you considered other factors such as comorbidity that might influence the dosing and monitoring?
6. Are you prepared to diagnose and manage any adverse event?
7. Have you counseled the patient on potential side effects, how to recognize early signs, and necessary actions?
8. When discharging patients, have you provided reasonable information about the risks of adverse reactions, precautions to be observed, and person to notify?

Patients who suffer from adverse effects may be willing to forgive a physician's failure to provide informed consent when that therapy is indicated. However, in situations where the treatments were not indicated, or of questionable value, then any adverse event would likely be unacceptable to the plaintiff or courts.

8.3 Case 3: Failure to Recognize Complications of Steroid

A 38-year-old male with steroid-dependent Crohn's colitis (diagnosed 6 years before) called his FP for advice regarding chickenpox from his young son who was recently diagnosed with it at a daycare center. The patient was experiencing retrosternal and epigastric pain on swallowing. The FP prescribed omeprazole 20 mg once daily and ibuprofen over the phone, without seeing the patient. Later in the night of the same day, the man presented to the emergency department of a local hospital. The ER physician noted that the patient was chronically on methylprednisolone 8 mg once daily for Crohn's disease, and that he had developed local pustules consistent with early varicella within the past 4 days. However, the main concern of the patient was severe retrosternal, mid-chest pain on swallowing and radiating through his back for 24 h. The recorded vital signs showed a temperature of 38.3°C, blood pressure of 155/110 mmHg, heart rate of 81/min, and respiratory rate of 20/min. The examination revealed scattered vesicles/pustules on the patient's face, soft palate, and pharynx. Treatment on discharge consisted of liquid bupivacaine swish and swallow (topical anesthetic), oxycodone-acetaminophen, and metoclopramide. An electrocardiogram was normal and the discharge diagnosis listed possible esophageal involvement with varicella.

Within 72 h, the subject returned to the same ER with worsening symptoms, and was seen by the same physician. The symptoms consisted of swelling of his face, fever, sweats, productive cough of blood-streaked sputum, and persistent chest pain. Examination reports revealed a very ill looking male with a temperature of 39.5°C, heart rate of 169/min, blood pressure of 131/87 mmHg, and respiratory rate of 30/min. His face was swollen and edematous with closure of the right eye, extensive vesicles and pustules on the face, soft palate with edema and inflammation of the gingivae, and numerous skin lesions over the trunk and proximal limbs. Oxygen saturation on room air was 92% and the chest radiograph was reported as normal.

Investigations revealed anemia, thrombocytopenia, liver disturbance, and evidence of disseminated intravascular coagulopathy. Intravenous acyclovir was started and the patient was transferred to the ICU of a tertiary care center, where he died within 38 h after the second presentation. Autopsy revealed disseminated varicella with involvement of the brain, lung, heart, liver, esophagus, and stomach.

8.3.1 Medico-legal Issues

The wife and family of the deceased man launched medical malpractice litigation against the FP, ER attending physician, and the local hospital. Charges against the FP were as follows: (1) substandard care reasonably expected of a general practitioner, (2) he should have advised or warned the patient and provided early treatment, especially since he knew that his son had chickenpox, (3) he knew, or ought to have known that the deceased was immunosuppressed from chronic steroids and therefore at increased risk, (4) he failed to provide medical assistance and prescribe the correct drug (acyclovir) on presentation, (5) he failed to make the patient aware of the potential complications of his long-term steroid use, and (6) he failed to refer the deceased to an appropriate specialist.

The accusations against the ER attending physician were similar: (1) his negligence was the direct cause of the deceased's death, (2) his medical care fell below the standard reasonably expected from an ER physician, (3) he ought to have known that the patient was immunosuppressed from steroids, and therefore at high risk for complications from chickenpox, (4) he failed to provide proper medical assistance and treatment, (5) he failed to appropriately admit the patient on initial presentation and institute intravenous acyclovir, and (6) he failed to consult an appropriate specialist (internist or infectious disease specialist).

Damages were sought by the plaintiffs for pain and suffering, deprivation of a husband and father, loss of economic benefit afforded to the family from potential employment earnings of the deceased over the next 27 years (assuming retirement at age 65).

Counsel for the defendants requested expert opinion on two key issues: (1) was the steroid dose the deceased received sufficient to cause immunosuppression? (2) If appropriate therapy with acyclovir were started at initial presentation with chickenpox, would the outcome have been any different?

8.3.2 Medical Aspects of Chickenpox and Immunosuppression

Chickenpox (varicella) has dramatically declined in all age groups, but most markedly in children since the introduction of varicella vaccine in 1995 in North America and developed countries. Since the introduction of the vaccine, the decline in varicella-related hospitalization in the US was greatest among 0–4 year-old children, but rates also declined in older youths (5–19 year) and adults.[24] In temperate regions, 90% of cases of varicella occur in children <10 years of age, 5% occur in individuals >15 years old, and adults (>20 year) only account for 2%. The risk of hospitalization and death is greater in young infants and adults than children, and most varicella-related deaths occur in previously healthy people.[25] Although varicella is much less common in adults than children, 47% of the deaths from complications occur in adults.[26] In tropical and subtropical countries, the mean age of patients with varicella is higher than in temperate regions, and up to 40% of immigrants from these areas are susceptible to varicella.

Healthy children rarely suffer from complications of varicella, with the most common one being secondary bacterial infection (*Streptococcus and Staphylococcus*) of the skin and soft tissue. Immunocompromised children are predisposed to more severe and progressive diseases (up to one third) with multiple organ involvement, lungs, liver, and central nervous system issues being the most frequent.[27] Mortality in these children range from 15% to 18% and those with lympho-proliferative malignancies on chemotherapy have the greatest risk.

Bone marrow transplant recipients also have a high risk of varicella zoster virus (VZV) infection, with a probability of VZV infection at 30% by 1 year after transplant.[28] In a series of 231 cases of VZV infection, 36 presented with chickenpox and 195 with herpes zoster. The overall VZV infection mortality was

9.7% (23 of 231) – all with disseminated infection in the first 9 months. However, the mortality in those with herpes zoster was only 6.6% versus 27.7% of those with varicella.[28]

High dose corticosteroids are also associated with significant complications of varicella and herpes zoster.[29] Immunosuppression is most commonly seen with high daily dose of ≥1 mg/kg of prednisone or moderate doses for prolonged periods. Rates of infectious complication were not increased in patients given a daily dose of less than 10 mg daily, or a cumulative dose of less than 700 mg prednisone in a meta-analysis of 73 controlled trials.[30] Many experts consider prolonged daily dose ≥15 mg prednisone or equivalent to be immunosuppressive. The US Food and Drug Administration (FDA) states that low doses of prednisone (or similar agents) for prolonged periods may also increase the risk of infection.[31] Corticosteroids can suppress several stages of the immune response that leads to inflammation, but the main immunosuppressive effect is on the cellular immunity. Thus, steroids can increase the risk and severity of a variety of infectious agents (virus, bacteria, fungi, and parasites). Most notable are agents that require intact cellular immunity for control and eradication, such as herpes viruses, mycobacteria, listeria, nocardia, pneumocystis, candida, cryptococci, toxoplasma, and strongyloides, etc., are increased in patients on prolonged corticosteroids.

The effect of corticosteroids on the inflammatory and immune responses is pleomorphic. An earlier study in guinea pigs demonstrated that similar levels of lymphocytopenia were induced by acute and chronic corticosteroid administration, but only chronic treatment was associated with depression of certain cell-mediated lymphocyte functions.[32] Chronic cortisone treatment resulted in marked decrease in both antigen-induced migration inhibitory factor (MIF) and proliferation, although mitogen responses remained normal. Over the last few decades, corticosteroids have been found to inhibit the function of various cell types: (1) macrophage/monocytes – inhibit cyclooxygenase-2 and phospholipase A_2 (interrupting prostaglandin and leukotriene pathways), and suppress cytokine production and release of interleukin (IL)-1, IL-6, tumor necrosis factor (TNF)-α, (2) endothelial cells – impair endothelial leucocyte adhesion molecule-I (ELAM-I), and intracellular adhesion molecule-I (ICAM-I), that are critical for leucocyte localization, (3) basophils – block histamine and leukotriene 4c IgE-dependent release, (4) fibroblast – inhibit arachidonic pathway (as with monocytes) and suppress growth factor-induced DNA synthesis and fibroblast proliferation, (5) lymphocytes – inhibit cytokines IL-1, IL-2, IL-3, IL-6, TNF-α, GM-CSF, and interferon γ production or expression.[33]

The association of steroid therapy and increased risk, severity and complications of VZV infections has been well established for decades.[34] Patients receiving high-dose corticosteroids are at risk for disseminated disease and fatality, whereas patients on low-dose schedules are not at increased risk.[34,35] Esophagitis and gastrointestinal involvement of VZV are distinctly rare and have been described in both immunocompromised hosts and apparently healthy subjects as complications of chickenpox or herpes zoster. Disseminated varicella in autopsy studies of children with acute lymphoblastic leukemia or lymphoma on chemotherapy had demonstrated involvement of the esophagus, small bowel, colon, liver, spleen, and pancreas.[36] Fulminant and fatal cases of varicella hepatitis have been described predominantly in immunosuppressed children and adults, but also in healthy people.[37] Rare cases of adult varicella on chronic steroids (for asthma) have been reported with small bowel involvement presenting with abdominal pain and gastrointestinal bleeding.[38] However, it appears that the patient may have been on moderately high dose of methylprednisolone (40 mg daily). In an immunocompetent young adult on inhaled steroids for asthma, varicella has been reported to cause diffuse abdominal pain and tenderness with hepatic, esophageal, and pulmonary involvement, with recovery after acyclovir therapy.[39]

Bullous and necrotic ulcerative lesions of the esophagus and stomach have been described in the pathology literature of fatal varicella as early as 1940.[40] Stomach and small bowel changes detected by radiological imaging has also been reported in a case of chickenpox.[41] Occasionally healthy adults with varicella may have mild symptoms of esophagitis that respond to antihistamine-H2 blockers, suggesting temporary esophageal reflux.[42] Shingles esophagitis have also been seen on endoscopy in patients without widespread dissemination of herpes zoster and benign course.[43]

8.3.3 Discussion of Medico-legal Aspects

The deceased patient (Case 3) was receiving 8 mg daily of methylprednisolone prior to his presentation with chickenpox. This dose is equivalent to 10 mg prednisone and normally would not be considered to be immunosuppressive. However, the course of the disease and widespread dissemination with fatality resembles that of an immunocompromised host. How can we explain this reaction? The possibilities include: (1) inaccurate history of the steroid dose provided by the patient, (2) rarely, dissemination and fatality can occur in healthy adults, (3) unrecognized immunocompromised state such as HIV infection or rare genetic mutations, or polymorphisms in genes involved in cellular immunity, and (4) higher free active concentration of the drug than would be expected. Methylprednisolone (Medrol) is 70% bound to protein, mainly albumin, and decrease in serum albumin by 30–50% could increase the active unbound drug almost to the same proportion. On admission to hospital, the patient's serum albumin was 15 g/L (lower limit of normal 35 g/L), 42% of the normal lower limit. Although the serum albumin can decrease in acute illness from varicella, the half-life of circulating albumin is 15 days and thus, even after 7 days of chickenpox, it should not decrease more than 25% below normal, even if his liver stopped producing any protein (which is not likely). Hence, the patient probably had a chronically low serum albumin from his chronic colitis. His free concentration of corticosteroid should have been greater than 50% of his expected active drug, which is equivalent to $\geq$15 mg prednisone/day.

Can this information absolve the defendants from responsibility of the patient's adverse outcome? It could be argued by the defendants that it is not common knowledge or usual practice to consider the protein binding effects of drugs on their toxicity. Furthermore, it would not be expected that the FP and ER physicians be cognizant of these facts. The defendants maintain that their management did not fall below the expected standard of care, and most reasonable physicians would not have considered the patient immunocompromised on such a low dose of prednisolone. The outcome was unpredictable and only in hindsight was it evident that the deceased was likely immunocompromised and susceptible to a higher risk for adverse outcome.

Experts' opinions for the plaintiffs' side argued that the involved physicians should have been aware that adults (even normal hosts) are at a greater risk of severe disease and complications than children from chickenpox are. Therefore, the FP and ER physician were remiss in not prescribing antiviral drug (acyclovir). The ER physician should have admitted the deceased at the first presentation and started intravenous acyclovir, as he suspected visceral dissemination (esophagitis) with varicella, irrespective of the immune state of the patient.

Previous randomized control trial (RCT) of oral acyclovir therapy for uncomplicated varicella in healthy adults have reported mild clinical benefit (decrease of symptoms, fever and time to cutaneous healing), but only in those initiating treatment within 24 h of the rash.[44] Late treatment (25–72 h) had no benefit. The low frequency of serious complications (pneumonia, encephalitis, or death) precluded any evaluation of acyclovir on these outcomes. In immunocompromised patients with VZV infection, later initiation of therapy ($\geq$72 h after onset of rash) may be of value.[45,46] Although there is no RCT to prove the benefit of intravenous acyclovir in normal adults with varicella complicated by visceral involvement, observational and cohort studies suggest benefit.[47] Thus, intravenous acyclovir continues to be the standard therapy for healthy adults and immunocompromised hosts with clinically significant visceral disease (pneumonia, encephalitis) or dissemination.

8.4 Precautions for Chronic Steroid Therapy

Chronic corticosteroid therapy can have numerous side effects and complications. It is important for physicians to counsel their patients on these potential adverse events, and provide a risk-benefit assessment. Many organs and systems in the body can be adversely affected by chronic steroid therapy (endocrine, bone, eyes, muscle, brain, immune system, skin, etc.). It is important to counsel on potential increased risk of infectious diseases and certain precautions should be taken before embarking on chronic therapy. These include a Mantoux skin test and treatment for latent tuberculosis in those with positive reactions and about to receive prednisone $\geq$15 mg/day for $\geq$30 days.[48] A baseline chest radiograph for active or inactive disease should be performed beforehand. It is also recommended that steroid dependent children should undergo VZV antibody test, and if this were negative, then varicella vaccination should be offered.[33] It seems prudent to apply these guidelines to adults as well on chronic steroid therapy. For patients with previous chickenpox or adequate antibodies, varicella zoster vaccine may be considered to reduce the risk and severity of shingles. This vaccine, a live attenuated vaccine has been found effective and is recommended for persons ≥60 years of age to reduce the burden of illness and incidence of postherpetic neuralgia.[49] Presently, this vaccine is not indicated in immunocompromised adults, so it should be administered before starting prolonged steroids. The product monograph of Zostavax™ (Merck), states that the varicella zoster vaccine is contraindicated in patients receiving high-dose corticosteroid, but not contraindicated for individuals on inhaled or low-dose steroids. The varicella vaccine has been found safe in children with moderate immune deficiency,[50] but it is contraindicated in those with substantial suppression of cellular immunity (as with high-dose steroid).[51]

What should have been the appropriate steps of action in this case? Once the FP was notified that the patient's child had chickenpox, he should have counseled the father and determined his previous past history or antibody level against VZV. For patients considered non-immune and severely immunosuppressed (moderate to high-dose corticosteroid (≥20 mg/day) VZV immune globulin should be offered and treatment with acyclovir should be instituted at the first sign of varicella.[33] Since the deceased was considered to be receiving a low dose of steroid, then it was more appropriate to offer treatment with acyclovir at the first sign of a typical rash, or provide a prescription to be filled within 24 h of onset of varicella.

References

1. McEvoy GK, Miller J, Kesler L, Welsch OH (eds), (2009). Isoniazid. In: AHFS Drug Information. Am Soc Health Syst Pharmacists, Bethesda, p. 580–585.
2. Geng E, Kreiswirth B, Driver C, Li J, Burzynski J, Della Latta P, La Pas A, Schluger NW, (2002). Changes in the transmission of tuberculosis in the New York City from 1990 to 1999. *N Engl J Med* 346:1453–1458.
3. Martinez-Lirola M, Alonso-Rodriguez N, Sanchez ML, Herranz M, Andries S, Penafiel T, Rogado ML, Cabezas T, Martinez J, Lucerna MA, Rodriguez M, Bonillo MDC, Bouza E, de Viedma DG, (2008). Advanced survey of tuberculosis transmission in a complex socioepidemiologic scenario with a high proportion of cases in immigrants. *Clin Infect Dis* 47:8–14.
4. Centers for Disease Control and Prevention, (2007). Trends in tuberculosis incidence – United States 2006. *MMWR Morb Mortal Rep* 56:245–250.
5. World Health Organization, (2006). The global plan to stop TB 2006–2015. World Health Organization, Geneva, Switzerland.
6. World Health Organization, (2009). Global tuberculosis control: epidemiology strategy, financing. World Health Organization Report, Geneva, Switzerland.
7. Ramkumar D, La Brecque DR, (2003). Drug-induced liver disease and environmental toxins. In: (eds) Zakin D, Boyer TD; Hepatology: a textbook of liver disease; 4th Edition, Saunders, Philadelphia, p. 755–838.
8. Altman C, Biour M, Grange J, (1993). Hepatic toxicity of antitubercular agents. Role of different drugs, 199 cases. *Press Med* 22:1212.
9. Dickinson, D, Bailey WC, Hinoschowitz BI, Soony SJ, Leslie MD, Hodgkin MM, (1981). Risk factor for isoniazid (INH)-induced liver dysfunction. *J Clin Gastroenterol* 3:271–279.

10. Raviglione MC, O'Brien RJ, (2008). Tuberculosis. In: (eds) Fauci AS, Braunwald E, Kasper DC, Hauser SC, Longo DC, Jameson JC, Loscalzo J. Harrison's Principles of Internal Medicine, 17th Edition, McGraw-Hill, New York, p. 1006–1020.
11. Canadian Pharmacists Association, (2009). Isoniazid. In: CPS 2009: Compendium of Pharmaceuticals and Specialties. Can Pharm Assoc, Ottawa, p. 1187–1190.
12. Tostmann A, Boeree MJ, Aanoutse RE, de Lange WL, van der Ven AJ, Dekhuijen R, (2008). Antituberculosis drug-induced hepatotoxicity: concise up-to-date review. *J Gastroenterol Hepatol* 23:192–2002.
13. Fong IW (ed.), (2009), Emergency issues in head and neck infections. IN: Emerging issues and controversies in infectious disease; Springer, New York, p. 27–46.
14. International Rhinosinusitis Advisory Board, (1992). Infectious rhinosinusitis in adults: classification, etiology and management. *Ear Nose Throat J.* 76 (72):1–19.
15. Finegold SM, Flyan MJ, Rose FK, Jousimies-Somer H, Jukielaszek C, McTeague M, Wexler HM, Bekowitz E, Wynne B, (2002). Bacteriologic finding associated with chronic bacterial maxillary sinusitis in adults. *Clin Infect Dis* 35:428–433.
16. Lawson DH, Price BJ, (1982). Adverse reactions to trimethoprim-sulfamethoxazole. *Rev Infect Dis* 4:429–433.
17. Valeyrie-Allanore L, Roujeau JC, (2008). Epidermal necrolysis (Stevens-Johnson Syndrome and toxic epidermal necrolysis). In: (eds) Wolff K, Goldsmith LA, Kutz SI, Gilchrist BA, Paller AS, Leffell DJ: Fitzpatrick's Dermatology in General Medicine, 7th Edition; McGrawHill, New York, p. 349–355
18. Roujeau JC, Kelly JP, Naldi L, Rzany B, Stern RS, Anderson T, Auquier A, Bastuji-Garin S, Correia O, Locati F, Mockenhaupt M, Paoletti E, Shapiro S, Shear N, Schöpf E, Kaufman DW, (1995), Medication use and the risk of Stevens-Johnson Syndrome or toxic epidermal necrolysis. *N Engl J Med* 333:1600–1608.
19. Bastuji-Garin S, Fouchard N, Bertocchi M, Roujeau J-C, Revuz J, Wolkenstein P, (2000). SCORTEN: a severity-of-illness score for toxic epidermal necrolysis. *J Invest Dermatol.* 115:149–153.
20. Guégan S, Bastuji-Garin S, Poszepczynska-Guigne E, Roujeau JC, Revuz J (2006). Performance of the SCORTEN during the first five days of hospitalization to predict the prognosis of epidermal necrolysis. *J Invest Derm* 126:272–276.
21. Brown KM, Silver GM, Halerz M, Walaszek, Sandroni A, Gamelli RL (2004). Toxic epidermal necrolysis: does immunoglobulin make a difference? *J Burn Care Rehabil* 25:81–88.
22. Bachot N, Revuz J, Roujeau JC (2003). Intravenous immunoglobulin treatment for Steven-Johnson syndrome and toxic epidermal necrolysis: prospective non-comparative study showing no benefit in mortality or progression. *Arch Dermatol* 139:33–36.
23. Berris K (2009). Adverse events – physician-prescried opioids. Risk identification for all physicians. April 2009; Ri0915E. CMPA website: www.cmpa-acpm.ca
24. Straus SE, Oxman MN, Schmader KE (2008). Varicella and Herpes Zoster. IN (eds): Wolff K, Goldsmith LA, Katz SI, Gilchrest BA, Paker AS, Leffell DJ: Fitzpatrick's Dermatology in General Medicine, 7th Edition, McGraw-Hill Medical, New York, p. 1885–1898.
25. Choo PW, Donahue JG, Manson JE, Platt R, (1995). The epidemiology of varicella and its complications. *J Infect Dis* 172:706–712.
26. Preblud SR (1986). Varicella complications and cost. *Pediatrics;* 78 (suppl): S728–S735.
27. Feldman S, Hughes WT, Daniel CB (1975). Varicella in children with cancer.: seventy seven cases. *Pediatrics*; 56:388–397.
28. Locksley RM, Flournay N, Sullivan KM, Myers JD (1985). Infection with varicella zoster virus after marrow transplantation. *J Infect Dis* 152:1172–1181.
29. Weller TH (1983). Varicella and herpes zoster: changing concepts of the natural history, control and importance of a not-so-benign virus (first of two parts). *N Engl J Med* 309:1362–1368.
30. Frey FJ, Speck RF (1992). Glucocorticoids and infection. *Schweiz-Med-Wochenschr* 122:137–144.

31. Arcuri LB, Cashman LD, Cassaro AT, Kostick JH, Hsiung JRS, Wolfe KR (2009). Hormones and synthetic substitutes: adrenals. In AHFS Drug Information Am Soc Health Sys Pharmacists. Bethesda, MD, p. 3036–3128.
32. Balow JE, Hurley DL, Fauci AS (1975). Immunosuppressive effects of glucocorticosteroids: differential effects of acute vs chronic administration on cell-mediated immunity. *J Immunol* 114:1072–1076.
33. Schimmer BP, Parker KL (2005). Adrenocorticotropic hormone, adrenocortical steroids and their synthetic analogs: inhibitors of synthesis and actions of adrenocortical hormones. IN: (eds) Brunton LL, Lazo JS, Parker KL: Goodman & Gilman's: The Pharmacological Basis of Therapeutics, 11th Edition, McGraw Hill, New York, p. 1587–1612
34. Weller TH (1983). Varicella and herpes zoster: changes in concepts of natural history, control and importance of a not-so-benign virus (second of two parts). *N Engl J Med* 309:1434–1440.
35. Zaia JA (1981). Clinical spectrum of varicella-zoster virus infection. IN: (eds) Nahmlas AJ, Dowdle WR, Schinazi RE: The human herpesviruses: an interdisciplinary perspective; Elsevier, p. 10–19.
36. Miliauskas JR, Webber BL (1984). Disseminated varicella at autopsy in children with cancer. *Cancer* 53:1518–1525.
37. Patti ME, Selvaggi KJ, Kroboth FJ (1990). Varicella hepatitis in the immunocompromised adult: a case report and review of the literature. *Am J Med* 88:77–80.
38. Sherman RA, Silva J Jr. Gandour-Edwards R (1991). Fatal varicella in an adult: case report and review of gastrointestinal complications of chickenpox. *Rev Infect Dis* 13:424–427.
39. Maillot C, Riachi G, Francois A, Ducrotte P, Lerebours E, Hemet J, Colin R (1997). Digestive manifestations in an immunocompetent adult with varicella. *Am J Gastroenterol* 92:1361–1363.
40. Johnson HN (1940). Visceral lesions associated with varicella. *Arch Pathol* 30:292–307, 13.
41. Marshak RH, Lindwern AE (eds), (1976), in: Radiology of the small intestine; Saunders, Philadelphia, p. 487–496.
42. Bardhan KD (1978). Cimetidine in "chickenpox esophagitis". *BMJ* 1:370 (letter).
43. Gill RA, Gebhard RL, Dozeman RL, Summer HW (1984). Shingles esophagitis: endoscopic diagnosis in two patients. *Gastrointestin Endosc* 30:26–27.
44. Wallace MR, Bowler WA, Murray NB, Brodine SK, Oldfield III EC (1992). Treatment of adult varicella with acyclovir. A randomized placebo-controlled trial. *Ann Intern Med* 117:358–363.
45. Balfour HH, Bean H, Laskin OL, Ambinder RF, Myers JD, Wade JC, Zaia JA, Aeppli D, Segreti AC, Keeney RF (1983). Acyclovir halts progression of herpes zoster in immunocompromised patients. *N Engl J Med* 308:1448–1453.
46. Shepp DH, Dandliker PS, Meyers JD (1986). Treatment of varicella-zoster virus infections in severely immunocompromised patients. *N Engl J Med* 314:208–212.
47. Haake DA, Zakowski PC, Haake DL, Bryson YJ (1990). Early treatment with acyclovir for varicella pneumonia in otherwise healthy adults: retrospective controlled study and review. *Rev Infect Dis* 12:788–797.
48. Escalante P (2009). In the clinic: tuberculosis. *Ann Intern Med* 150: ITC 6-1-14.
49. Oxman MN, Levin MJ, Johnson GR, Schmader KE, Strauss SE, Gelb LD, Arbeit RD, Simberkoff MS, Gershon AA, et al, for the Shingles Prevention Study Group (2005). A vaccine to prevent herpes zoster and postherpetic neuralgia in older patients. *N Engl J Med* 352 V:2271–2284.
50. Yeung CY, Liang DC (1992). Varicella vaccine in children with acute lymphoblastic leukemia and non-Hodgkin lymphoma. *Pediatr Hematol Oncol* 9:29–34.
51. Kroger AT, Atkinson WL, Marcuse EK, Pickering LK (2006). General recommendations on immunization. Recommendations of the Advisory Committee on Immunization Practices (ACIP). *MMWR*:53(RR15);1–48.

Part IV
Surgical Disciplines

Chapter 9
Orthopedic Complications

9.1 Case 1: Early Post-operative Complication

An orthopedic surgeon performed an elective second metatarsal wedge osteotomy (day surgery procedure) on a middle-aged healthy female for recurrent painful plantar callus. Pre-operative counseling was provided, including mention of uncommon possible complications of swelling and infection that may delay healing but "can always be corrected with a few days of appropriate antibiotics." The procedure was performed without any obvious complications. However, 1 day post-operatively, the patient called the surgeon complaining of intense pain in the foot. Over the phone, the surgeon advised cutting the bandage and elevating the foot. On the fourth day, the patient called the office again complaining of worsening, intense pain, redness and swelling of her foot, and hot and cold spells.

The following day in the physician's office, pus was expressed from the wound; it was lavaged with saline, and then closed partially with steri-strips. The patient was discharged with oral ciprofloxacin 500 mg twice daily and acetaminophen 300 mg/codeine 30 mg every 4 h. Later, on the fifth post-operative day, the patient went to the ER, as her symptoms were not improved. She was admitted to the hospital, where the wound was opened under local anesthetic, and debrided of necrotic tissue and placed on intravenous cloxacillin. A radiograph showed gas in the tissues and cultures were reported 2 days after admission as growing Clostridium perfringens.

The patient was then transferred to a tertiary care hospital for further management. Examination at the center revealed evidence of cellulitis, extending from the foot to the knee, with marked swelling and greyish, necrotic tissue surrounding the wound. Surgery was performed under general anesthesia, which included debridement of extensive necrosis of the interosseous muscles, subcutaneous tissue and devascularized bone of the second digit. The antibiotic therapy then was changed to penicillin G 18 million units/day, clindamycin 1,200 mg/day in four divided doses, and tobramycin 240 mg/day. The patient also received hyperbaric oxygen therapy, was hospitalized for about 2 weeks, and required skin grafting.

I.W. Fong, *Medico-Legal Issues in Infectious Diseases*, Emerging Infectious Diseases of the 21st Century, DOI 10.1007/978-1-4419-8053-3_9,

9.1.1 Medico-legal Issues

The patient initiated medico-legal actions against the surgeon for negligent care and failure to provide early and adequate treatment of her post-operative complication. Specifically, he failed in his obligation to provide meticulous care by not examining the patient and arranging for exploration of the wound by the first post-operative day when she called his office. It was his responsibility to ensure adequate and proper management of any post-operative complications, and his substandard care of the plaintiff resulted in pain and suffering, prolonged hospitalization and deformity of her foot.

The defendant and his lawyer countered that the plaintiff was made aware of these possible complications before surgery, and was provided informed consent. Pain and swelling post-operatively within 72 h are very common and are rarely secondary to infection. Thus, his advice over the phone on the first post-operative day represented standard practice and responsible care. When the plaintiff called his office on the fifth day, he was informed by his secretary that the plaintiff called to request more pain medications (analgesics). He was unaware that the plaintiff was having fever, chills and redness of the wound. This is why he decided to reassess her the following morning at the first pre-arranged follow-up visit. At the examination for discovery, the defendant stated that the patient did not appear very ill and the wound appeared to be mildly infected with no evidence of crepitus or gas in the soft tissue. Hence, his initial management was appropriate, and most likely the infection progressed later that day to the more severe stage when she was seen in the emergency department.

The defendant's team also argued that clostridial infection in this setting is very rare, as the patient had no traumatic wound, evidence of diabetes, or peripheral vascular disease. This infection was unexpected and unusual. Most wound infections that occur usually appear 5–12 days after surgery, and are easily treated with oral antibiotics. Hence, the defendant's medical care was appropriate and met the standard of his peers.

9.1.2 Medical Aspect

In clean surgery, as performed in this case, the chance of post-operative wound infection is very low, less than 2–3% in a normal host. Most cases of infection usually manifest after 5–14 days of surgery; and the predominant pathogens are *Staphylococcus aureus* and *Streptococcus* species. Wound infections can occasionally appear within 72 h of surgery and are usually caused by *Streptococcus pyogenes* and *Clostridia* species. Although early wound infections are very uncommon, patients with symptoms should be examined to exclude this possibility. There are two possibilities in this case to explain the course and progression of the patient's illness: (1) the initial pain and swelling on the first post-operative day was likely due to early infection by *C. perfringens*. This would be unlikely unless there was pre-existent poor blood supply and anaerobic environment for proliferation of the organism, or heavy inoculation of the wound at the time of the surgery. (2) More likely would be the development of post-operative local wound hematoma, which caused the initial intense pain and swelling. This acted as a good media for proliferation of a few spores of *Clostridia* that may have contaminated the wound at surgery, probably from the patient's shoes and the callus of her foot. An inflammatory reaction with edema, tissue necrosis, and swelling then progressed over the next few days.

What should be the appropriate therapy for surgical wound infection? This depends on several factors, which are determined by the physician's clinical assessment and judgment. Foremost, would be the severity of infection or illness and the patient's general condition and presence or absence of significant co-morbid illness. If the wound infection were judged to be mild to moderate in severity, it would be quite appropriate to open the wound for drainage and treat with anti-staphylococcal beta-lactam oral antibiotics. In this situation, the patient should be reassessed in a few days and advised to return sooner or attend an ER for any worsening of the patient's condition. For infected wounds that are opened for drainage, the wound should be left open and parked with saline soaked gauze or wick. Closure of an infected wound, even with steri-strips, is not recommended and defeats the purpose of opening the wound.

It is quite possible that a wound infection may appear mild initially and become much worse within several hours. This is not uncommon, even in normal hosts with group A *Streptococci* or *Clostridia* infections. In cases of infected wounds that are deemed severe or extensive, or if the patient appears seriously ill or septic, immediate admission to a hospital is warranted. Appropriate intravenous antibiotics should be started (for staphylococcal and streptococcal coverage, but broader spectrum antibiotics are indicated for severe sepsis and septic shock) as soon as possible. Initial drainage of the wound could be attempted in the emergency department, but definitive drainage and debridement should be further arranged under general anesthetic. Blood cultures and wound drainage cultures should be obtained under aseptic technique, preferably before starting antibiotics. For wounds with fluctuation and clinical abscess formation, cleansing the wound with chlorhexidine, then aspiration of the pus with a needle and syringe is the best method to obtain a sample for gram stain, aerobic and anaerobic culture. The antibiotics should then be adjusted according to the results of culture and susceptibility.

9.1.3 Medico-legal Discussion

Was the physician negligent in not examining the patient on the first post-operative day? In hindsight, the answer obviously should be yes, but this may not be simple in a busy medical practice. The physician should best speak to the patient directly (rather than indirectly through a secretary), in order to assess the severity of the patient's symptoms. For mild to moderate pain and swelling, it is reasonable to give verbal advice over the phone for the most common complications. However, the patient should be counseled to come in for an appointment the next day or visit an ER if the symptoms worsen or do not improved. If the symptoms of the patient were considered severe, intolerable, or out-of-keeping with the usual post-operative pain and common complications, then it would be advisable for the physician to assess the patient the same day or arrange assessment in a hospital. For uncertain situations (as to the severity of the patient's condition), it would be best to give the patient the choice of being assessed in his office (or clinic) the same day or soon after.

It has been my experience in several medico-legal cases of this type, that often, the physician fails to speak to the patient directly, and he or she makes a clinical judgment and provides advice secondhand (through a secretary or receptionist). This is a bad practice and should be avoided by physicians in their office practice, emergency department, or clinics.

9.2 Case 2: Complication After Hip Surgery

A healthy 45-year-old male suffered an intertrochanteric fracture with sub-trochanteric extension of his left hip after a fall on a skating ice rink. A closed reduction with insertion of a dynamic hip screw was performed that night at a community hospital. The following morning, the patient complained of soreness of the hip, but had no fever, and was treated with analgesics (morphine) after assessment by the surgeon. Later that afternoon, fever was recorded from 4 p.m. to 10 p.m., with temperatures of 39.2–39.7°C. The surgeon was informed of the patient's temperature at 10 p.m. that night. No new orders were made and the surgeon interpreted the fever as a normal post-operative reaction.

On the morning of the second post-operative day, the physician assessed the patient and thought that he was not particularly ill. The vital signs at 6 a.m. showed temperature of 38.2°C, blood pressure of 100/60 mmHg, pulse of 145/min and respiratory rate of 20/min. Examination of the wound revealed crepitus at the edges of the wound and upon opening the wound, there was a watery discharge (with no odor) and minor necrosis of the adjacent muscle. Intravenous broad-spectrum antibiotics for mixed infection were ordered and plans for debridement later in the day were made. The surgeon then went to his office practice outside the hospital, but not before blood tests and wound cultures were obtained. An hour later, the surgeon was notified of the results of the patient's investigation. The complete blood count revealed hemoglobin of 116 g/dL and leukocyte count of 23,600 cells/mm^3 and gram stain of the wound swab showed no pus cells, but a few large gram-positive bacilli. The surgeon then called a microbiologist at a university hospital for advice. After discussion of the clinical status of the patient, it was the opinion by the microbiologist that the gram-positive bacilli seen on the gram stain could represent a diphtheroid bacteria rather than clostridia, as he would expect the patient to be more seriously ill with a clostridial infection. The patient was placed on ampicillin, clindamycin, and gentamicin that morning.

The surgeon was called to reassess the patient by noon on the same day, as the patients' condition appeared to be worsening. Vital signs then revealed a temperature of 39.8°C, heart rate of 150/min, blood pressure of 98/68 mm\Hg and a respiratory rate of 28/min. Crepitus was palpable, extending down the thigh to the knee, and upwards above the inguinal ligament. Urgent surgery was requested which commenced at 1:50 p.m. that day. Extensive muscle involvement involving the whole lower limb and extending to the lower trunk was evident and amputation of the entire limb was performed. He was then transferred to a tertiary center for hyperbaric oxygen therapy. Eventually, he was cured of the infection but required hospitalization for over a month. Cultures of the wound grew C. perfringens and pathology confirmed extensive gas gangrene with widespread muscle necrosis.

9.2.1 Medico-legal Issues

The patient sued the surgeon and the hospital for medical malpractice and negligent post-operative care. Expert opinion provided by the plaintiff's lawyer's consultation indicated several areas of substandard care provided by the physician and the nursing staff. The surgeon was blamed for not assessing the patient in the evening when the plaintiff developed fever on the first post-operative day. Moreover, the plaintiff's team charged that the surgeon should have arranged urgent surgery on the morning of the second day when he recognized crepitus of the wound.

Nurses' charting on the first post-operative day was assessed as being inadequate by the consultant – that between 6 p.m. and 2 a.m., there was no blood pressure, pulse or respiratory rate recorded, besides temperature. The nurses should have recorded the state of the wound. It was also charged that the nurse on duty should have notified the attending physician at 2 a.m. when the blood pressure had fallen to 100/60 mmHg. The baseline blood pressure on the day of operation was 120/60–70 mmHg.

Furthermore, the plaintiff charged that if early and appropriate management were instituted on the morning of the second post-operative day, he would not have required amputation of his lower limb, and surgery could have been limited to debridement and drainage of the wound.

Counsel for the defendant requested an independent medical opinion to assess these charges against his client. Specific questions were: (1) what is the significance of post-operative fever in the first 24–48 h? (2) Should the surgeon have seen the patient in the evening when fever was initially reported? (3) Would the outcome be any different if surgery were performed in the morning of the second post-operative day?

9.2.2 Medical Aspects

There are many causes of fever in the post-operative patient, and although infection is a major concern, it has been recognized by surgeons for decades that fever in the first 72 h is usually benign and rarely secondary to infection. It was a general belief and teaching for many decades that early post-operative fever was caused by pulmonary atelectasis. However, very few large prospective studies have been published that address this issue.

In a series of 537 consecutive patients undergoing major gynecologic surgery, 211 (39%) developed post-operative fever.[1] Investigations and clinical findings were not very revealing with all blood cultures (77) reported negative, 11 of 106 (10%) urine cultures positive, and 5 of 54 (9%) chest radiographs were abnormal. A pathologic process was detected only in 8% of febrile patients. In a pediatric prospective study[2] of post-operative fever following tonsillectomy in 100 children, there was no association with positive blood cultures, core, or surface of the tonsils and incidence of severity of fever. Fever (>38°C) occurred in 30%. The investigators concluded that post-operative fever within 24 h was rarely the result of infection. A retrospective review[3] of fever response after 200 cases of total knee or hip arthroplasty, found that fever (≥38°C) was almost universally present by day 1 or 2 post-operatively. Maximum daily post-operative temperature occurred in most patients on day 1 and gradually leveled off towards normal by the fifth day. Nineteen percent of patients (38 of 200) had a maximum temperature of ≥ 39°C. None of the patients had evidence of atelectasis (clinically or radiologically) and the presence of a positive urine culture (11 patients) had no effect on the fever, none had symptoms of urinary tract infections, and most of the positive results were reported after the temperature had returned to normal. The authors concluded that early post-operative fever is a normal inflammatory response to surgery, and that investigations for sepsis were not indicated unless other signs or symptoms were present.[3]

The mechanisms of post-operative fever appear to be similar to other causes of pyrexia such as infections or inflammatory conditions. The tissue injury caused by surgery itself can stimulate the pro-inflammatory cytokines and prostaglandins to produce fever by the neurohormonal pathway involving the thermal centers in the hypothalamus.[4,5] Thus, post-operative fever within the first 48 h is a physiological response to tissue injury. When should a physician be concerned about early post-operative fever? Although infections in the early post-operative period are rare (except for pre-existing infection or gross contamination at surgery), they do occur and often with catastrophic consequences because of the virulence of the organisms (group A *Streptococcus* and *Clostridia* species). Furthermore, there are other causes of fever during this period that should be considered, such as blood products reaction, drug fever, thromboemboli (especially those confined to bed for days before surgery), malignant hyperthermia post-anesthetic (temperature ≥ 42°C), and other rare disorders (i.e., neuroleptic malignant syndrome, associated with altered mental state and muscle rigidity).

There are currently no guidelines with respect to approach to clinical assessment and investigation of early post-operative fever. The expected standard of a surgical team would be clinical assessment and should include review of the patient's vital signs and symptoms, examination of the wound, skin, and chest; and further examinations or investigations depending on the site of surgery and specific organ related symptoms.

9.2.3 Discussion of Medico-legal Issues

One of the charges against the defendant was his failure to assess the plaintiff within hours or soon after fever was reported by the attending nurse on the night of the first post-operative day. It should be noted that the surgeon did assess the patient earlier that day (in the morning). Thus, his response that "the fever was a normal reaction to surgery" was appropriate and within accepted standards, provided there was no other significant abnormality or symptoms (besides the expected post-operative pain). It would then be appropriate to delay clinical reassessment until the following morning.

A criticism of the nursing care (by the expert witness for the plaintiff) was failure to notify the surgeon in the late night (or early the next morning of the second day) of a low blood pressure (100/60 mmHg). It was argued by the defendants that this does not represent true hypotension or worrisome low blood pressure (normally defined as a systolic blood pressure $<$90 mmHg, or a fall $\geq$40 mmHg of the usual baseline systolic blood pressure). Thus, it was not expected that the nurse should notify the attending physician simply for this observation. There are several benign physiologic responses that can explain a low blood pressure post-operatively. The systolic blood pressure normally falls up to 20 mmHg during sleep,[6] and the baseline blood pressure of the plaintiff was 120–130/60–70, thus a systolic blood pressure of 100 mmHg would not be of concern. Other factors that cause vasodilatation can lead to lowering of the blood pressure including fever, tissue injury from surgery,[7] morphine, and other narcotics. Morphine in usual therapeutic doses does not cause significant hypotension, but hypovolemia and blood loss at surgery combined, may cause symptomatic hypotension.[8]

A major issue in this case is the failure of the surgeon to arrange immediate or urgent surgery in the morning of the second post-operative day, when he found evidence of wound infection with crepitus. It has been argued by the plaintiff that a delay in surgery by several hours resulted in adverse outcome with loss of his lower limb. This statement is probably correct. When wound infection was diagnosed in the morning of day 2 post-operatively, the patient was stable with no evidence of widespread tissue necrosis. Thus, earlier emergency surgery with debridement and opening of the fascial planes was likely to have resulted in better outcome and preservation of the limb.

The defendant argued that in hindsight, immediate surgery should have been done, but at the time of assessment, the patient did not appear to be very ill, as he would have expected for clostridial myonecrosis or gas gangrene. In this hospital and others, there is a classification system used by surgeons to denote the degree of urgency for surgery that has to be communicated to the operating room personnel to gain access to an operating room, anesthetist, and staff. The classification was as follows: "red case" – immediate threat to life or limb, and the patient needs surgery right away or less than 2 h, "blue case" – urgent but not immediate life or limb threatening, and surgery can be performed within 6–8 h (initial classification of this case in the morning), and "relatively urgent" – needs surgery within 24–72 h.

Was the judgment error of the surgeon between classifying the patient as "blue case" rather than "red case," negligent or substandard care? This is a difficult decision and the judicial outcome may not be the same in different courts or trials. When a surgeon is confronted with these circumstances, what should be recommended? There are several factors that need to be considered and each case has to be individualized. Although the physician's main concern is the patient's safety and outcome, he or she has to balance the risk of delaying surgery for several hours and the practicality of performing immediate surgery. Some factors to be considered are the availability of an operating room, anesthesiologist, and supporting staff. It may be much easier to arrange for immediate surgery at 8 a.m. in the morning than at midnight or 1–2 a.m. In circumspection, whenever there is evidence of necrotizing or crepitant soft tissue infection, it is best to assume the worst and arrange for immediate surgery. It is not uncommon for the patient to appear relatively well ("not septic") in the early stages of these infections then in several hours later become critically ill and hemodynamically unstable (the calm before the storm). Thus, in most cases, the best decision would be to arrange immediate surgery.

The gram stain of aseptically collected fluid aspirate or debrided tissue can be of value in the decision-making. Typically, in clostridial soft tissue infection, the drainage is not frankly purulent and is usually described as "dish water" in appearance. On gram stain, leukocytes are sparse, large, "brick-shaped," gram-positive bacilli that can be seen. The gram stain report in this case was typical for clostridial infection, and unfortunately, the microbiologist consulted was from a different center and did not have access to the slide for review. In these situations of crepitant soft tissue infection, an urgent ("stat") gram stain is useful in differentiating between primary clostridial infection versus mixed synergistic infection (with combination of anaerobes, streptococci, staphylococci and coliforms), when the gram stain reveals a mixture of different phenotypes and gram stain appearance. As a general rule, clostridial and streptococcal necrotizing infections are more rapidly progressive and require more urgent surgery than in mixed anaerobic fasciitis, where surgery can be delayed for 6–8 h in stable patients.

9.3 Case 3: Infection After Total Hip Replacement

An elderly woman, age 71, underwent a right total hip replacement at a regional hospital in a small town. She had a history of severe osteoarthritis of multiple large joints with previous prosthetic knee replacements performed by the same surgeon. Other past medical conditions included a history of degenerative lumbar disc disease with previous sciatica, asthma, hypertension, and hyperlipidemia. Surgery was performed with perioperative cefazolin prophylaxis and was uncomplicated. The first post-operative follow-up visit occurred 2 weeks after surgery, and bloody discharge was noted from the non-inflamed wound. Incision and drainage were performed after local anesthetic, with removal of about 100 mL of blood, which was not sent for culture. A week later the patient returned for another visit, and this time there was localized swelling and pointing at the lower end of the wound, but no evidence of redness or drainage. After aseptic cleansing and incision, drainage of multiple clots took place and cultures were sent out. Twelve days later the patient returned for follow-up visit, and the wound appeared infected, with purulent wound discharge. Admission to hospital and subsequent surgery was facilitated later that day. The surgical staff then noted that the wound culture taken 12 days before was reported as growing Pseudomonas aeruginosa, resistant to ciprofloxacin and ceftazidime but susceptible to tobramycin, piperacillin, and imipenem. Imipenem was started soon after admission before surgery.

At surgery, deep wound drainage and debridement was performed, with extraction of a retained foreign body consisting of a small gauze pad. The prosthesis was left in place. Intravenous (IV) imipenem was continued after hospital discharge for home parenteral therapy. Ten days after hospital discharge the patient returned to the emergency department (ER) with severe lower backache radiating down her leg with numbness. A low-grade fever of 38°C was documented, but no focal neurological signs. Blood cultures tested negatively, but piperacillin was added nonetheless. Computerized tomography (CT) of the lumbar spine revealed evidence of degenerative disease, but MRI scan a week later showed evidence of discitis and vertebral osteomyelitis with an adjacent phlegmon. A disc aspirate between L4-5 was performed which failed to grow any bacteria, and the neurosurgical consultant recommended continued antibiotics with no spinal surgery. After 2 months of parenteral antibiotics, the patient gradually improved, but she was left with residual weakness of hip flexion, was only able to walk very slowly with a cane, and had a limp 2 years after the event.

9.3.1 Medico-legal Issues

The patient requested legal counsel to consider and initiate medico-legal suit for medical malpractice and negligence against the surgeon/his assistants and the hospital. The lawyer requested expert medical witness to review the case and provide an independent medical opinion. The questions posed by the counsel were: (1) was the presence of retained gauze the cause of the post-operative wound infection? (2) Does the presence of an unintentional foreign body in the wound represent an act of medical negligence? (3) Was the hip wound the source and cause of the spinal infection? (4) Could earlier recognition and treatment of the hip wound infection have prevented seeding to the spine? (5) Was the initial treatment appropriate for the prosthetic joint infection?

9.3.2 Medical Aspects

Infection of a prosthetic joint causes substantial morbidity and health care expenditure, with an estimated average cost for combined medical and surgical treatment of

US $30,000 for an infected prosthesis.[9] Since the advent of routine perioperative antimicrobial prophylaxis and laminar airflow surgical operating rooms, the risk of post-operative infection is presently less than 1% for hip and less than 2% for knee replacement.[10] However, certain high risk conditions have higher rates of infection, such as underlying diabetes mellitus, rheumatoid arthritis, repeat surgery on the same joint, previous infection of the prosthesis, presence of wound hematoma, and nasal colonization with *Staphylococcus aureus*. Retained foreign body (other than the intended prosthesis) dramatically increases the risk of infection in any surgical procedures. It is generally accepted that the bacteria causing infection are usually colonizing skin flora introduced at the time of surgery. Bacteria are occasionally introduced post-operatively following aspiration or drainage of fluid collections, or reoperations. Hematogenous seeding to a prosthetic joint usually presents as late infection after a year. The risk of hematogenous seeding in a prospective study over 6 years was only 0.3%,[11] but *S. aureus* bacteremia have greater risk of infecting prosthesis than other bacteria. The most common recovered bacteria from infected joint prosthesis are coagulase-negative staphylococci (CONS) in 30–43%, *S. aureus* (12–23%), mixed bacteria (10–11%), *Streptococci* (4–10%), gram-negative bacilli (3–6%), enterococci (3–7%), and anaerobes (2–4%).[12]

The diagnosis of infected prosthesis is often challenging, as the clinical presentation can be subtle without the obvious signs of infection (fever, redness, and purulent drainage) and diagnostic tests are inaccurate. The surgeon or physician has to have a high index of suspicion as any wound drainage or fluid (i.e., hematoma) can be infected early without the usual signs of wound infection. Chronic, low-grade prosthetic joint infection often presents with just joint pain and loosening of the prosthesis on imaging. Thus, draining wounds or hematoma in the wound should be cultured on evacuation (irrespective of fever or redness), but after cleansing the skin with chlorhexidine-alcohol solution. Patients with chronic joint pain, with or without joint loosening, should undergo aseptic aspiration of the joint under ultrasound guidance for aerobic and anaerobic culture. An elevated erythrocyte sedimentation rate (ESR) and C-reactive protein (CRP) are highly sensitive for prosthetic joint infections, but would be non-specific within the first few weeks of surgery.

Early prosthetic joint infection (as in this case) usually occurs within the first 3 months of surgery, as a local wound complication. Typical signs of wound infection are commonly seen with virulent organisms such as *S. aureus*, β-hemolytic *Streptococci*, and some gram-negative bacilli. However, low virulent organisms (normal skin flora) and opportunistic bacteria (*Pseudomonas* species in the normal host) may present in the early stages with non-purulent discharge, pain, and fluid collection or infected hematoma (personal experience).

The management of an infected orthopedic prosthesis in the early stage is a combination of antibiotics and surgical intervention. Drainage, debridement, and irrigation with retention of the prosthesis has been recommended if symptoms of infection are less than 14 days,with salvage of the prosthesis in ≥70%, in the absence of loosening of the prosthesis or sinus.[12] However, this actually depends on the recovered organisms. With *S. aureus*, early surgery within ≤48 h is warranted, and for gram-negative bacilli, less than 4 days after signs/symptoms of infection results in reasonable rates of cure and salvage of the prosthesis.[13] There is also evidence that multi-resistant bacteria such as methicillin resistant *S. aureus* (MRSA) or *P. aeruginosa* usually will require total excision of the prosthesis before cure can be obtained.

Should combinations of antibiotics be used for serious *P. aeruginosa* infection? Although some experts recommend a combination of the two antibiotics to which the organism is susceptible, based on reviews of the literature, there is no good evidence to support this.[14]

However, single use of imipenem or meropenem for pseudomonas infection is commonly associated with development of resistance, thus combined therapy in that situation is reasonable (especially in the presence of chronic therapy for retained prosthesis). However, it is unknown whether combined imipenem and piperacillin (or with tobramycin) would be any more effective than piperacillin itself.

9.3.3 Medico-legal Discussion

It is highly likely that the retained gauze played a direct role in leading to the prosthetic hip wound infection. However, this may not represent negligence if the usual precautions were taken at the end of the procedure to do an instrument and sponge/towel count to guard against such an accident. Moreover, human error can occur despite these precautions, which may not be considered an act of negligence by the courts. In the case under discussion, the count sheet for surgical paraphernalia was noted to be correct. This raises an issue of an incorrect count or the possibility of the gauze being left in place after the first drainage of the hematoma.

Although there was no confirmation that the microorganism causing the prosthetic hip infection was the same as that in the lumbar spine, it can be surmised that the two conditions were directly related. The probability that the same organism (*P. aeruginosa)* metastatically seeded to the lumbar disc/vertebra would be much greater than the chance of an unrelated infection developing at the same time. Failure to grow any bacteria from the disc space was likely due to the combination of antibiotics the patient had received before and during the aspirate. The chance of seeding to the spine would have been greatest when the subject was not receiving any effective antibiotic (before recognition of the infection), but could have occurred during the second surgical procedure for drainage and debridement. Thus, it could be argued by the plaintiff's legal counsel that if recognition of the infection had been made 10 days before re-admission when the wound culture reported a pathogen, then the adverse outcome of the vertebral osteomyelitis could have been averted.

Was the initial management appropriate and accepted standard for the prosthetic hip infection? Neither imipenem nor meropenem is considered first line therapy for pseudomonas infection, but the carbapenems can be used as alternatives (second line) when there is microbial resistance or previous allergies to first line agents.

Although many experts may have chosen piperacillin alone or combined with tobramycin (initially) for serious pseudomonas infection (based on the susceptibility in this case), it is very doubtful that the outcome would have been any different. It is even more contentious whether the surgical drainage (as performed) versus removal of the entire prosthesis would have changed the outcome.

9.4 Case 4: Post-arthroscopy Infection

In January of 1998, a 24-year-old male consulted an orthopedic surgeon for a painful right knee. He had suffered a hockey injury to the knee the year before, and as a result was unable to run, crouch, kneel, or play his sport. Besides having knee pain and intermittent swelling, his knee would give way on occasion. The physician suspected a lateral meniscus injury and recommended arthroscopy. The procedure was performed a few months later at a community hospital. At the operation, a crack with adjacent softening of the knee cartilage traversing the anterior-posterior diameter was found. This was treated by mechanical removal of the damaged and softened area of the cartilage. The procedure itself was uncomplicated, but the following day the man was seen in the hospital with severe knee pain. There was minimal swelling of the knee, with no signs of inflammation, so he was discharged on analgesics.

The young man returned to the ER 4 days later with continued severe knee pain, fever, and progressive swelling of the knee. He was then admitted for septic arthritis under the care of another orthopedic surgeon. Arthroscopic drainage and debridement was performed and he was started on parenteral antibiotics (cefazolin). Culture of the drainage grew S. aureus susceptible to beta-lactams. He was immobilized in a splint for 3 days. The patient continued to have fever and a swollen, erythematous knee joint. A second arthroscopic drainage was performed 4 days later and the culture was still positive from the drainage. Fever persisted until the evening just before discharge 4 days later, when he was sent home on a week's supply of oral cloxacillin.

A week later the patient was re-admitted with pain, swelling, and limitation of flexion of the knee, as well as persistent low grade fever (38.2°C). Intravenous antibiotic was reinstituted and physiotherapy performed, but no further surgery or aspiration. Subsequent imaging investigation revealed osteomyelitis of the femoral condyle. Parenteral antibiotics were continued at home for a total of 6 weeks course. The subject was left with residual stiff knee and was reevaluated by a consulting orthopedic surgeon at a tertiary center about 14 months later. On examination, there was limitation of movements with only 20° of flexion. He then underwent two arthroscopic debridements with subsequent improvement in mobilization of the knee. The prognosis was considered guarded, future surgery would likely be required, and permanent damage with osteoarthritis could be expected.

9.4.1 Medico-legal Issues

Medico-legal action was taken by the young man against the surgeons involved and the institution for medical malpractice. Specific charges were failure to take proper precautions to prevent the complication of septic arthritis, and substandard therapy for the infection leading to chronic disability that required further surgery. The defendants claimed that proper aseptic surgical technique was undertaken, and that he was appropriately treated for septic arthritis. The very low risk of a post-arthroscopic infection was discussed with the plaintiff before the procedure, but this risk is never zero, even under the best circumstance or technique.

9.4.2 Medical Aspects

The type of articular cartilage injury suffered by the plaintiff could be classified as a type I lesion (Bain and Jackson classification), with a linear crack and softening of the area.[15] This represents the mildest type of cartilage injury and the recommended treatment is initial non-operative management which involves decreasing the load on the joint (by weight loss), altering activities, strengthening of the supporting muscles, orthosis or braces, and nonsteroidal anti-inflammatory agents (NSAIDS) for pain control.[15] If the patient fails to improve on medical management then arthroscopic debridement and lavage should be recommended.

Should prophylactic antibiotic be used before arthroscopy to prevent post-operative infection? Current guidelines do not recommend antibiotic prophylaxis for arthroscopy as the infection rate is very low (<1%), unless it involves implantation of foreign materials.[16] Moreover, a previous prospective, randomized, double-blind trial of prophylactic antibiotics in arthroscopic surgery found no benefit, as the rate of infection was so low.[17] Besides recent joint surgery, the plaintiff had no other known risk factors for septic arthritis, although he may have had undetected chronic nasal colonization with *S. aureus*. This alone may have increased the risk for post-operative wound infection in clean surgery.[18] Up until very recently, decolonization with nasal mupirocin had not been demonstrated to reduce wound infection rates in *S. aureus* nasal carriers.[19] However, in a very large multicenter, randomized trial just recently published, intranasal mupirocin with daily baths with chlorhexidine soap for 5 days reduced *S. aureus* surgical site infection by 60%.[20] A previous non-randomized study had shown that *S. aureus* nasal decolonization in joint replacement surgery reduced the rate of infection.[21] Interestingly, just using chlorhexidine-alcohol instead of povidone-iodine for surgical site antisepsis can reduce *S. aureus* wound infections by 50% and overall surgical site infection rate by 40%.[22]

When should septic arthritis be suspected? The clinical picture of a hot, red, tender, and swollen joint with fever is the classic presentation of acute bacterial arthritis, but this presentation is not highly sensitive. A recent review of the clinical manifestations of septic arthritis showed that fever occurs in only about 57%, and that joint pain and swelling are the two most sensitive (but non-specific) symptoms.[23] Also, risk factors identified for septic arthritis include age older than 80 years, diabetes mellitus, recent joint surgery, hip or knee prostheses, skin infection, and HIV infection. The physical examination was not found to be predictive of bacterial arthritis, and thus a low threshold for arthrocentesis for white blood cell count and culture should be considered for anyone with monoarthritis.

The management of an infected joint involves a combination of antibiotics and drainage to control and eradicate the infection, pain and inflammation control, and subsequent physiotherapy. The aims of therapy are to eradicate infection, restore joint function, and limit damage to the articular cartilage. Experimental infectious arthritis in the rabbit model with *S. aureus* demonstrate rapid destruction of cartilage.[24] The quantity of cartilage lost from the joint as a result of infection in untreated animals after 3 weeks was about 55%. When antibiotic was started 1 day after the infection, the loss of cartilage averaged 30% and increased to approximately 50% when treatment was delayed for 7 days.[25] Thus, it is imperative to rapidly clear the infectious material from articular cartilage, even before control of the infection, as the bacterial products through pro-inflammatory cytokines and metallomatrix enzyme generation lead to irreversible cartilaginous damage. Drainage can be accomplished through daily aspiration by arthrocentesis, arthroscopy, or surgical procedure. There is a paucity of prospective studies and there is no randomized prospective trial to evaluate medical versus surgical drainage. Review and evaluation of the available evidence indicates that regular aspiration appears to be as effective as arthroscopic or surgical drainage.[26] Indications for surgical or arthroscopic drainage have been recommended for hip joint, loculation of pus or when the fluid is too thick for aspiration, failure of medical therapy, soft tissue extension, and all prosthetic joints. Failure of medical therapy has been defined as positive synovial fluid culture after 72 h of appropriate antibiotics, and persistence of symptoms of infection or inadequate clinical response after 5 days.[27] Surgical drainage should be considered initially for patients with underlying damaged joints due to rheumatoid arthritis, because of the poor outcome with these joints.[28]

Medical management of septic arthritis depends on the microorganism isolated and the presence of complications and prosthesis. Gonococcal arthritis is usually easily treatable and most textbooks or experts recommend parenteral therapy for about 2–3 days, then oral antibiotics for another 7–10 days. The initial empiric therapy for non-gonococcal bacterial arthritis (native joint) should be broad spectrum for streptococci, staphylococci (including MRSA) and gram-negative bacilli. Treatment should then be adjusted for the specific isolate susceptibility. Unfortunately, there is very little solid evidence to guide therapy on the type and duration of parenteral therapy, or the optimal duration of the total course of treatment. Thus, recommendations in various textbooks and guidelines have been numerous and based largely on experience and retrospective data. For instance, for non-gonococcal bacterial arthritis the recommendations by Harrison's Principles of Internal Medicine varies with the bacteria; for *S. aureus,* 4 weeks duration (β-lactams for MSSA and vancomycin for MRSA), for susceptible *Streptococcus pneumoniae* and other streptococci, 2 weeks of intravenous (IV) penicillin, and for gram-negative, 3–4 weeks of IV cephalosporin (or carbapenem or broad spectrum penicillin) or oral quinolone.[29] In a recent Rheumatology textbook,[30] 2 weeks total antibiotics is recommended for uncomplicated susceptible organisms (streptococci, *Hemophilus influenzae* or gram-negative cocci); duration of parenteral not indicated, but oral can be used if adequate blood levels can be achieved, and for *S. aureus* 3–4 weeks antibiotics (mainly parenteral), and gram-negative bacilli or *S. pneumoniae* should be treated for a total of 4 weeks. The authors also recommended that extensive infection and immunosuppressed subjects be treated for 4–6 weeks.[30] In an authoritative Infectious Diseases textbook,[31] it was recommended that gonococcal arthritis usually responds to 2 weeks course, and *S. aureus* and gram-negative bacilli should be treated for 4 weeks (with 2–4 weeks parenteral for *S. aureus*).

A guideline recently published by the British Society of Rheumatology and other British health professionals, recommended 2 weeks IV antibiotics for non-gonococcal septic arthritis (or until signs improve), and then orally for around 4 weeks.[32] However, as noted by a systematic review and meta-analysis of antibiotic therapy for bone and joint infections, the lack of high quality data and heterogeneity of the study populations preclude inferences from the available data.[33] There are some recent, prospective, randomized trials in children with bacterial arthritis that may provide some direction for the adult management.

Previous studies published almost 4 years ago on the duration of treatment for bacterial arthritis in children (based on clinical response and decreasing ESR) found the mean duration of therapy for MSSA septic arthritis was 3 weeks, and this was then used as standard guideline.[34] In a small proof-of-concept study, 33 children with acute hematogenous bone or joint infection were treated with short-term (7 days for septic arthritis, 10 days for osteomyelitis) or long-term (14 days for joints, and 21 days for bone) IV antibiotics after surgical drainage.[35] No difference in outcome was found between short-term versus longer-term therapy, but the study was underpowered to show a difference. In a more recent larger, prospective, randomized trial for childhood septic arthritis in Finland, short-term therapy appeared just as effective as long-term treatment.[36] This multicenter trial enrolled 130 cases (88% caused by *S. aureus* [>50%], *H. influenzae* or *Streptococcus pyogenes*) and 63 patients received short-term (10 days) while 67 received long-term (30 days) treatment. Intravenous was used for the first 2–4 days, and then followed by a well-absorbed oral agent, only one joint aspiration was performed for most cases, and surgical procedures in 12% of patients. The primary end point was full recovery by clinical response and the treatment continued until the CRP decreased to $\leq$20 mg/L without further need for antibiotics (for osteoarticular indication) 12 months after therapy.[36]

For patients with severe pain, rest and immobilization for the first few days (with or without a splint) are recommended, and then once clinical improvement occurs, passive physiotherapy and non-steroidal anti-inflammatory agents (NSAIDS) can be used to improve function. Initial NSAIDS should be avoided, as it may provide a false sense of clinical response to antibiotics. Most textbooks recommend daily joint aspiration until the joint is dry, or there is only a non-inflammatory effusion. In children, a short course of dexamethasone for 4 days has been shown to improve residual dysfunction and shortened duration of symptoms at 6 and 12 months in a double blind, randomized, placebo-controlled trial.[37] This is compatible with the findings in animal experimentation.

9.4.3 Medico-legal Discussion

In this patient (Case 4), standard pre-operative preparation was used, antibiotic prophylaxis was not indicated, and the fact that a post-operative wound infection occurred does not imply medical negligence or substandard care. The management of the plaintiff's infection (septic arthritis) was sub-optimal and could be considered below accepted medical standards. Current guidelines and textbooks recommend at least 2 weeks of parenteral therapy followed by oral antibiotics for *S. aureus* septic arthritis, and the patient only received 7–8 days of parenteral therapy. Is this by itself sufficient to be considered substandard care or medical negligence? Unfortunately, the court's yardstick for assessing standards of medical care often depend on commonly used textbooks and medical societies' recommendations and guidelines, despite the lack of solid evidence to support them. Although trials in children are not directly applicable to adults, they often result in the same conclusions when similar studies are performed. For instance, corticosteroid was first demonstrated to be of benefit in childhood bacterial meningitis, and is now an accepted adjunct in adult pneumococcal meningitis. A short course of parenteral therapy followed by a longer course of oral antibiotics was first demonstrated to be as effective as the full course of parenteral antibiotic in children with acute osteomyelitis. This approach is now an accepted practice in adults with osteomyelitis, supported by cumulative, retrospective, and prospective collected data from non-randomized studies. Thus, it could be argued by the defendants that there is no good evidence that the shortened IV antibiotic course can lead to a worse outcome in septic arthritis.

The other aspect in the management of the plaintiff's infection may be harder to defend. Expert medical witnesses contended that the surgeon discharged the patient on oral antibiotics prematurely, as there was evidence of persistent uncontrolled infection with fever on the evening before hospital discharge. Moreover, with a persistent positive culture and fever after the second arthroscopy, there was indication to repeat the arthroscopy and aspirate the joint or perform open surgery to ensure improvement in cell count and culture. At this stage, it would have also been recommended by most experts to perform MRI of the knee to determine extending into the bone. It was also the opinion of the medical experts reviewing the case that on re-admission, repeat arthroscopy, or open surgery should have been performed. This negligence contributed to further joint damage and adhesions, which led to poor function and recovery; and resulted in the need for further surgeries later on.

9.5 Case 5: Infection After Arthroscopic Knee Surgery

An orthopedic surgeon recommended arthroscopic surgery to a 48-year-old female with chronic osteoarthritis of the knee complicated by pigmented villonodular synovitis. The patient was obese and taking metoprolol for unspecified cardiac arrhythmia. Debridement, synovectomy, and removal of the nodules were performed without any intra-operative complication, and the patient was sent home on analgesics for routine post-operative pain on the same day. On post-operative day 2, the patient experienced intense pain, which progressed over the next 2 days (over the weekend) with gross swelling and increased warmth, plus drainage of reddish colored fluid. She called the surgeon's office on the morning of day 5 with her complaints, and was told by the secretary that this was a normal reaction after the surgery. Her symptoms were no better the following day and she made arrangements to see the surgeon on post-operative day 6.

The patient claimed that at that visit, her knee and upper leg were grossly swollen, extremely painful, hot, tender, and oozing a discharge that soaked the bandages. According to the surgeon's clinic note however, the wound was healing well, there was no evidence of inflammation or significant discharge. A minor post-operative bleed was considered to be the cause of the patient's symptoms, so he prescribed stronger analgesics and a knee-immobilizer to rest the joint. Despite this, the patient's symptoms continued to worsen and she was admitted to the hospital emergency department on day 10 after surgery. Her complaints were increasing knee pain and swelling, purulent wound discharge for 3–4 days, and night sweats, but no definite fever or chills. Examination by the ER physician recorded a normal temperature of 37°C, tachycardia of 100 bpm and a swollen, erythematous, warm, tender left knee with limited flexion. A knee aspirate performed revealed thick, purulent fluid. A diagnosis of septic arthritis was made, intravenous antibiotics were instituted, and the consulting orthopedic surgeon on call performed an emergency arthroscopic drainage and synovectomy.

She was hospitalized for 10 days and received 3 weeks of intravenous cefazolin for S. aureus infection. Residual pain, stiffness, and limitation of flexion resulted, plus there was radiological evidence of loss of the knee cartilage. Two years later, the patient required a total knee replacement with prosthesis.

9.5.1 Medico-legal Issues

Medico-legal litigation was subsequently initiated by the patient against the original orthopedic surgeon for negligence in post-operative care. The plaintiff charged that the defendant failed to provide adequate medical care by: (1) not ordering prophylactic antibiotics to prevent the infection, (2) failure to diagnose and treat the infection at an early stage, (3) minimizing her symptoms and complaints, and his negligence resulted in pain, suffering, and loss of function in her knee.

Furthermore, if he had instituted proper treatment 4 days earlier, when she was seen in his clinic, the outcome could have been averted.

Counsel for the defendant requested an independent medical opinion to determine the validity of the plaintiff's claims, and to provide any reasonable answers or rebuttal.

9.5.2 Medical Aspects

Most orthopedic procedures are considered clean surgery, unless there was infection present in or around the joint or an open wound, and the rate of post-operative infection is normally very low (<2%). Clean surgery does not involve entry into the alimentary, respiratory, or genito-urinary tract and the colonizing burden of bacteria is low, thus, aseptic surgery with proper pre-operative skin decolonization are highly effective in preventing infection. Since most clean operations have very low rates of post-operative infection, prophylactic use of antimicrobials in most clean procedures is not usually justified. However, procedures involving insertion of prosthetic materials may justify the prophylactic use of antimicrobials. While the risk of infection may be low, the consequences of infection can be serious and catastrophic. Moreover, some randomized controlled trials (RCT) have shown the value of prophylactic antibiotics in some clean surgical procedures (e.g., total hip replacement, hip fracture repair with internal fixation, and cardiac bypass surgery).[38,39] In arthroscopic surgery of the knee not involving implantation of foreign material, antibiotic prophylaxis is not indicated and is of no proven value.

Besides the colonizing bacteria at the site of surgery, many other factors may influence the risk of surgical site infection, such as aseptic surgical technique, length of the procedure, operating room environment, comorbid conditions (diabetes mellitus, obesity, malnutrition, immunosuppression, smoking, preoperative nasal colonization with *S. aureus,* etc.), presence of infection elsewhere, surgeon's experience, and frequency of performing the procedures, emergency conditions, perioperative blood transfusions, and others.

Would appropriate treatment instituted 4 days earlier for septic arthritis have made a difference in the outcome? It has been the traditional teaching that infected joints (septic arthritis) should be diagnosed quickly and receive urgent aspiration and antibiotic therapy, because if left undetected, can lead to rapid joint obstruction.

Infection of the synovial space occurs either by hematogenous seeding or by direct invasion or inoculation at surgery/intervention (as in this case). The synovium is extremely vascular and contains no protective basement membrane, thus, there is easy access to the synovial space. Within 24–48 h of inoculation or bacterial or invasion, there is marked granulocyte infiltration, vascular congestion and proliferation of cells lining the joint space.[40] During the week following bacterial invasion, there is marked synovial proliferation, persistent purulent effusion, subsequent infiltration by mononuclear cells, and granulation tissue and abscess (loculated pus) develop. Cytokine induced proteolytic enzymes are released by the inflammatory cells, which lead to cartilage and bone destruction in as early as 10 days.[40] Delay in treatment can result in joint destruction and systemic sepsis.

Bacterial arthritis has been reported to result in loss of joint function in 25–50% of patients.[41,42] Factors that have been reported to be associated with adverse outcome include: extremes of age, persistent joint disease (notably rheumatoid arthritis), comorbidity, immunosuppression, hip joint infection, adjacent osteomyelitis, virulent organisms (i.e., *S. aureus* and coliforms), presence of bacteremia, a long duration of symptoms, and persistent synovial fluid cultures during treatment. However, the information on these factors has largely been derived from retrospective studies with small sample sizes. Retrospectively collected data suggest that poor outcome (on joint function) was associated with symptoms of infection for 7 days or more before treatment.

The largest prospective community-based study on outcome of bacterial arthritis in 154 patients found no association between a poor outcome and treatment delay.[43] Bacterial arthritis had a poor outcome in almost 50% of patients; factors associated with the poorest prognosis were older age, preexisting joint disease, and presence of synthetic material.[43] The median number of days with symptoms of infection before treatment was 4 days in adults and 2 days in children, but the range was quite wide. In this study, *S. aureus* infection did not predict a poor outcome as reported in retrospective studies,[44] or animal models, which showed rapid cartilage destruction.[45]

In rabbit models of acute septic arthritis with *S. aureus* degradation of cartilage, as reflected by quantified analysis of glycosaminoglycan and collagen, starts within 24 h of infection.[45] Macroscopic quantitative loss of cartilage averaged between 30% and 50% when treatment was delayed for 1–7 days, respectively.[46] In retrospective studies of septic arthritis, significant mortality (11.5%) has been associated with confusion at presentation, age ≥65 years, multiple joint involvement, poor functional outcome with delay in presentation of greater than 3 days, presence of prosthetic and arthroscopic material, and open surgical drainage.[47] The poorer outcome found with surgical intervention may well be due to reflection of the more severe and complicated cases in retrospective series. In a review of surgical drainage versus daily needle aspiration for bacterial arthritis, the conclusion was that there is no compelling evidence to recommend surgical lavage for the initial management of uncomplicated septic arthritis.[48] Surgical drainage should be reserved for more complicated cases, such as those associated with prosthetic joints or soft tissue extension, or joints poorly accessible to needle aspiration, like the hip.

Choice of initial antibiotics will depend on the most likely microorganisms, and then tailored according to culture results and susceptibility. The optimal duration of therapy with antibiotics has been based on retrospective studies and expert opinions. For native joint infection, 2 weeks of antibiotic therapy is usually adequate for *Neisseria* and sensitive *Streptococcus* infection, but *S. aureus,* pneumococci and gram-negative bacillus infections are treated for 3–4 weeks, with intravenous therapy for the first week.[49]

In the randomized trial in children with septic arthritis (previously mentioned), the number of S. *aureus* septic arthritis was similar in the two groups (35/63 and 41/60).[36] The response rate between short-term and long-term therapy was similar, and all patients recovered without sequelae.

9.5.3 *Discussion of Medico-legal Issues*

To address the charges by the plaintiff against the surgeon, the following responses were provided by the expert witness for the defendant:

1. Prophylactic antibiotic for routine arthroscopy surgery is not recommended and therefore was not indicated in this case.
2. Based on the surgeon's clinical notes, when the plaintiff was assessed in the clinic on post-operative day 6, there was no indication to perform a knee aspiration to exclude infection, as there were no clinical signs of inflammation. Moreover, the risk of infection was low (<2%), whereas post-operative hematoma would be more common, as pigmented villonodular synovitis is very vascular, and could account for the defendant's symptoms and signs.
3. There is no good clinical evidence from prospective studies that an earlier diagnosis of infection and initiating antibiotics 4 days earlier would have affected the outcome.

9.6 Case 6: Post-operative Knee Pain

A healthy young man (20 years of age) had suffered from a knee injury during athletic activities. He subsequently underwent an arthroscopic anterior cruciate ligament reconstruction (bone-tendon-bone graft) at a community hospital. The procedure was uncomplicated and he was seen 1 day post-operatively in the clinic for removal of a hemovac drain.

On post-operative day 8, the patient presented to a hospital emergency department at 3 a.m. with increasing left knee pain starting after physiotherapy. He was assessed by the attending nurse who spoke to the ER physician on-call by phone. The patient was provided with oxycodone-acetaminophen tablets with advice to see his FP the following day. This was never done as the patient felt better, with less pain. Three days later (11 days post-operatively) the patient returned to the same hospital emergency department with persistent pain and oozing of blood from the surgical wound. The ER physician noted the patient's temperature was 37.9°C with no signs of inflammation or purulent drainage. A hematoma from the wound was drained (no cultures sent), stronger analgesics were prescribed, and the patient was advised to follow-up with his surgeon in 3 days as scheduled.

The patient was assessed by the surgeon 13 days post-operatively and was found to have a grossly swollen, inflamed knee with purulent drainage from the incision site. He was then admitted to hospital, underwent arthroscopic drainage and debridement of the knee, and started on intravenous antibiotics. The graft was found to be intact and the pus grew S. aureus, which was treated with intravenous, followed by oral cloxacillin for 4–6 weeks. Although the infection resolved, the patient was left with residual stiffness of the knee.

A year later, the patient was found to have a swollen knee with instability, and failure of the tendon graft was suspected. This was confirmed by MRI. He underwent revision and reconstruction of the graft using a patella tendon as autograft. The screw from the original surgery was completely loose, and thus removed. Cultures from the site of the screw grew S. aureus (again) and he was retreated with a 6-week course of cloxacillin.

9.6.1 Medico-legal Issues

The patient initiated medical malpractice litigation against the ER physician for failure to properly diagnose and treat infection of his knee in a timely manner. The lawyer for the plaintiff charged that the ER physician was negligent in not sending drainage from the knee for culture when the plaintiff was first assessed in the rural hospital emergency department. Furthermore, his negligence resulted in pain, suffering, and the delay in diagnosis and treatment, which was the direct cause of the following surgery.

Expert witnesses for the plaintiff (an orthopedic surgeon and an ER physician) provided opinions that the defendant was negligent and provided substandard care. They opined that the defendant should have suspected a bacterial infection when the plaintiff first presented to the emergency department (post-operative day 8) and that if a knee aspirate were sent for gram stain and culture, and appropriate drainage and antibiotics instituted, the adverse outcome could have been avoided.

9.6.2 Medical Aspect

Infection of a joint with a graft or foreign material carries a worse prognosis than a native joint. However, most of the data in the medical literature relates to large prosthesis such as in hip and knee arthroplasties, and internal fixation of fracture sites with rods, screws, and plates. There is inadequate information of the rates of infection, outcome, and prognosis of anterior cruciate ligament grafts. Since most of the graft consists of autologous tendon and bone, it is probably accurate to surmise that the rates of infection and outcome would be better than those of larger prosthesis. However, even the presence of a small screw could compromise the results when compared to native joints, as the pathogenesis of infection would be similar to any foreign-body related infection. It would be fair to estimate that in this case, the incidence of infection after surgery and prognosis should be worse than native joints, but better than total joint arthroplasties.

In the past decade or more, there have been great strides made in our understanding of the mechanisms and biology of foreign-body related infections. The majority of orthopedic foreign-body infections are caused by bacteria that normally colonize the skin, such as coagulase-negative *Staphylococcus* and diphtheroids, which are of low virulence and often present in an insidious manner. *S. aureus* is also a significant pathogen, more likely to infect those who are chronic carriers, and clinically present earlier (within 2 weeks of procedure).

Insertion of a prosthesis increases the risk of infection (by lowering the infectious dose or inoculum required to produce infection), but probably also by increasing the duration of the surgical procedure. Thus, even an extremely low number of bacteria introduced at the time of aseptic surgery can multiply on and around the foreign material, before onset of clinical manifestations, varying from days to months. The hallmark of foreign-body related infection (irrespective of the size and material) is the development of biofilm phenotype. This phenotype may appear in <24 h to days later. Biofilm bacterial colonies are exceedingly difficult to eradicate without removal of the implant for several reasons. The biofilm represents complex colonies of microbes within a physio-chemical barrier of "slime" (a thick, hydrated, polyanionic-gelled polysaccharide/glycoprotein complex). This slime barrier impedes phagocytosis of neutrophils, impairs humoral and cellular immunity, slows the penetration of antimicrobials, and greatly reduces their activity.[49] In vitro studies have demonstrated that the killing activity of antibiotics is greatly reduced (10–1,000 fold) in this environment.[51] Thus, routine testing of recovered bacteria is not useful in predicting response unless the device is removed.

A recent review of orthopedic prosthetic infections indicate that cure can be achieved in a significant proportion of selected cases (due to low virulence, very sensitive organisms) in early post-operative infection, with prompt surgical drainage, debridement, with prolonged antibiotics and salvage of the prosthesis.[50] This approach is more commonly accepted in Europe than North America, as studies in the United States have found less favorable results, but usually less stringent criteria were used for selecting the cases. Generally, retention of the prosthesis will result in a less favorable outcome for *S. aureus*, gram-negative bacillary infection (especially more resistant strains), and in these cases when there is a delay in surgical drainage and debridement >48 h after onset of symptoms.[51]

9.6.3 Medico-legal Discussion

Prophylactic antibiotic (a single dose of first generation cephalosporin) would normally be recommended with this type of arthroscopic surgery with a graft. However, this was not an issue that was raised by the plaintiff and the surgeon was not included in the malpractice litigation suit.

The main issues raised by the plaintiff against the ER physician for providing negligent care were: (1) failure to clinically examine the patient on the first visit (day 8) at the emergency department, and (2) failure to diagnose and initiate treatment on the second emergency department visit (day 11). Furthermore, his failure to recognize and treat the infection earlier resulted in failure of the anterior cruciate graft and created the need for further surgery.

To make a valid judgment, one has to assess the availability of healthcare personnel and standard operations of the small community emergency department. It is apparent that an ER physician is not physically present in the department 24 h a day. After certain hours, the nurse on duty assesses the patients then calls the physician to discuss the issues and need for immediate medical attention. The ER physician decided in the morning of the first emergency department visit that based on the information provided, this was not a true emergency requiring immediate medical attention. His decision was that the plaintiff could wait a few more hours to be assessed by his FP later in the day. This appears to be a reasonable decision, considering the patient presented with only increased knee pain after physiotherapy. Increased joint pain after surgery and physiotherapy is considered very common and usually of a benign nature. However, a visit to the hospital emergency department at 3 a.m.is worrisome and suggests that if the patient is in enough pain to warrant a visit at that time, it could be a signal of a more sinister condition than usual aggravation of post-operative pain from physiotherapy.

The severe knee pain on that first visit likely represented an early manifestation of septic arthritis (<24 h). Hence, it could be argued that if the plaintiff were assessed that morning, knee aspirate performed, and aggressive treatment by arthroscopic drainage and antibiotics within 24 h of the emergency visit instituted, then it would be more likely than not have resulted in a better outcome. A similar outcome may have occurred if the plaintiff went to see his FP or orthopedic surgeon later that day, as advised by the ER physician. Thus, the failure of the patient to follow the advice of the ER staff places the responsibility on the plaintiff himself.

A decision the court would have to make is whether or not the ER physician's failure to assess the patient clinically for severe knee pain represented reasonable medical care, with the expectation that the patient would be assessed by his FP or operating surgeon several hours later that day. Since this was not a limb or life-threatening emergency medical condition, it may be considered that even 12–24 h delay would not have affected the patient adversely. Obviously, the judge or jury would consider the medical opinion of other ER physicians in the community to determine the accepted standard of care.

On the second emergency department visit 3 days later, the same ER physician assessed the patient but failed to send a knee aspirate for bacterial culture and never considered the diagnosis of septic arthritis. This was considered negligent care by expert medical witnesses for the plaintiff. In hindsight, this was clearly a judgment error. The defendant's defense was that the patient did not present with the usual manifestations of septic arthritis (redness, fever, increased warmth, or purulent discharge from the wound). Moreover, there was clinical evidence of wound hematoma, which would account for the patient's symptoms. Despite these arguments, expert witnesses opined that infection should still need to be excluded, and that joint aspirate and wound hematoma should have been sent for immediate gram stain and culture. Would earlier treatment at this stage result in a different outcome?

If we assume that the infection in the subject's knee was likely inoculated with *S. aureus* into the joint at the time of the surgery, and first clinically manifested 8 days post-operatively (first emergency treatment visit), then even if appropriate arthroscopic drainage, debridement and several weeks of antibiotics were instituted at the second emergency visit, the chance of a better outcome may not be greater than 50%. Data from infected prosthetic joints due to *S. aureus* suggest that debridement and drainage 2 days or later after onset of symptoms (even with long term antibiotics) was associated with higher probability of treatment failure than those debrided within 2 days of onset (relative risk = 4.2).[50,51] Although the plaintiff did not have a prosthetic joint or arthroplasty, there is reason to believe that his outcome and course may be similar due to the presence of a screw.

9.7 Lessons to Be Learned

The six cases described in this chapter can provide physicians with insight in avoiding errors in diagnosis and management of orthopedic related infections. Even though clinical signs of infection may not be evident, physicians should be vigilant to exclude this post-operative complication, as the outcome can be devastating to the patients. A summary of the lessons to be learned from each of the cases described is listed below.

1. Surgeons should never guarantee outcome of any surgical procedure and avoid the term "can always" and use "usually" or "most instances" to estimate outcomes.
2. Although post-operative pain is very common and represents pain from surgical injury in most instances, severe pain uncontrolled by usual analgesics should warrant a clinical assessment as soon as possible.
3. Physicians should avoid making decisions from second hand information, but instead speak directly to the patient.
4. In most cases, early post-operative fever within 72 h of surgery is usually benign and represents immune response to tissue injury from the surgery, but requires a clinical assessment to exclude infections and other causes.
5. Wound infection with crepitus (gas in soft tissues), or suspicion of necrotizing fasciitis should best have urgent surgical debridement (within 2 h if feasible).
6. After joint surgery, the surgeon should be vigilant to exclude early post-operative infection for worsening pain and swelling of the joint, with or without fever or purulent drainage. Any drainage or aspirated fluid (even hematomas) should be sent for gram stain and culture. Remember that local collection of serum (seroma) or blood (hematoma) provides an ideal median for bacterial growth and proliferation.

7. Orthopedic procedures with prosthesis or foreign body are high-risk procedures, therefore early recognition of infection, prompt surgical drainage, and aggressive antibiotic therapy are necessary to preserve the function of the joint and offer possible salvage of graft or prosthesis. Always warn patients that despite the best medical care, infections can occur and are difficult to cure and frequently necessitate removal of the prosthesis or graft.

References

1. Fanning S, Neuhoff FJ, Brewer JE, Casteneda T, Marcott MP, Jacobson RL (1998). Frequency and yield of post-operative fever evaluation. *Infect Dis Obstet Gynecol* 6:252–255.
2. Anand VT, Phillips JJ, Allen D, Joynson DHM, Fielder HMP (1999). A study of post-operative fever following pediatric tonsillectomy. *Clin Otolaryngol* 24:360–364.
3. Shaw JA, Chung R (1999). Febrile response after knee and hip arthroplasty. *Clin Orthoped Related Res* 367:181–189.
4. Lin E, Lowry SF (1999). Inflammatory cytokines in major surgery: a functional perspective. *Intensive Care Med* 25:255–257.
5. Lin E, Calvano SE, Lowry SF (2000). Inflammatory cytokines and cell response in surgery. *Surgery* 127:117–126.
6. Ganong WE (ed) (2005). Dynamics of blood and lymph flow. In: Review of Medical Physiology, 22nd Edition, Lange-McGraw Hill, New York, p577–596.
7. Ganong WE (ed) (2005). Cardiovascular regulatory mechanisms. In: Review of Medical Physiology, 22nd Edition, Lange-McGraw Hill, New York, p597–610.
8. Gutstein HB, Akil H (2005). Opioid analgesics. In: (eds) Brunton LL, Lazo JS, Parker KL. Goodman & Gilman's: The pharmacological basis of therapeutics, 11th Edition, McGraw-Hill, New York, p547–590
9. Darouiche RO, (2004). Treatment of infections associated with surgical implants. *N Engl J Med* 350:1422–1429.
10. Harris WH, Sledge CB, (1990). Total hip and knee replacement. *N Engl J Med* 323: 801–809.
11. Ainscow DA, Denham RA, (1984). The risk of hematogenous infection in total joint replacement. *J Bone Joint Surg Br* 66:580–582.
12. Zimmerli W, Trompuz A, Ochsner PE, (2004). Prosthetic joint infections. *N Engl J Med* 351: 1645–1654.
13. Fong IW (ed), (2009). Current concepts of orthopedic implants and prosthetic joint infections. In: Emerging issues and controversies in infectious diseases. Springer, New York, p299–326.
14. Fong IW (ed), (2009). Combination antimicrobial therapies. In: Emerging issues and controversies in infectious diseases. Springer, New York, p327–377.
15. Miller III RH, Azar FM, (2008). Knee injuries. In (eds): Canale ST, Beatty JH: Campbell's Operative Orthopedics 11th Edition. Mosby/Elsevier, Philadelphia, p2395–2660.
16. Bratzler DW, Houck PM, for the Surgical Infection Prevention Writers Workgroup, (2004). Antimicrobial prophylaxis for surgery: an advisory statement from the National Surgical Infection Prevention Project. *Clin Infect Dis* 38:1706–1715.
17. Wieck JA, Jackson JK, O'Brien TJ, Lurate RB, Russell JM, Dorchack JD, (1997). Efficacy of prophylactic antibiotics in arthroscopic surgery. *Orthopedics* 20:133–134.
18. Wertheim HF, Mellers DC, Vos MC, van Leeuwan W, van Belkun A, Verbrugh HA, Nouwen JL, (2005). The role of nasal carriage in *Staphylococcus aureus* infection. *Lancet Infect Dis* 5:751–762.
19. Perl TM, Cullen JJ, Wenzel RP, Zimmerman B, Pfaller MA, Sheppard I, Twombley J, French PP, Herwaldt LA and the Mupirocin and the Risk of *Staphylococcus aureus*

Study Team, (2002). Intranasal mupirocin to prevent post-operative *Staphylococcus aureus* infection. *N Engl J Med* 346:1871–1877.
20. Bode LGM, Kluytmans JAJ, Wertheim HF, Bogaers D, Vandenbroucke-Grauls CMJ, Roosendaal R, Troelstra A, Box ATA, Voss A, van der Tweel I, van Belkum A, Verbrugh HA, Vos MC, (2010). Preventing surgical site infections in nasal carriers of *Staphylococcus aureus*. *N Engl J Med* 362:9–17.
21. Hacek DM, Robb WJ, Paule SM, Kudrna JC, Stamos VP, Petersen LR, (2008). *Staphylococcus aureus* nasal decolonization in joint replacement surgery reduces infection. *Clin Orthop Relat Res* 466:1349–1355.
22. Darouiche, RO, Wall MJ, Itani KMF, Otterson MF, Webb AL, Carrick MM, Miller HJ, Awad SS, Crosby CT, Mosier MC, Al Sharif A, Berger DH, (2010). Chlorhexidine-alcohol versus povidone iodine for surgical-site antisepsis. *N Engl J Med* 362:18–26.
23. Margaretten ME, Kohlwes J, Moore D, Bent S, (2007). Does this adult patient have septic arthritis? *JAMA* 299:1478–1488.
24. Smith RL, Schurman DJ, Kajiyama G, Mell M, Gilkerson E, (1987). The effect of antibiotics on the destruction of cartilage in experimental infectious arthritis. *J Bone Joint Surg* 69-A:1063–1068.
25. Manadan AM, Block JA, (2004). Daily needle aspiration versus surgical lavage for the treatment of bacterial septic arthritis in adults. *Am J Therapeut* 11:412–415.
26. Stimmler MM, (2000). Infectious arthritis. In (eds): Humes HD, Dupont HC, Griffin JW Harris Jr ED, Hoggardd WA, King TE, Loriaux DC, Nabel ED, Todd III RF, Traber PK. Kelly's Textbook of Internal Medicine. Lippincott, Williams and Wilkins, Philadelphia, p1418–1424.
27. Mandell BF, (2007). Septic arthritis. In (eds): Dale DC, Federman D. ACP Medicine, 3rd Edition, WebMDInc, New York, p1533–1541.
28. Gardner GC, Weisman MH, (1990). Pyarthrosis in patients with rheumatoid arthritis: a report of 13 cases and review of the literature from the past 40 years. *Am J Med* 88:503–511.
29. Maddoff LC, (2008). Infectious arthritis. In (eds): Fauci AS, Braunwald E, Kasker DL, Hauser SL, Longo DL, Jameson JL, Loscalzo J. Harrison's Principles of Internal Medicine, 17th Edition, McGraw-Hill, New York, p2169–2175.
30. Ho Jr G., Siraj DS, Cook PP, (2009). Bacterial arthritis. In (eds): Firestein GS, Budd RC, Harris Jr ED, McInnes IB, Sergent JS, Kelby's Textbook of Rheumatology, 8th Edition, Saunders/Elsevier, Philadelphia, p1701–1714.
31. Ohl CA, (2005). Infectious arthritis of native joints. In (eds): Mandell GA, Bennett JE, Dolin R. Principles and Practice of Infectious Diseases, 6th Edition. Elsevier/Churchil/Livingstone, Philadelphia, p1311–1322.
32. Weston V, Coakley G, (2006). Guideline for the management of the hot swollen joint in adults with a particular focus on septic arthritis. *J Antimicrob Chemother* 58:492–493.
33. Stengel D, Bauwens K, Sehouli J, Ekkernkamp A, Porzsolt F, (2001). Systemic review and meta-analysis of antibiotic therapy for bone and joint infections. *Lancet Infect Dis* 1:175–188.
34. Bradley JD (2009). What is the appropriate treatment course for bacterial arthritis in children? *Clin Infect Dis* 48:1211–1212.
35. Jaberi FM, Shahcheraghi GH, Ahadzadeh M, (2002). Short-term intravenous antibiotic treatment of acute hematogenous bone and joint infection in children: a prospective randomized trial. *J Pediatr Orthop* 22:317–320.
36. Peltola H, Pääkkonen M, Kallio P, Kallio MJT, for the Osteomyelitis-Septic Arthritis (OM-SA) Study Group (2009). Prospective, randomized trial of 10 days versus 30 days of antimicrobial treatment, including a short-term course of parenteral therapy, for childhood septic arthritis. *Clin Infect Dis* 45:1201–2110.
37. Odio CM, Ramirez T, Arias G, Abdelnour A, Hidalgo I, Herrera MC, Bolaños W, Alpizar J, Alvarez P, (2003). Double blind randomized, placebo-controlled study of dexamethasone therapy for hematogenous septic arthritis in children. *Pediatr Infect Dis J.* 22:883–888.

38. Abramowicz M, Zucotti G (eds), (2006). Treatment Guidelines from the Medical Letter: antimicrobial prophylaxis for surgery. *Med Letter* 4:83–87.
39. Bratzler DW, Houck PM, for the Surgical Infection Prevention Guidelines Writers Workgroups, (2004). Antimicrobial prophylaxis for surgery: an advisory statement from the national surgical prevention project. *Clin Infect Dis* 38:1706–1715.
40. Goldenberg DL, Reed JL (1985). Bacterial arthritis. *N Engl J Med* 312:764–771.
41. Cooper C, Cawley MID (1986). Bacterial arthritis in an English health district: a 10-year review. *Ann Rheum Dis* 45:458–463.
42. Youssef PP, York JR (1994). Septic arthritis: a second decade of experience. *Aust NZ J Med* 24:307–311.
43. Kaandorp CJE, Krijnen P, Moens HJB, Habbema JDF, van Schaardenburg D (1997). The outcome of bacterial arthritis. A prospective community-based study. *Arthritis Rheum* 40:884–892.
44. Dubost JJ, Fis I, Soubrier M, Lopitaux R, Ristori JM, Bussiere JL, Sauvezie B (1994). Septic arthritis in patients with rheumatoid arthritis. *Rev Rheum* 61:143–156.
45. Riegels-Nielsen P, Frimodt-Moller N, Jensen JS (1987). Rabbit model of septic arthritis. *Acta Orthop Scand* 58:14–19.
46. Smith RL, Schurman DJ, Kajiyama G, Mell M, Gilkerson E (1987). The effect of antibiotics on destruction of cartilage in experimental infectious arthritis. *J Bone Joint Surg* 69:1063–1068.
47. Weston VC, Jones AC, Bradbury N, Fawnthrop F, Doherty M (1999). Clinical features and outcome of septic arthritis in a single UK Health District. 1982–1991. *Ann Rheum Dis* 58:214–219.
48. Manadan AM, Block JA (2004). Daily needle aspiration versus surgical lavage for the treatment of bacterial septic arthritis in adults. *Am J Therapeut* 11:412–415.
49. Fong IW (ed) (2009). Device-related infections. In: Emerging issues and controversies in Infectious Disease. Springer, New York, p299–326.
50. Fong IW (ed) (2009). Current concepts of orthopedic implants and prosthetic joint infections. IN: Emerging issues and controversies in infectious disease. Springer, New York, p299–326.
51. Brandt CM, Sistrunk WW, Duffy MC, Hassen AD, Stecklberg JM, Ilstrup DM, Osreon DR (1997). *Staphylococcus aureus* prosthetic joint infection treated with debridement and prosthesis retention. *Clin Infect Dis* 24:914–919.

Chapter 10
Neurosurgical Medico-legal Issues

10.1 Case 1: A Comedy of Errors Leads to a Catastrophic Outcome

A 58-year-old female presented to the ER of a small community hospital with symptoms of fever, chills, and right-sided back pain of 1-day in duration. There was a past history of kidney infection 3 years before, type II diabetes diagnosed several months prior, and chronic smoker's cough, which was somewhat increased in the last 2 days. The vital signs and examination were reported to be normal except for occasional crackles of the right lower lung. The only investigation performed was a urinalysis, which showed 1+ blood, but no other abnormality. The ER physician diagnosed probable mild pneumonia. There was a history of penicillin allergy, so treatment was initiated with azithromycin for 5 days and she was advised to follow-up with her family physician (FP). A mid-stream urine culture was requested of the patient to be performed the following morning.

After 2 days, the patient returned to the same ER with symptoms of persistent fever, chills and worsening right-sided back and flank pain. Vital signs were normal except for a temperature of 38°C, pulse of 100/min and she appeared unwell. Clinical examination revealed no focal findings of infection, urinalysis showed 2+ ketones but was otherwise normal; complete blood count including leucocyte count and differential counts were normal and the chest radiograph revealed no pulmonary infiltrates. She was admitted for fever of unknown cause and maintained on her usual medications, including azithromycin 250 mg daily, and hypoglycemics. She was started on intravenous saline and given a single dose of morphine, followed by acetaminophen 325 mg with codeine 30 mg every 4 h. Blood cultures were obtained and a computerized tomography (CT) of the abdomen was performed (at another hospital). The patient remained febrile for 24 h with a maximum temperature of 39.5°C. A urine culture taken the day before admission was reported by the second hospital day as growing Staphylococcus aureus >10^6 CFU/mL, resistant only to penicillin. Blood cultures taken on admission (two sets) were reported as growing staphylococcal species (preliminary report) on the second

I.W. Fong, *Medico-Legal Issues in Infectious Diseases*, Emerging Infectious Diseases of the 21st Century, DOI 10.1007/978-1-4419-8053-3_10,

hospital day. CT scan of the abdomen was reported as normal with no evidence of hydronephrosis, abscess, or nephrolithiasis.

The patient was discharged on the third day of admission with advice to complete one more day of the azithromycin and to make an appointment for follow-up with her FP. The discharge diagnosis was listed as urosepsis. Within a week of hospital discharge, the subject was seen by the FP and oral ciprofloxacin for 7 days was prescribed, presumably for urosepsis or pyelonephritis, although the patient had no urinary symptoms nor any further investigations. Over the next several weeks, the patient was assessed at regular intervals by the FP.

She continued to have severe, excruciating, right-sided mid-back and flank pain extending to the anterior sub-costal region, aggravated by movements and worst on bending forward at the waist. The FP tried various medications, which only partially relieved the pain, including ketorolac tromethamine (Toradol), oxycodone-acetaminophen (Percocet) and a local intramuscular injection of corticosteroid. Repeat investigations showed a mild normocytic anemia (hemoglobin 112 g/dL), high erythrocyte sedimentation rate (ESR) of 105/h, high C-reactive protein (CRP) of 40 mg (>10 mg indicative of inflammation), normal abdominal ultrasound, and radiograph of the lumbar-sacral spine revealed degenerative disc disease at multiple levels. A general surgeon consultation was requested, yet no firm diagnosis was made. The opinion was that the patient did not have an intra-abdominal process accounting for the pain. Despite a normal creatine kinase and negative investigations for a collagen vascular disorder, the FP tried the patient on oral prednisone 30 mg daily with a working diagnosis of some form of myositis or rheumatological disorder. A rheumatology consultation was requested and magnetic resonance imaging (MRI) of the muscles (lower thorax and upper abdomen) was arranged at a university hospital.

The initial MRI revealed no abnormality of the muscles but there was increased intensity and abnormality of the thoracic spine. The patient was recalled for a second MRI with specific spinal views, which showed evidence of T8-9 discitis/osteomyelitis and inflammatory mass compressing the spinal cord. Later that day, after interpretation of the MRI by the radiologist, the subject was notified to attend the ER of the university hospital.

She was admitted later that night to the internal medicine department for conservative management. The neurosurgical team considered no need for urgent surgical intervention, since there was no definite neurological deficit found on examination. Medical management consisted of intravenous vancomycin (after obtaining blood cultures), and morphine for pain. Over the next 12 h, her condition deteriorated with inability to move her lower limbs and urinary retention. Immediate neurosurgical reassessment was done and surgical drainage and decompression were performed in the morning of the next day (6 weeks after her initial presentation to the rural hospital). Repeat surgery was required, but the patient was left with residual paraparesis and inability to walk independently. She required 3 months of in-patient rehabilitation at a specialized center.

10.1.1 Medico-legal Issues

Malpractice litigation was initiated by the patient and her husband (plaintiffs) against the attending physicians of the community hospital and teaching hospital, the institutions themselves, and the FP. Claims against the FP, the community hospital, and the attending ER physician (a GP), were negligence in management of *S. aureus* bacteremia, failure to investigate and diagnose a spinal infection, failure to refer the patient to an appropriate specialist, misdiagnosis, and inappropriate treatment. Furthermore, their negligence in making an early diagnosis led to severe pain and suffering and the development of an epidural abscess that caused her permanent disability.

The charges against the teaching hospital and neurosurgeon were failure to operate in a timely fashion before the plaintiff developed paralysis of her lower limbs. It was charged that the physicians were aware of an epidural abscess on admission and should have made arrangements for immediate surgery. Moreover, they demonstrated negligence in their management by transferring the patient to a medical service. Expert witness for the plaintiff opined that surgery should have been arranged on admission, or at least hourly neurological monitoring on the neurosurgery service with expectant surgery at the earliest sign of neurological impairment.

10.1.2 Medical Aspect

Expert witnesses for the plaintiffs strongly criticized the management and investigations performed during the first admission to the community hospital as being grossly inadequate and below the standard of care. It was noted that specific therapy for *S. aureus* bacteremia was never instituted, and that the patient's symptoms of radicular pain were misinterpreted and never properly investigated. Moreover, the physician should have consulted an internist or infectious disease specialist (if available) before discharging the patient from hospital.

The first mistake made by the admitting physician was to assume that the patient's pain, fever, and bacteremia were due to pyelonephritis or urosepsis. *S. aureus* is a most unusual cause of pyelonephritis in the community, and only occasionally causes urinary tract infection in hospitals, mostly in those with urethral catheterization. Usually, when *S. aureus* is recovered from blood and urine, the source is not the urinary tract, but rather the blood stream. Infection can seed to the kidney from another source. The attending physician should also have been aware that in symptomatic pyelonephritis, it is expected that the urinalysis would show increased leucocytes and +/− nitrates, unless there is complete ureteric obstruction on the affected side or very severe neutropenia.

Furthermore, antibiotic therapy for *S. aureus* bacteremia was never instituted, and the only antibiotic the plaintiff received was azithromycin started before admission for possible community-acquired pneumonia. Although the *S. aureus* was susceptible to the macrolides, this class of antimicrobials is not considered adequate for systemic blood stream infection with this bacterium. Their potency against *S. aureus* is modest and resistance can readily occur with a one-step mutation. Also, for primary *S. aureus* bacteremia (with two or more separate positive blood cultures), the main sources of complications to consider (in absence of skin and soft tissue infection) are bacterial endocarditis and bone or joint infection. For two or more decades, it has been standard practice and recommendation to administer intravenous beta-lactam antistaphylococcic agents for 2 weeks for *S. aureus* bacteremia, to assess clinically for septic arthritis/osteomyelitis, and perform trans-esophageal echocardiogram (TEE) to exclude endocarditis.[1] Treatment would be terminated for those without any obvious source or complications at the end of 14 days, and then repeat blood cultures taken several days later. In patients with allergy to penicillin, the best choice of antibiotics would be a first generation cephalosporin (cefazolin) for mild cutaneous reaction, or vancomycin for those with history of a severe reaction.

It was very likely in this case, that the right-sided back and flank pain was originating from the spine at the first admission and the *S. aureus* bacteremia either represent asymptomatic seeding to the kidney, or contamination of the urine in a subject colonized with the organism. Even 2 weeks of intravenous anti-staphylococcal therapy would be inadequate in this case, as the clinical symptoms of radicular pain indicated vertebral osteomyelitis with nerve root compression from the onset. It is quite likely that the course of azithromycin and ciprofloxacin had some effect in moderating the infection, causing it to become more sub-acute-chronic without curing the infection. It is common for sub-acute-chronic osteomyelitis to not manifest symptoms such as recurrent or persistent fever.

It is quite evident that neither the FP nor the surgeon, who assessed the plaintiff later in the course of her illness, gave consideration to the *S. aureus* bacteremia and the connection to the symptoms and abnormal blood tests. The anemia of chronic inflammatory disease and high markers of systemic inflammation (ESR and CRP) were quite typical for chronic osteomyelitis or abscess, and should have raised red flags to signal the causation of the subject's symptoms.

10.2 Case 2: Too Little, Too Late, for Backache

An unemployed 52-year-old male presented to the ER of a university teaching hospital with worsening of lower backache for a month. There was a past history of back trauma 10 years before which resulted in episodic mild chronic backache, usually aggravated by lifting. However, this present pain was more constant and severe in nature. The patient had no recent constitutional symptoms of fever, chills, night sweats, or weight loss. However, he did recall having a febrile 'flu-like' illness about 6 weeks before, which subsided when he was prescribed oral antibiotics for a week, when he visited a 'walk-in clinic'. No investigation or diagnosis were ever made. The past history was not significant except for 'binge-drinking' of five to six bottles of beer on weekends.

The vital signs, including temperature were normal and the examination revealed a moderately obese male with some mild tenderness of the lumbar spine on percussion or bending over. A radiograph of the lumbar-sacral spine showed evidence of degenerative disc disease. The patient was reassured that this is a common benign condition and he was prescribed naproxen 500 mg every 8 h. The patient returned to the same ER 5–7 days later with unrelenting backache, not adequately relieved by the naproxen.

No investigation was performed and stronger analgesics were prescribed – acetaminophen 325 mg with codeine 30 mg, two tablets every 4 h. A week later, the subject revisited the ER with persistent pain, nausea, constipation, some difficulty climbing stairs, and also rising from the sitting position. The pain was present at night, not relieved by lying supine and awakened the patient from sleep. Cursory examination performed demonstrated some spinal tenderness as before, minimal weakness at hip flexion. Again, no investigations were performed and the physician attributed the constipation and nausea to the codeine and the mild weakness on elevation of his limbs secondary to the pain. This time oxycodone-acetaminophen two tablets every 6 h, lactulose for constipation, and dimenhydrinate for nausea were prescribed. He was advised to follow-up at the walk-in clinic closest to his home, as he did not have an FP.

Ten days later, the patient was brought to the ER with inability to walk, urinary retention and constipation. Examination revealed a temperature of 38°C and abnormalities confined to the neurological system, with paralysis of both lower limbs with slight movement from side to side, distended bladder, and lax anal sphincter with impacted stools. Blood tests revealed an anemia of 108 g/dL, leucocyte count of 13,500/uL with 90% neutrophils, and an ESR of 98 mm/h. An urgent MRI of the spine revealed discitis/osteomyelitis and an epidural abscess extending from T10 to L2. Urgent surgery was performed for drainage, decompression, and laminectomy. Despite surgery, prolonged course of antistaplylococcal antibiotics (intravenous, then oral cloxacillin), the patient remained paralyzed and required a wheelchair for ambulation.

10.2.1 Medical-legal Issues

A lawsuit was launched by the patient against the hospital and ER physicians for medical malpractice. Claims were made by the plaintiff that the ER physicians were negligent in not investigating his severe backache earlier, and that delay in diagnosis of his spinal infection led to his paralysis and severely affected his future income and enjoyment of life. Charges were also made that the ER physicians should have performed blood tests, CT or MRI of the spine and detailed neurological examination at least by the second ER visit, if not at the first assessment. Moreover, the physicians were careless and negligent in not requesting a consultation earlier to a back specialist (orthopedic or neurosurgeon).

The defendants countered that chronic low back pain is very common in the population, and in most cases, it is of benign nature and can be managed conservatively with analgesics and back exercises. Furthermore, the ER physicians followed the current guidelines for management and investigation of chronic low back pain. Thus, the standard practice for patients presenting similar to the plaintiff does not include investigations with CT or MRI of the spine at the outset, unless there is evidence of infection (such as fever) or neurological deficit.

10.2.2 Medical Aspects of Chronic Lower Backache

Low back pain is a very common ailment among the population that will be presented to FPs or ER physicians. Non-specific backache with no underlying pathology accounts for 85% of backache, and 80–90% of subjects with this benign condition will have resolution of their symptoms in 4–6 weeks.[2] Back pain may arise from the muscles, ligaments, discs, vertebrae and nerve roots of spinal cord, but can also be referred from visceral organs in the abdomen (kidney, pancreas, aorta, etc.). Although backache is commonly attributed to muscle strain, the pathogenesis is often unclear and can be related to stress, and in viral infections (i.e. influenza) it is probably related to pro-inflammatory cytokine release.

It is estimated that nearly everyone at some point in their lifetime has back pain that interferes with work and routine daily activities; it is the most common cause of job-related disability and the leading contributor to missed work.[3] Each year Americans spend at least $50 billion for low back pain. Back pain is categorized into three groups: acute or short term low back pain (<6 weeks) which generally lasts a few days to a few weeks, sub-acute back pain with symptoms from 6 to 12 weeks, and chronic back pain for more than 12 weeks.[4]

Acute low back pain most often is mechanical in nature and a result of trauma to the lower back or a rheumatological disorder. Pain from trauma may be caused by sports injury, a sudden jolt in a motor vehicle accident, work around the home or garden, or other stress on the spine and surrounding tissues. The symptoms can vary from dull muscle ache to sharp stabbing or shooting pain, limited flexibility (such as bending over) or inability to walk straight.

Chronic low back pain may be stable with intermittent exacerbations, but it is often not progressive in nature. It occurs most often between the ages of 30 and 50, of equal frequency in men and women, and in part related to the aging process, sedentary lifestyle, and being overweight. Spinal degenerative disc disease increases with age and is commonly the attributable cause of chronic backache.

Since low back pain is such a frequent complaint and the majority are benign in nature, the physician has to be selective in considering more serious underlying conditions, and the need for more extensive (and expensive) investigations.

Table 10.1 Risk factors in back pain

History	Risk
Age <18 years	Congenital anomaly
Age >50 years	Tumor
Major trauma	Fracture
Minor trauma (>60 years)	Osteoporosis – fracture
Duration >6 weeks	Tumor, infection
Fever, chills	Infection
Weight loss	Tumor, infection
IVDA	Infection
Immunocompromised	Infection
Unremitting pain	Tumor, infection
Incontinence	Epidural compression
Numbness saddle area	Epidural compression
Weakness of lower limbs	Epidural compression
Fever	Infection
Writhing in pain	Infection
Positive straight leg raise test	Herniated disc
Lax anal sphincter	Epidural compression
Sensory loss – perianal/perineum	Epidural compression
Motor weakness lower limbs	Epidural compression

Often, a detailed history and physical examination is sufficient to decide on management and need for investigations. Table 10.1 outlines risk factors on history and physical examination that may guide investigations to exclude important conditions.

The history is very important to define the onset, site, severity, duration, and precipitation of the pain, and any limitations of movement. Dull aching pain that worsens with movement but improves with rest and lying supine is most often a benign condition (non-specific backache). Pain worst on coughing or straining (Valsalva maneuver), or by sitting and relieved by supine position is suspicious for disc herniation. Back pain that occurs at night, awakens the patient from sleep, or is unrelenting despite rest and appropriate analgesics should raise the alarm for serious pathology, such as tumor or infection.[4] Sciatica is a well-recognized complication of lumbar disc herniation by most physicians, but only occurs in about 1% of subjects with back pain. Bilateral sciatica worsened by walking, prolonged standing or back extension and relieved by rest and flexion is indicative of spinal stenosis.

Although fever should raise the suspicion of spinal infection in localized backache, it can be non-specific in viral illnesses such as influenza. Moreover, fever is present in only 27% of subjects with tuberculosis of the spine, 50% in pyogenic vertebral osteomyelitis and about 80% in spinal epidural abscess.[5] Thus, absence of fever is common in patients with spinal infection. Some of these patients may have received antibiotics for febrile illness in the prior 3 months that could modify the course and presentation of the infection. A history of preceding infection, especially with bacteremia and candidemia (within the past 3–6 months) should alert the physician of possible vertebral osteomyelitis.

In subjects with severe, prolonged back pain, a detailed physical examination is important not only of the spine, back and lower extremities, but also of the heart, lungs, abdomen, skin, peripheral joints, and lymphatic system for any signs of systemic disease that may provide a clue to the diagnosis. Tenderness of the paravertebral muscles may indicate a muscle strain, but can be non-specific in systemic viral infections, and fibromyalgia (characterized by multiple tender joints over the neck, shoulders, spine, and hips). Localized point tenderness on percussion over the spine is very suspicious of pathology in the vertebra. It is commonly present in fractures and infections of the spine, with a sensitivity of 86% and specificity of 60%.[5] A positive straight leg raise test is about 80% sensitive for a herniated disc, most frequently L4-5 and L5-S1.

There is a wide variation in appropriate clinical evaluation, investigation, and management of patients with acute or chronic low back pain by primary care physicians or between specialties. Attempts to standardize the approach to evaluation and management have been made by various guidelines published through medical societies of several countries. More than 85% of patients with low back pain cannot attribute it to any specific disease or spinal abnormality (non-specific low back pain). Only a minority of patients seen by the FPs or ER physicians with low back pain have an underlying specific disorder (not associated with major trauma). These include underlying cancer ($\approx$0.7% of cases), compression fracture (4%), or spinal infection (0.01%).[6] However, delay in diagnosis or misdiagnosis in these conditions can result in catastrophic outcome. Herniated discs accounts for about 4% of patients with low back pain, spinal stenosis about 3%, and ankylosing spondylitis range from 0.3% to 5%.[7]

The American College of Physicians and American Pain Society guidelines[8] for initial assessment of a patient with low back pain is to do a focused history and physical examination to determine the likelihood of specific underlying conditions (including neurologic involvement). Patients can then be classified into one of these three categories: non-specific low back pain, back pain that might be associated with radiculopathy or spinal stenosis (presence of sciatica or pseudoclaudication), and back pain potentially associated with another spinal cause. The small group of patients with potentially rapid neurological deterioration such as those with infection, tumor, or cauda equina syndrome falls into the third category, and these patients need prompt evaluation to diagnose and initiate treatment.

Although predictive factors for vertebral infection have not been well studied, they include fever, chills, night sweats, recent infection within the prior 3 months (especially blood stream infection), and intravenous drug abuse (IVDA). However, the absence of fever does not preclude the diagnosis of spinal infection, and should not deter the pursuit of an investigation to exclude infection. Risk factors for cancer as a cause of low back pain in a large prospective study include a history of cancer, unexplained weight loss, failure to improve after a month, and age >50 years.[9] All patients should be evaluated for neurological deficits including motor weakness at different levels, fecal incontinence, or urinary retention. Urinary retention is the most frequent finding of a cauda equina syndrome, with a sensitivity of 90%.[5] In the absence of urinary retention, the probability of a cauda equina syndrome is very low, about 1 in 10,000.

Besides major trauma, risk factors for vertebral fracture are older age, history of osteoporosis, steroid use, and unexplained "stress fractures" peripherally. Ankylosing spondylitis should be suspected in younger age, history of morning stiffness, improvement with exercise, alternating buttock pain, awakening with pain in the early morning,[10] chronic diarrhea (inflammatory bowel disease) and rash (psoriasis).

Routine plain radiography (or other imaging) of the spine is not recommended for the vast majority of patients with back pain, if the clinical findings are most consistent with non-specific low back pain.[8] Plain radiography as initial evaluation should be performed for suspected compression fracture (steroid use, osteoporosis), symptoms of sciatica, radiculopathy or suggestive symptoms of spinal stenosis. Prompt investigation should be performed on patients with severe or progressive neurologic deficits or suspected infection with MRI or CT scan of the spine. Emergency imaging to evaluate spinal cord compression or cauda equine syndrome is best done by MRI (most sensitive),[6] but CT scan can be performed if MRI is not readily available.

For patients with no evidence of neurological deficit (or radiculopathy), but serious underlying pathology is possible (cancer or infection), an initial screening with plain radiography and blood tests such as ESR, CRP and complete blood count is reasonable and recommended in part.[8] An ESR $\geq$20 mm/h is associated with a sensitivity of 78% and specificity of 67% for cancer. In vertebral infection, the ESR and CRP should be elevated in most patients, and normal values of both are strongly against spinal infection. The leukocyte count, however, is less predictive and may or may not be elevated. An MRI should be performed on patients with abnormalities on initial testing.[8]

Routine radiograph of the spine is often normal in early infections, as definitive bone destruction may not become evident until 8 weeks after onset.[12] The first radiographic abnormality in infection is very subtle and may be overlooked – irregularity of the end plate. Erosion of the end plate and widening of the paravertebral spine (displacement of the paravertebral line on routine frontal radiographs) may be seen with progression. Unfortunately, after a variable period of time (8–12 weeks), bone regeneration occurs with visible sclerosis which is often interpreted as due to degenerative disc disease with Charcot-type spine (similar radiographic appearance).[13]

MRI is the preferred imaging for spinal infection (gold standard) and allows assessment for epidural and paravertebral abscesses and spinal cord or nerve root compression. In cases of contraindications or unavailability of MRI, high resolution reformatted CT scanning is still an excellent investigative tool for vertebral infection[13] and would be the next best imaging technique.

10.2.3 *Epidural Abscess and Spine Infection*

Spine infections can be a postoperative complication of disc surgery or after invasive spinal procedures, with an incidence in about 1–4% of spinal operations.[14]

Spontaneous hematogenous vertebral osteomyelitis (non-tuberculous) is relatively rare and the incidence is estimated to be five cases per million people per year.[15] For spinal epidural abscesses, the annual incidence is much lower at about 0.2–2 cases per 10,000-hospital admissions.[14]

Although healthy subjects with no risk factors can be affected, recognized risk factors for spinal infections include advanced age, diabetes, human immunodeficiency infection (HIV), IVDA, chronic renal or hepatic disease, long-term steroid use, severe trauma, malignancy, chemotherapy and previous surgery.[14] In adults, spontaneous vertebral osteomyelitis/discitis most commonly affects males and is predominantly caused by *S. aureus* (60–65%), *Streptococcus* species (10%), and gram-negative bacilli (especially in the elderly with urinary tract infections as the source).[14] Hematogenous seeded spinal infection commonly affects two adjacent vertebrae and the intervertebral disc, as there is a common blood supply by the segmental artery. The paraspinal venous plexus may contribute to the spread of infection, particularly in sources from the genito-urinary tract. The lumbar spine is more commonly affected than the thoracic or cervical spine, and the anterior vertebral area is the most commonly involved. However, fungal infection and actinomycosis may involve the posterior sections of the spine. Mycobacterial vertebral osteomyelitis has the same pathogenic mechanisms as pyogenic involvement, but with a more indolent course, and the disc space is affected later in the disease, after vertebral destruction.[16]

Delayed or misdiagnosis is a common problem in the management of spinal infection[14] that can predispose to litigation. In a series of 101 patients, misdiagnoses occurred in 33.7%, and the average delay between clinical manifestations to diagnosis was 2.6 months.[17] Although neurological deficit has been estimated to occur in about 6–17% of pyogenic vertebral infections,[18] it can be much higher in: (1) subjects with subacute or chronic infection, especially in the elderly where fever is often absent, (2) those presenting with atypical chest and abdominal pain (where radicular pain is frequently misinterpreted), (3) the immunocompromised patients as a subtle clinical presentation is common, (4) indolent infections by specific pathogens (i.e., mycobacteria or fungi),[14] or (5) partial treatment with antibiotics of common bacterial pathogen (personal experience).

In children, acute presentations with fever, back pain and local spinal tenderness are commonly found in vertebral osteomyelitis. However, in adults, this triad is frequently absent and often accounts for delayed or misdiagnosis. The clinical presentation varies according to the location of the involved vertebra. Localized back pain (aggravated by activities and relieved by rest) with percussion tenderness appears to be the most common presentation (83–90%) in spinal infection in some studies.[19,20] However, fever in these studies was present in only 61–65%, leukocytosis in 57–61%, elevated ESR in 76–95%,[19,20] and elevated CRP in 97%.[19] Other studies have also found that CRP levels may be more sensitive than the ESR in spinal infection, but both can be used to assess response to therapy.[14,21,22]

Epidural infections can affect the spinal cord or cauda equina with direct compression by an abscess, inflammatory mass (phlegmon), or indirectly as a result of vascular occlusion caused by septic thrombophlebitis or invasive arteritis.

The clinical features can be variable depending on the stage of progression or presentation. In Stage 1 (the early phase), back pain is localized to the area of the spine affected. Stage 2 is associated with nerve root pain radiating from the involved spinal area (radicular pain). Stage 3 represents onset of motor weakness, sensory deficit, and bladder or bowel dysfunction. Stage 4, the final phase, progresses to paralysis.[23] Epidural abscesses are more common in posterior than anterior areas of the spine, and more common in thoraco-lumbar than cervical vertebrae because of larger spaces.

The clinical manifestations of an epidural abscess depend on the site and stage of the disease. In Stage 1, the presentation is that of an uncomplicated discitis/vertebral osteomyelitis. At Stage 2, a cervical abscess usually presents with neck pain with radiation down the arms, while lumbar abscess may present with low back pain radiating down the legs. Thoracic abscess (Stage 2) with chest and abdominal (or hip) pain is often misinterpreted as arising from organs in the thorax and abdomen.[23] A comprehensive review of the international literature collected on 915 patients with spinal epidural abscess found that 71% had back pain as the initial symptom and 66% had fever.[24] Paralysis (Stage 4) affected 34% of the patients and the mortality was 15%, even in the 1990s. Nerve root initiation or radicular pain occurred in 20%, muscle weakness and urinary dysfunction in 26%, and fecal incontinence in 24%.[24] Surprisingly, in this meta-analysis only 17% of the patients had local spine tenderness on percussion in the initial stage. The inflammatory markers (ESR and CRP) were elevated in almost all patients with spinal abscess, but are non-specific findings.[23] However, a normal ESR and CRP is strongly against the diagnosis of spinal infection and epidural abscess. The ESR was found to be elevated (>20 mm/h) in 94% of 117 patients with epidural abscess, and a leukocyte count >10,000 cells/uL in 78% of 218 patients in the review.[24]

Bacteremia has been detected in about one-third of patients with pyogenic vertebral osteomyelitis/discitis overall,[14] but appears to be higher in those complicated by epidural abscess (about 60%), especially in those infected with *S. aureus*.[23]

Delayed diagnosis and sub-optimal early management are the main factors responsible for poor outcome in spinal epidural abscess. It is estimated that nearly 50% of cases are initially misdiagnosed,[23] and the most important predictor of outcome is the neurological status immediately before surgical intervention and decompression. Patients with paralysis for up to 24–36 h are likely to have some neurological improvement, but left with permanent neurological deficit. Surgical intervention after 36 h of paralysis will likely result in little or no neurological improvement.[23,24] Surgical intervention in patients without clinical neurological findings or just radicular symptoms (Stage 1 or 2) have the best prognosis for neurological recovery and usually have no residual neurological deficit. Prompt earliest surgery in patients with rapid progression and virulent infection may result in better outcome than any delay in intervention.[23] Delay in surgical intervention while waiting for an accurate diagnosis or trial of medical therapy after admission with neurological deterioration can lead to a poorer outcome.[25,26]

In the meta-analysis of spontaneous spinal epidural abscess (N = 599), there was a mortality of 16% and complete recovery only in 43%, with 26% left with residual neurologic deficit (other than paralysis or paresis).[24] The conclusion of this review is that the main problem in the management of epidural abscess lies in the necessity of early diagnosis and timely surgical treatment to avoid or reduce permanent neurological deficits.[24]

The antibiotic treatment should be tailored to the recovered organism and susceptibility, and is usually administered for 6 weeks for acute vertebral osteomyelitis, and 3 or more months in those with chronic osteomyelitis (intravenous and oral combined). Surprisingly, there is no data on the value of dexamethasone in reducing permanent neurological deficits.

10.3 Discussion of Medico-legal Issues

The two cases described in this chapter represent the varied clinical presentation of spinal epidural abscess, which resulted in delayed diagnosis and surgical management culminating in catastrophic outcome.

In case 2, the patient presented to the ER with persistent, severe backache without fever, and appears to have been misdiagnosed as being due to back strain or benign non-specific back pain. The diagnosis of vertebral osteomyelitis that led to the spinal epidural abscess was not considered until the patient returned with paralysis. Unfortunately, misdiagnosis is common in vertebral osteomyelitis. Although most patients with low back pain have no serious underlying pathology, some patients have conditions that require remedial surgical therapy such as disc protrusion, abscesses, or cancer and a careful diagnostic evaluation is important.

The decision by a GP or ER physician to perform diagnostic imaging or blood tests are usually based on the history and physical examination to determine low from high risk conditions. A previous review of the literature on the accuracy of history and physical examination in diagnosing low back pain was published in 1995.[27] Thirty-six studies were reviewed, but only 19 were of good methodological quality, scoring $\geq$55 points out of a maximal score of 100. The combined history and the ESR had relatively high sensitivity for vertebral cancer. The straight leg-raising test had high sensitivity and low specificity for lumbar radiculopathy, but other neurological tests (reflexes, paresis, and impaired sensibility) had very variable sensitivity and specificity for radiculopathy.[11] There were insufficient numbers of patients in the studies to assess diagnostic accuracy in vertebral osteomyelitis and epidural abscess.

Misdiagnosis of vertebral osteomyelitis is fairly common (nearly 34%), and the average delay in diagnosis is almost 3 months (range 0.2–12 months).[17] Factors that have been associated with misdiagnosis (or delay) include older age, absence of fever, and a positive straight leg-raising test.[17] Backache is present in almost all patients, but localized spinal tenderness can be elicited on percussion in 83–90%.[19,20] When the diagnosis is uncertain and the persistence and severity of pain (especially with lack of response to standard NSAIDS) is not in keeping with benign low backache, then simple tests such as a complete blood count, ESR, and

CRP should be performed as normal results for all would be strongly against pathological spinal conditions in the presence of a normal examination. Any abnormality would suggest the need for more specific investigations, i.e., CT or MRI of the spine.

In an epidural spinal abscess, the same problem exists in delayed or misdiagnosis as in vertebral osteomyelitis, as both are closely related. However, there is an additional problem in delaying surgical intervention until it is too late to achieve full neurological recovery. This delay can occur even after the diagnosis is evident on MRI, and poor outcome may happen due to trial of medical therapy. Although there have been scattered reports of successful outcomes with medical management of epidural abscesses,[27] sudden neurological deterioration of patients receiving medical therapy (resulting in permanent neurological deficits) has also occurred[28] and poses risk for medico-legal litigation.

Review of the Canadian Medical Protective Association (CMPA) closed files from 1996 to 2006 revealed a total of 33 cases of spinal epidural abscesses that resulted in civil legal actions (30) or the College regulatory actions (3).[29] The emergency department was the most common location for the cases concluded in favor of the plaintiff (8 of 16 cases).

To heighten awareness of physicians to the diagnosis of epidural abscess, the CMPA had provided four risk management considerations[29]:

- Consider the diagnosis of spinal epidural abscess in patients presenting with back pain and unexplained fever.
- Physicians should be familiar with conditions associated with spinal epidural abscess (intravenous drug abuse, diabetes, recent spinal surgery or spinal anesthesia, recent bacteremia).
- When a spinal epidural abscess is suspected, act quickly to facilitate timely diagnosis and treatment (MRI or CT scan, neurological consultation, admission to hospital).
- Once a spinal epidural abscess is confirmed – arrange for urgent neurosurgical or orthopedic consultation.

These considerations, however, are broad generalized guidelines, that do not assist the physician in differentiating other causes of back pain and fever; such as seen with influenza (where the pain is not well localized), or with pyelonephritis where the pain and tenderness is lateral to the spine. Moreover, it does not provide assistance to the surgeons with respect to timing of surgical decompression. Some authors recommend surgical intervention for the following indications[28,30]:

1. deteriorating neurological deficit, or progression beyond Stage 2
2. persistent severe pain
3. increasing or persistent fever or leukocytosis
4. spinal deformity or instability
5. failure to identify causative organism
6. MRI showing >50% compression of the thecal sac
7. lack of availability of serial MRI

8. lack of availability of emergency spinal surgery facilities
9. failure to improve or resolve on intravenous antibiotics
10. presence of immunosuppression

In the algorithm outlined by Darouiche in 2006[23] surgical decompression is recommended in most cases of spinal epidural abscess unless declined by the patient, contraindicated because of high operative risk, or is unlikely to improve paralysis that existed for >24–36 h – but it still may be needed to control epidural infection and sepsis. In this recent review, conservative medical management is also considered reasonable for neurologically intact patients, with an identified organism (i.e., from blood cultures), and the patient can be closely monitored clinically and by serial MRI.[23]

A wait-and-see approach for non-surgical management of spinal epidural abscess still carries a risk of rapid deterioration in neurological deficit, and despite surgical decompression within 24 h, may result in permanent neurological impairment. This sudden deterioration can be the result of thrombosis of the anterior spinal artery, rather than mechanical compression of the spinal cord.[31]

In case I of this chapter, this conservative approach in a patient with confirmed epidural abscess with radicular symptoms (Stage 2) resulted in permanent paresis of the lower limbs, despite surgery within 24 h of paralysis. This has resulted in litigation against the neurosurgeon involved in the tertiary center. The argument by the plaintiff and legal representatives was that emergency surgical compression should have been done before waiting for the patient to develop signs of paralysis. Moreover, once paralysis occurred, it took over 12 h for surgery to be performed.

What can physicians learn from these cases and literature reports? The common pitfalls in the diagnosis and management of spinal epidural abscess are outlined in Table 10.2. To avoid these pitfalls, physicians first need to consider the possibility of spinal osteomyelitis and epidural abscess in their differential diagnosis of back pain, even in the absence of fever. Back pain failing to respond to NSAIDS or requiring strong narcotics should raise a red flag for serious spinal pathology. These

Table 10.2 Pitfalls in the diagnosis and management of vertebral osteomyelitis and epidural abscess

- Attributing back pain to muscle strain – despite severe, persistent pain on NSAIDS
- Misinterpretation of radicular pain arising from the abdomen or chest
- Misuse of antibiotics for treatment of fever before determining cause
- Failure to determine source of bacteremia (i.e., *S. aureus*)
- Failure to recognize deep wound infection after spinal surgery
- Imaging studies of vertebral level not affected
- Ascribing clinical and laboratory findings to non-infectious cause
- Failure to recognize need for urgent surgery
- Failure to plan for urgent surgery and inadequate monitoring for patients receiving medical therapy
- Failure to discuss options of medical versus early surgical interventions with the patient
- Failure to consult an appropriate specialist in a timely fashion

patients and those with history of fever, chills or night sweats (with no respiratory symptoms suggestive of viral or influenza infection), should initially have complete blood counts taken, ESR, CRP, plus a spinal x-ray. Blood cultures (at least two sets) should be taken in those with fever or elevated white blood count, ESR or CRP (in the inflammatory range). CT scan can be performed for those with an abnormality, spinal percussion tenderness, or straight leg-raising positive test. MRI is preferable for those with any neurological deficits and radicular pain, and should be arranged as an urgent investigation in these situations.

Antibiotics should be delayed (unless the patient is septic) after obtaining blood cultures or aspiration under imaging of infected appearing disc, vertebra, or abscess. Patients with any neurological signs or radicular symptoms (other than sciatica from suspected disc herniation) should have urgent neurosurgical or orthopedic consultation. Patients with strongly suspected or proven spinal epidural abscess without neurological deficit should also have non-urgent surgical consultation. The pros and cons of watchful waiting on antibiotics should be discussed with the patient. For epidural abscesses involving the cervical and thoracic spines, urgent surgical intervention should be strongly considered, even in stage 2 (radicular symptoms), as the chance of cord injury may be too catastrophic for conservative management. Other means of partial drainage, such as aspiration or catheter drainage under imaging may be another option. Standardized guidelines for monitoring neurological signs in patients with spinal epidural abscess need to be developed and adopted for patients on antibiotics alone, with evidence of epidural abscess. Currently, in many institutions, patients that are not considered for early surgical intervention are transferred to the medical service for antibiotic management with the proviso that the neurosurgery service be informed for any deterioration in neurological function. Unfortunately, this often results in unacceptable delays in surgical intervention before a point of irreversible damage.

References

1. Parsonnet, J, Dersiewicz, RL, (2001). Staphylococcal infection. IN (eds): Braunwald E, Fauci AS, Kasper DC, Hauser SL, Longo DL, Jameson JL. Harrison's Principles of Internal Medicine, 15th Edition, McGraw-Hill, New York: p889–901.
2. Deyo R, Weinstein J, (2001). Primary care. Low back pain. *N Engl J Med* 344:363–370.
3. National Institute of Neurological Disorders and Stroke. (2009). Low back pain fact sheet. National Institutes of Health http://www.ninds.nih.gov/disorders/backpain/detail_backpain.htm
4. Della-Giustina D, Coppola M, (2004). Thoracic and lumbar pain syndromes. In (eds): Tintanalli JE, Kelen GP, Stapczynski JS. Emergency Medicine: A Comprehensive Study Guide, 6th Edition, McGraw-Hill Co. Inc., New York: p1773–1779.
5. Deyo RA, Rainville J, Kent DL, (1992). What can the history and physical examination tell us about low back pain? *JAMA* 268:760–765.
6. Jarvik JG, Deyo RA, (2002). Diagnostic evaluation of low back pain with emphasis on imaging. *Ann Intern Med* 137:586–597.
7. Underwood MR, Davies P. (1995). Inflammatory back pain in primary care. *Br J Rheumatol* 34:1074–1077.

8. Chou R, Qaseem A, Snow V, Casey D, Cross JT Jr, Shekelle P, Owens DK for the Clinical Efficacy Assessment Subcommittee of the American College of Physicians and the American College of Physicians/American Pain Society Low Back Pain Guidelines Panel, (2007). Diagnosis and treatment of low back pain: a joint clinical practice guideline from the American College of Physicians and the American Pain Society. *Ann Intern Med* 147:478–491.
9. Deyo RA, Diehl AK, (1988). Cancer as a cause of back pain: frequency, clinical presentation, and diagnostic strategies. *J Gen Intern Med* 3:230–238.
10. Rudwaleit M, Metter A, Listing J, Sieper J, Braun J, (2006). Inflammatory back pain in ankylosing spondylitis: a reassessment of the clinical history for application as classification and diagnostic criteria. *Arthritis Rheum* 54:569–578.
11. Van den Hoogen HM, Koes BW, van Eijk JT, Bouter LM, (1995). On the accuracy of history, physical examination and erythrocyte sedimentation rate in diagnosing low back pain in general practice. A criteria-based review of the literature. *Spine* 20:318–327.
12. Dina TS, (1992). Infection of the spine. *Semin Spine Surg* 4:98–110.
13. Rothman SLG, (1996). The diagnosis of infections of the spine by modern imaging techniques. *Orthop Clin North Am* 27:15–31.
14. Tsiodras S, Falagas ME, (2006). Clinical assessment and medical treatment of spine infection. *Clin Orthop Related Res* 44:38–50.
15. Krogsgaard MR, Wagn P, Bengtsson J, (1998). Epidemiology of acute vertebral osteomyelitis in Denmark: 137 cases in Denmark 1979–1982, compared to cases reported to the National Patient Register 1991–1993. *Acta Orthop Scand* 69:513–517.
16. Tekkok IH, Peh WC, (2004). Skeletal tuberculosis in children. *Pediatr Radiol* 34:853–860
17. Buranapanitkit B, Lim A, Geater A, (2001). Misdiagnosis in vertebral osteomyelitis: problems and factors. *J Med Assoc Thai* 84:1743–1750.
18. Hsieh PC, Wienecke RJ, O'Shaughnessy BA, Koski TR, Ondra SL, (2004). Surgical strategies for vertebral osteomyelitis and epidural abscess. *Neurosurg Focus* 17:e4–7.
19. Kapeller P, Fazekas F, Krametter D, Koch M, Roob G, Schmidt R, Offenbacher H, (1997). Pyogenic infectious spondylitis: clinical laboratory and MRI features. *Eur Neurol* 38:94–98.
20. Chelsom J, Solberg CO, (1998). Vertebral osteomyelitis at a Norwegian University Hospital 1987–97: clinical features, laboratory findings and outcome. *Scand J Infect Dis* 30:147–151.
21. Mayer B, Schaller K, Rohde V, Hassler W, (1995). The C-reactive protein for detection of early infection after lumbar microdiscectomy. *Acta Neurochir (Wien)* 136:145–150.
22. Brown EM, (2002). Infections in neurosurgery: using laboratory data to plan optimal treatment strategies. *Drugs* 62:909–913
23. Darouiche RO, (2006). Spinal epidural abscess (current concepts). *N Engl J Med* 355:2012–2020.
24. Reihause E, Waldbaur H, Seeling W, (2000). Spinal epidural abscess: a meta-analysis of 915 patients. *Neurosurg Rev* 232:175–204.
25. Darouiche RO, Hamill RJ, Greenberg SB, Weathers SW, Musher DM, (1992). Bacterial spinal abscess: review of 43 cases and literature survey. *Medicine (Baltimore)* 71:369–385.
26. Danner RI, Hartman BJ, (1987). Update of spinal epidural abscess: 35 cases and review of the literature. *Rev Infect Dis* 9:265–274.
27. Wheeler D, Keiser P, Rigamonti D, Keay S, (1992). Medical management of spinal epidural abscesses: case report and review. *Clin Infect Dis* 15:22–27.
28. Harrington P, Millner PA, Veale D, (2001). Inappropriate medical management of spinal epidural abscess. *Ann Rheum Dis* 60:218–222.
29. Robertson J, (2008). Spinal epidural abscess: a rare, insidious and potentially catastrophic infection. Risk Identification for all Physicians. CMPA;R10811E: www.cmpa-acpmca.
30. Baker AS, Ojemann RJ, Baker RA, (1992). Editorial response to Wheeler et al: To decompress or not to decompress – spinal epidural abscess. *Clin Infect Dis* 15:28–29.
31. Baker AS, Ojemann RG, Swartz MN, Richardson EP Jr, (1975). Spinal epidural abscess. *N Engl J Med* 293:463–468.

Chapter 11
Litigations in General Surgery

11.1 Case 1: Laparoscopic Cholecystectomy

A 65-year-old female underwent laparoscopic cholecystectomy at an outlying small community hospital for cholelithiasis and history of biliary colic. She had a past history of mild hypertension that was well controlled on anti-hypertensive therapy, and no other medical illness.

The procedure itself was uncomplicated except that adhesions from the duodenum to the gallbladder were dissected and excised during the operation. The subject was kept overnight with the intention of discharge the next morning. Just before release from the hospital, the patient felt weak and dizzy and fainted. She was found to be hypotensive with a blood pressure (BP) of 80/60 mmHg, with evidence of hypovolemia, abdominal distention, and tenderness. An emergency laparoscopy was performed the same day and a bile leak with bile peritonitis was found, but the direct source of the leak was never identified. Intra-abdominal drains were inserted; the subject was started on broad-spectrum antibiotics, and transferred to the intensive care unit (ICU).

The patient was rehydrated and stabilized hemodynamically, but her course was complicated by renal impairment, pulmonary edema, sepsis syndrome with disseminated intravascular coagulopathy, and evidence of cerebrovascular accident. She was transferred to a university medical center 2 days after the initial laparoscopy. A laparotomy was performed and a perforated duodenum was found which was oversewn duodenum was found. Her course in the hospital was further complicated by deep vein thrombosis requiring insertion of a vena cava filter, further surgery to repair duodenal and jejunostomy leak, seizures, enterococcal bacteremia, and paroxysmal atrial fibrillation.

She had a long hospital course, required prolonged rehabilitation and was left with partial cortical blindness and mild residual weakness from the stroke.

11.1.1 Medico-legal Issues

The patient and her husband (plaintiffs) instigated legal action for medical negligence against the general surgeon and medical institution. Medical expert witnesses for

I.W. Fong, *Medico-Legal Issues in Infectious Diseases*, Emerging Infectious Diseases of the 21st Century, DOI 10.1007/978-1-4419-8053-3_11,

the plaintiff were critical of the surgeon, and said that he fell below the standard of care by not performing an immediate laparotomy the first post-operative day when there was evidence of peritonitis. At the second laparoscopy when the surgeon recognized a biliary leak with no obvious source, there was serious error in judgment by not converting the procedure to an exploratory laparotomy at that time.

11.1.2 Medical Aspects

Laparoscopic cholecystectomy (LC) was introduced in 1990 and it has become the preferred procedure for cholecystectomy, rather than laparotomy. This approach results in four less days in hospital, less operative pain, and fewer weeks of missed work. Serious complications and deaths related to the operation itself are rare. The operative mortality associated with LC has been estimated to be about 0.1% in patients under age 50 and about 0.5% in subjects over age 50.[1] In a large, population-based study in Western Australia from 1988 to 1994, the results of LC were compared to open cholecystectomy (OC).[2] Compared with OC, the laparoscopic technique carried a nearly 2-fold risk of major bile leak, vascular and bowel complications. It was found in the study, that the risk factors for bile leak or intraoperative injury were 2.3-fold greater in men than women, 2.6-fold greater in teaching hospitals than non-teaching hospitals, and were 3.47-fold greater in the presence of pancreatitis, jaundice, or cholangitis.[2]

From a review of five series assessing LC, the complications varied from 2.0% to 6.87% and the associated mortality range from 0.04% to 0.23%.[3] The six surgical complications related to LC were: (1) bile duct injury (0.11–0.24% since the 1990s), (2) bile spillage (most commonly from gall bladder injury), (3) bleeding from injury to cystic or hepatic artery (rarely from vena cava or aorta), (4) stones in the peritoneal cavity or residual in the common bile duct, (5) wound infection (rarely cholangitis if stones migrate to the common bile duct), and (6) injury to other organs (i.e., intestines, as in this case).[3]

In the present case, LC was indicated and a suitable procedure of choice. However, adhesions and the need for dissecting the gallbladder from tethered adhesions to the duodenum likely predisposed to the small bowel injury and subsequent biliary peritonitis. A contentious issue in this case was the choice of repeat laparoscopy over laparotomy for a bowel leak. Is the presence or suspicion of acute duodenal or gastric perforation an absolute indication for immediate open laparotomy? This issue is not clear-cut, and should depend on the individual patient's clinical status. Non-operative management of perforated duodenum is now an accepted option in selected patients.[4] It has been recognized over the past decade that over 40% of perforated duodenums may seal spontaneously without surgical intervention.[5] These are presumably small openings or leaks. Furthermore, in a randomized controlled trial of selected patients with perforated intestine, there was no difference in outcome (morbidity or mortality) between those receiving open surgery versus non-surgical medical management.[6]

11.1.3 Medico-legal Discussion

The decision by the surgeon (defendant) to perform a repeat laparoscopy is a reasonable and valid choice of surgical management for presumed peritonitis. This should not be considered negligence or sub-standard therapy. The most important aspects of surgical management in peritonitis from a bowel leak are to adequately drain the contents collected in the peritoneum and source control. In this case, source control could not be adequately achieved because the duodenal perforation could not be visualized. At this junction, it may have been more appropriate to convert to an open laparotomy to identify and correct the underlying intestinal perforation. Another option would be to institute naso-gastro-duodenal continuous drainage with abdominal drains in place. The latter procedure can be effective in the presence of small bowel tears as subsequent spontaneous closure may occur.

It could be argued, that although the defendant made a judgment error in not converting to an open laparotomy (as he could not identify the source of the bile leak) the outcome probably would not have been more favorable by converting to open surgery. Laparoscopy has been found in randomized, controlled trials to be of similar benefit as open surgery in perforation of the stomach or duodenum.[7–9] The most urgent need in perforation of an intra-abdominal viscus is drainage of the infected fluid or abscess, and it is common practice with many types of perforated intestines to insert a percutaneous drainage first, combined with antibiotic therapy, then do definitive surgery later.[10]

Supporting evidence in this case that initial laparotomy after the first signs of peritonitis following LC would not have affected the outcome, is the fact that even after corrective open surgery at the tertiary center, the plaintiff continued to manifest signs of sepsis and the intestinal leak persisted. Thus, she underwent a second laparotomy to repair the leak in the duodenum and jejunum.

Expert medical witnesses for the plaintiffs contend that a prompt laparotomy would have resulted in quicker clinical response and prevented her stroke, which they attribute to the development of disseminated intravascular coagulopathy, (DIC) as a result of uncontrolled sepsis. They argued that the surgeon's failure to do definitive surgery resulted in significant morbidity and residual weakness (limiting her daily function) resulting from the cerebrovascular accident. The development of stroke was likely indirectly related to the sepsis, either as a result of paroxysmal atrial fibrillation, which resulted in cerebral emboli, or thrombus in the cerebral vessels resulting from DIC.

The atrial fibrillation may have been precipitated by pulmonary emboli; the plaintiff suffered from. Both DIC and venous thrombo-embolism are complications of sepsis, which induce a procoagulant state by stimulating the clotting cascade and inhibiting the fibrinolytic system. Prophylactic heparin is indicated in these patients unless there is evidence of bleeding. Once sepsis has begun, these events can occur and the procoagulant state may persist for several days after treatment and adequate source control. It is likely, however, that persistent sepsis from inadequate source control would pose a greater risk for thrombosis and DIC. Even after laparotomy and over-sewing of the bowel perforation at the tertiary center, persistence of sepsis continued and source control was still not obtained. Thus, it can be argued that initial laparotomy instead of laparoscopy for peritonitis would not have made any difference in the outcome.

11.2 Case 2: Abdominal Pain Post-partum

Three months after delivering a healthy baby, a 33-year-old female attended the emergency department (ER) of a suburban community hospital. Her symptoms were nausea, vomiting, and lower abdominal pain for 12 h. The physical examination noted normal vital signs with a temperature of 36.9°C, and the abdomen was noted to be soft by the ER physician. She was treated with intravenous saline and dimenhydrinate, and then discharged. Over the next 3 days, the patient's symptoms improved, but on the fifth day, there was worsening of the abdominal pain involving the supra-pubic area and right lower quadrant with radiation to the vagina. She reported on the same day onset of fever, chills, and a few loose stools. The subject returned to the ER about 6 days after the initial visit complaining of severe peri-umbilical pain .Physical examination revealed a temperature of 39°C, pulse of 100 bpm, with marked tenderness, and guarding of the right lower abdomen. Investigations revealed significant leukocytosis and abdominal ultrasound suggested a pelvic abscess. A general surgeon and the gynecologist were consulted. Broad-spectrum antibiotics were instituted and a computerized tomography (CT) scan of the abdomen demonstrated a complex fluid and gas filled mass in the right lower quadrant of the abdomen, extending into the pelvis. Later that evening the patient underwent a laparotomy for appendectomy, right salpingo-oophorectomy, and drainage of an abscess. Abdominal drains were removed on the second post-operative day and she was discharged home on the third day.

Eight days later, she was re-admitted in the ER with worsening abdominal pain, fever, and feeling quite unwell. A CT scan of the abdomen then revealed a recurrence of the large pelvic abscess. Repeat laparotomy was performed, intravenous antibiotics instituted, and then oral antibiotics. She was discharged after 9 days in the hospital.

11.2.1 Medico-legal Issues

The patient and her husband initiated a lawsuit for medical negligence against the initial ER physician for missing the diagnosis of appendicitis, and against the surgeon and gynecologist for premature discharge from hospital with inadequate treatment of appendiceal rupture and abscess after the first laparotomy.

Medical expert witnesses for the plaintiffs opined that the ER physician fell below the standard of care. It was implied that the recorded history was lacking in sufficient detail with regards to the patient's main complaint of abdominal pain. Specifically, there was no description of the intensity, quality, duration, or radiation of the pain. Besides noting that the abdomen was soft, there was no notation of abdominal tenderness or rebound pain. Moreover, the ER physician did not generate a differential diagnosis or order investigations, but just treated the patient symptomatically for a probably viral infection. It was the opinion of the expert witness, an earlier diagnosis of acute or sub-acute appendicitis could have been made at the first ER visit and that a simple appendectomy would have prevented the rupture of the appendix and pelvic abscess.

A surgical expert witness for the plaintiff was critical of the management of the ruptured appendix and pelvic abscess. He charged that the duration of antibiotics was too short after the initial laparotomy (24 h), and that repeat white blood cell count and ultrasound should have been performed before hospital discharge. Furthermore, if appropriate management was instituted, the pain and suffering and second operation could have been avoided.

11.2.2 Medical Aspects

The clinical manifestations and differential diagnosis of acute appendicitis depends on four major factors: (1) the anatomic location of the inflamed appendix (retrocecal and pelvic appendix are more often misdiagnosed and present with atypical symptoms or signs), (2) the stage of the process (early, acute, simple inflammation or perforation), (3) age of the patient (atypical presentations are more common at the extremes of age), and (4) the patient's sex (false negative appendectomies being highest in young females 15–45 years of age [32–45%]).[11]

A diagnostic clinical accuracy of about 85% is considered reasonable and optimal in clinical practice for acute appendicitis.[12] Below 80% suggests an over diagnosis and above 90% indicates under diagnosis and not considering the diagnosis of appendicitis soon enough. Data suggests that the delay in presentation and making the diagnosis are responsible for the majority of perforated appendix, which is associated with greater morbidity and complications.

The most frequent conditions mimicking symptoms and signs of acute appendicitis in children are acute mesenteric adenitis (secondary to viral infection, often with upper respiratory tract symptoms), and acute gastroenteritis, which may be present in adults in those with retrocecal appendix. In young women (as in this case), the common conditions that should be considered in the diagnosis include acute pelvic inflammatory disease, twisted ovarian cyst, ruptured Graafian follicle, acute pyelonephritis, ruptured ectopic pregnancy, and endometriosis.[12]

A scoring system has been developed (Alvarado Scale) to assist and improve the diagnosis of acute appendicitis.[12] Relative weight to specific clinical manifestations (see Table 11.1) provide a total score of 10. A score of 9–10 indicate almost certainly appendicitis surgery should be performed without further investigations; 7–8 score presents a high likelihood of appendicitis, and 5–6 represents compatible

Table 11.1 Diagnostic scale for appendicitis (Data obtained from Jaffe and Berger[12])

	Manifestation	Value
Symptoms	Migration of pain	1
	Anorexia	1
	Nausea/vomiting	1
Signs	RLQ tenderness	2
	Rebound tenderness	1
	Fever	1
Laboratory values	Leukocytosis	2
	Left shift (↑ bands)	1
Total score		10

Abbreviations: *RLQ* right lower quadrant of the abdomen

possibility, but not a diagnosis of appendicitis. A CT scan of the abdomen is indicated for scores of 5–6 and possibly 7–8. A score of 0–4 indicates that the diagnosis of appendicitis is extremely unlikely and a CT scan is not justified.[12]

The incidence of ruptured appendix averages 25–27% in most series,[12] but is greatest in children under 5 years (45%) and adults above 65 years (51%) of age. The mean duration of symptoms at the time of perforation of the appendix is about 5–7 days, and the majority of the ruptures are localized and patients display local rebound guarding. Generalized peritonitis from ruptured appendix is uncommon, but more frequent in young children and the immunosuppressed, particularly those on corticosteroids.

Besides surgery, antibiotics are routinely used for the management of acute appendicitis. There is no single agent or combination of agents of choice, but several regimens are available with similar efficacy. The choice of a regimen may vary from center to center, depending on acquisition cost and toxicity. The chosen agent or agents should have adequate spectrum of antibacterial activity against the majority of colonic microbiota – i.e., coliforms, anaerobes, and streptococci. The duration of treatment depends on the presence or absence of perforation of the appendix. For uncomplicated acute appendicitis, it is reasonable and common practice to discontinue the antibiotics the following day after appendectomy.

For complicated cases with perforation, abscess and peritonitis, a longer course of antibiotics is needed (as with other types of intra-abdominal infections). Current guidelines[13] recommend a minimum of 5–7 days depending on the clinical status of the patient. The patient should be clinically stable, afebrile for at least 2 days, and have a normal white blood cell count. In an analysis of 2,567 patients from 11 prospective clinical trials of antibiotic therapy for surgical infection, reliable predictors of sepsis eradication were identified.[14] Of these patients, 1,419 had some form of intra-abdominal infection. The findings of this study are the basis for current guidelines for duration of antibiotics in intra-abdominal infections. In subjects with elevated white blood count (WBC) and normal temperature, cessation of antibiotics resulted in reappearance of infection in 19% of patients. In those patients with normal WBC and temperature, the chance of recurrence of the original infection was 6%. For patients who were afebrile with a normal WBC, with

3% or less immature granulocytes (3% bands), the likelihood of infection was only 0.2%.[14] If fever was present, however, 65% of the patients had recurrent sepsis.

11.2.3 Medico-legal Discussion

The charge of medical negligence against the ER physician was based on his failure to diagnose or consider a differential diagnosis of acute appendicitis in the plaintiff. Expert medical witness (ER physician in a community hospital) for the plaintiff described his history and examination as cursory and inadequate based on the medical records. Moreover, a differential diagnosis was never generated, nor a plan of investigations to exclude more serious conditions than "acute viral illness." These are valid criticisms that are commonly found from review of medico-legal cases resulting from missed or delayed diagnoses in ERs.

The defendant's counter argument was that at the time of the first ER visit, there was little or no evidence to suggest appendicitis or serious intraabdominal infection, as the patient was afebrile without any localized abdominal tenderness. On further examination, investigations such as imaging would not be indicated if the maximum diagnostic score for appendicitis was only three (pain, anorexia, nausea or vomiting). However, a complete blood count (CBC) and differential was not performed, and if there were leukocytosis or left shift, the score could have been 5–6, which would indicate the need for a CT scan. Furthermore, localized right lower quadrant abdominal tenderness would be often absent for a pelvic appendix. Thus, a rectal and pelvic examination should have been performed to determine localized tenderness.

Based on the current medical evidence, the plaintiff received inadequate duration of antibiotics, which likely predisposed to recurrent sepsis. The antibiotics course should have been for a minimum of 5–7 days, and only if the leukocyte count were normal with no significant bands. The subject should have been sent home on antibiotics (a suitable choice would have been oral amoxicillin-clavulanic acid) with a follow-up visit and repeat complete blood count in 5–7 days. A repeat pelvic ultrasound should be performed for subjects with persistent symptoms, fever, or leukocytosis. However, most physicians would consider a follow-up ultrasound despite absence of symptoms of persistent infection.

In this case, an out of court settlement was reached, but it was clear from the weight of the evidence that the defendants were not in a favorable position of winning the lawsuit.

11.3 Case 3: Post-gastroplasty Sepsis

Five years before, a morbidly obese female at age 50 underwent vertical banded gastroplasty which resulted in weight loss, but was complicated by dysphagia of solid foods and intolerable retrosternal heartburns. She therefore requested a reversal of the vertical gastric banding. The procedure was performed at a university teaching hospital by the original general surgeon. Perioperative prophylaxis was administered in the form of two doses of 1 g cefazolin, and the procedure consisted of reversal of the vertical gastric banding and repair of a hiatus hernia. Her body weight at the time was 90 kg. Post-operatively she developed a low-grade fever (38°) for 48 h and investigations failed to reveal any infections.

Because of problems with intravenous access and the need for maintaining IV fluids, a central line was inserted 4 days post-operatively (with some difficulty) by the surgical resident. Her temperature was recorded as being normal over the subsequent 48 h. Three days after insertion of the subclavian central venous line, the patient developed high fever ranging from 38.6°C to 39.9°C. She also developed right shoulder pain and there was evidence of redness and pus around the insertion site of the subclavian venous line. Blood cultures, swab of the exit site were obtained, and the IV line was promptly removed. Intravenous cefazolin was administered via a peripheral vein for 48 h, but the catheter became dislodged. Cultures from the blood and swab grew Staphylococcus aureus (sensitive to beta lactams). However, due to poor venous access, the antibiotic was changed to oral cloxacillin 2 g/day for a week. During this time, the patient remained afebrile.

Three days after completion of the oral antibiotic, fever recurred with evidence of pain, tenderness, and swelling of the right upper chest wall at the base of the neck. A peripheral vein IV was started and intravenous cloxacillin restarted when a rash appeared after the second dose. Cefazolin was tried, but discontinued (due to worsening of the rash), then vancomycin was instituted, but venous access was lost soon after. Oral erythromycin was prescribed to be taken after discharge for a total of 18 days, but was prematurely stopped 4 days before discharge because of reappearance of another rash.

About 12 days after hospital discharge, she was seen by her family physician with pain and swelling of the right sterno-clavicular joint and fever. Imaging studies demonstrated evidence of septic arthritis and osteomyelitis of sterno-clavicular bone and joint and she was readmitted to the same hospital. She eventually required a prolonged course of IV vancomycin, multiple surgical debridements, and was not able to return to work until 1½ years later.

11.3.1 Medico-legal Issues

The patient and her husband brought suit against the surgeon, alleging negligence in her management, in particular with respect to placement of the central venous line and subsequent infection. The specific issues in this case were: (1) What caused the osteomyelitis, and could it have been prevented? (2) When the patient was found to have *S. aureus* bacteremia, was she adequately treated? (3) Was there a significant delay in the diagnosis of her osteomyelitis and subsequent appropriate treatment?

Medical expert witness for the plaintiffs opined that the surgeon and his assistants provided substandard post-operative care and listed several limitations of their management. These included: (1) failure to consult interventional radiology to obtain venous access for continuation of IV antibiotics, (2) misdiagnosis of the plaintiff's shoulder pain and signs of inflammation over the sternoclavicular joint as being due to cellulitis, (3) failure to recognize the seriousness of *S. aureus* bacteremia, (4) failure to obtain an appropriate consultation for management of the infection (i.e., infectious disease specialist), (5) inappropriate choice and dose of antibiotics for treatment of bacteremia, (6) failure to expedite readmission for investigation and treatment of septic arthritis and osteomyelitis after hospital discharge (they allowed a week to go by before any treatment was instituted).

11.3.2 Medical Aspects

It is estimated that each year >150 million intravenous devices are inserted in hospitals in the United States for various purposes. In addition, about 80,500 central venous catheter (CVC) related blood stream infections occur in ICUs each year in the United States.[15] Most of the CVC infections arise from the insertion site, hub, or both. The four most common microbes causing blood stream infection from these catheters are coagulase-negative staphylococci, *S. aureus,* gram-negative enteric bacilli, and *Candida* species. Multiple factors can influence the development of CVC infections, such as experience of the operator and degree of difficulty of insertion, type and site of catheter inserted, aseptic technique, underlying host factors (diabetes, immunosuppression, colonization with *S. aureus*, renal failure, etc), and probably obesity, which may predispose to technical difficulty and hematoma at the site of insertion, duration of CVC, purpose, and frequency of catheter access.[16,17]

It is standard procedure to remove non-tunneled, infected CVC, but the optimal treatment of those with *S. aureus* bacteremia has never been proven by randomized controlled trials (RCT). However, guidelines have been published by expert panels based largely on retrospective studies, clinical experience, and empiricism. The major complications of *S. aureus* CVC blood stream infection are the risks of developing endocarditis and metastatic seeding to the joints and bones (25–30%).[18] It is crucial to fully assess patients for these complications, especially after a week, and within a month of the bacteremia. Patients with metastatic complications are usually treated for 4–6 weeks with anti-staphylococcal agents. Since clinical endocarditis is a serious complication which may go unrecognized for the first few weeks, studies were implemented to screen for valvular vegetations after the first week of treatment for *S. aureus* bacteremia (two or more positive blood cultures taken at separate times). Several studies had found the incidence of valvular vegetations in groups of these patients without obvious clinical endocarditis in 25–32% on transesophageal echocardiography (TEE).[18] A subsequent, cost-effective analysis of routine TEE to determine the duration of therapy for *S. aureus* bacteremia related to CVC was performed.[19] The results of this study demonstrated that performing a TEE after the first week of therapy was the most cost-effective strategy; compared to treatment of all patients with intravenous antibiotics for at least 4 weeks, or treating patients for 2 weeks intravenously, then repeating blood cultures post-therapy and readmitting patients with relapse of bacteremia or development of clinical endocarditis for further therapy of 4–6 weeks intravenously.[19] Patients with prompt removal of CVC and resolution of fever and bacteremia within 72 h may not require TEE if there are no physical signs of metastatic infection, such as persistent fever at 72 h and bacteremia at 48–96 h after initiation of antibiotics and removal of CVC.[20]

There is recent evidence from retrospective collective data that initial, empiric, inadequate therapy for *S. aureus* bacteremia may not affect the 30-day mortality,[21] but inadequate course of therapy may predispose to greater risk of metastatic complications.[22]

11.3.3 Medico-legal Discussion

It is not standard practice of surgeons to discuss the risk of CVC blood stream infection when obtaining informed consent pre-operatively. The individual risk per person is small for short-term CVC (4.4% per 100 device or 2.7 per 1,000 catheter days).[16] Hence, patients who suffer from this complication are upset and incensed by the unexpected, particularly when there is metastatic seeding and significant morbidity or mortality.

There was no doubt that the development of septic arthritis and osteomyelitis of the sterno-clavicular joint was a direct result of the CVC infection and bacteremia. Furthermore, by current standards,[18] the plaintiff never received adequate therapy for the bacteremia or joint and bone involvement until the subsequent hospital readmission.

It could be argued by the defendant that the standard guidelines for treatment of *S. aureus* CVC related bacteremia are not well known to surgeons. However, this defense would not be acceptable in a teaching hospital or urban center, where there is usually access to an infectious disease consultant. Failure to consult a specialist can be considered substandard care.

It was evident in this case that the choice of antibiotics after the first few days was sub-optimal because of the poor venous access and development of drug reactions (rash). Optimum therapy should have consisted of at least the 2 weeks intravenous antibiotics and further course of an oral agent for 2–4 weeks if there was clinical evidence of bone and joint infection. The defendant countered there was no peripheral venous access and an attempt at another CVC could be associated with several complications including bleeding (hematoma), clotting, or thrombosis of the subclavian vein, pneumothorax, and secondary infection of the new catheter. Therefore, choosing continuation of therapy with an oral agent was a reasonable alternative.

Expert medical witness for the plaintiff argued that erythromycin was not an adequate treatment for *S. aureus* bacteremia or bone/joint infection. The macrolides are considered inadequate or sub-optimal for staphylococcal infections because of low activity and high incidence of developing resistance by one-step mutation.

Suitable alternatives for patients allergic to penicillin or cephalosporin for oral therapy include clindamycin, trimethoprim-sulfamethoxazole, doxycycline, and the more expensive linezolid. Lack of knowledge by the surgeon would not be an acceptable excuse, as even an oral opinion could be readily obtained form a suitable specialist or consultant.

Misdiagnosis of sternoclavicular arthritis and osteomyelitis before hospital discharge was also charged by the plaintiff. The surgical team made a diagnosis of metastatic skin and soft tissue infection without any further investigations, and therefore, interruption of an already inadequate oral treatment could have allowed progression to bone destruction. It would have been prudent for the physician to investigate the plaintiff's symptoms with a CT scan or magnetic resonance image (MRI) of the sterno-clavicular joint, which would have confirmed bony involvement before hospital discharge.

11.4 Case 4: Acute Abdominal Pain and Melena

A 50-year-old male presented one night to the ER of a small community hospital with acute, colicky, lower abdominal pain and frequent black stools of a few days duration. The patient had no significant past medical illness, except a week before he was started on oral cloxacillin for possible cellulitis of the leg, and celecoxib for lower backache with tenderness over L4–5 spine by his FP. The ER physician noted a temperature of 38.5°C, pulse of 90/min and a blood pressure of 130/90 mmHg. There was no diffuse left lower quadrant tenderness, no rebound or percussion tenderness, and the bowel sounds were normal or active. Blood count revealed a leukocyte count of 10,700 cells/mm^3 (with a left shift), and a normal serum creatinine and amylase. Radiograph of the abdomen (three views) revealed no free air or evidence of obstruction. A provisional diagnosis of diverticulitis was made and intravenous saline, narcotics, and antibiotics were instituted. The patient was to be assessed and managed by the surgeon on call the next morning (a long holiday weekend).

At 8 a.m. the following morning, the on-call general surgeon assessed the patient and found clinical evidence of peritonitis with marked rebound tenderness and guarding over the left lower quadrant of the abdomen. A computerized tomography (CT) of the abdomen was requested at a tertiary center over 40 miles away. A CT scan was performed around 4:30 p.m. that day which revealed intestinal perforation with a fluid collection. Attempted percutaneous drainage obtained fecal content. The patient was transferred back to the admitting hospital with stable vital signs. On his return, the attending surgeon decided not to do surgery, as there would be no available anesthesiologist from the next day for 48 h. Arrangements were made with a consulting general surgeon at the tertiary center to accept the patient for surgical management. His antibiotics at that time included a triple combination of intravenous ampicillin, gentamicin, and metronidazole.

The patient arrived in the ER of the tertiary center at 7:40 p.m. with normal vital signs and with signs of generalized peritonitis being worst in the lower abdomen. However, surgery was not performed until early the next morning at 5:15 a.m. A large hole in the distal sigmoid colon was found with fecal contamination of the peritoneum and a large presacral abscess. Partial colectomy, creation of a colostomy, drainage of the abscess, and irrigation of the abdominal cavity was performed. Pathology revealed areas of a thickened, inflamed colon with a perforated diverticulum and multiple diverticuli. His post-operative course in the ICU was complicated by persistent septic shock, disseminated intravascular coagulopathy and obtundation, weakness of the lower limbs, and multi-organ failure (including ARDS and renal failure). The patient eventually died about 9 days post-operatively.

11.4.1 Medico-legal Issues

The family of the deceased launched litigation proceedings for medical malpractice against both institutions and the surgical teams involved in his care. Charges against the initial admitting health care team and surgeon were: (1) failure to diagnosis-perforated intestine in a timely manner, and (2) delay in arranging immediate surgery or transfer for surgery at another center. Their delay in performing surgery and urgent transfer was the direct result of the patient's death. If surgery were performed on the day of, or soon after admission to the initial hospital, the outcome could have been avoided.

Accusations against the medical personnel of the tertiary care center were: (1) failure of the radiologist performing the CT scan to consult a general surgeon for immediate admission to their center instead of transferring the patient back to the community hospital, and (2) delay by the surgeon accepting the patient from the first hospital to arrange for immediate laparotomy on his arrival, (instead the delay of 10 h contributed to sepsis and death). The surgeon and his assistants knew the diagnosis and presence of generalized peritonitis, and should have been aware of the seriousness of the patient's condition and the need for immediate surgery.

11.4.2 Medical Aspects

Diverticulosis of the colon is a very common condition and occurs in about half of the population >50 years of age in Europe and North America.[23] Diverticulitis is the inflammation and infection associated with diverticulosis and occurs in about 10–25% of subjects with diverticulosis. The manifestations and presentation vary from mild, uncomplicated diverticulitis to free colonic perforation with diffuse peritonitis that requires emergency laparotomy.

Uncomplicated diverticulitis usually presents with left lower quadrant abdominal pain and tenderness, and CT scans generally reveal pericolic soft tissue swelling, colonic wall thickening and/or phlegmon (inflammatory mass).[23] The patient may or may not have mild diarrhea. In mild cases of diverticulitis, patients can be treated with a course of oral antibiotics for 7–10 days and a low residue diet as an outpatient. Patients with more severe pain and leukocytosis should have imaging (CT scan) to exclude an abscess and be treated in hospital with parenteral antibiotics, bowel rest, and narcotics. Deterioration of the patient's clinical status or development of peritonitis is an indication for laparotomy. About 57–70% of these patients recover without further episodes of diverticulitis, and a sigmoidoscopy or colonoscopy is recommended 4–6 weeks after recovery.

Complications of diverticulitis may include abscess, obstruction, lower gastrointestinal hemorrhage, free perforation with diffuse peritonitis, and fistulas between the colon and adjacent structures (colovesical, colovaginal, and coloenteric). The Hinchey staging system for severity of complicated diverticulitis is useful for diagnosis and management, as follows: Stage 1 – colonic inflammation with an associated pericolic abscess; Stage 2 – colonic inflammation with a retroperitoneal or pelvic abscess; Stage 3 – associated with purulent peritonitis; Stage 4 – associated with fecal peritonitis.[23] Management depends on the subject's overall clinical condition and degree of contamination/infection. Small abscess (<2 cm in diameter) may be treated with parenteral +/− oral antibiotics plus bowel rest. Larger localized abscess can be treated with CT guided percutaneous drainage with antibiotics, and the majority will require one-stage resection as an elective procedure later.

Inaccessible abscess to percutaneous drainage should undergo laparotomy, deterioration or failure to improve requires urgent surgery, and free air on abdominal radiograph or diffuse peritonitis (Hinchey stages 3–4) also requires emergent surgery.[23,24] In Hinchey, stages 1–2 require sigmoid colectomy and primary anastomosis and stages 3–4 or large abscess most commonly undergo sigmoid colectomy with colostomy and a Hartmann pouch.[23] The mortality rate from generalized peritonitis associated with diverticulitis ranges from 12% to 26%.[24,25] Early identification of free perforation is critical and CT scan can be used to confirm the diagnosis in ambiguous cases but either free air on plain radiograph or strong clinical suspicion with signs of diffuse peritonitis are sufficient to justify urgent laparotomy.[24]

11.4.3 Medico-legal Discussion

At presentation to the ER, the patient may well have had Stage 1–2 complicated diverticulitis and imaging was indicated. Since CT scan was not available, an abdominal ultrasound should have been requested (available in the center). At the time of assessment in the ER, there was no good evidence for free perforation or diffuse peritonitis. However, plain radiograph may not have detected free intra-abdominal air (pneumoperitoneum) and the CT scan is more sensitive. In one study, free air was detected in only 5 of 13 (38%) patients with intraperitoneal air by plain radiograph, whereas all were detected by CT scan.[26] At the time of assessment on the following morning, the clinical findings were consistent with severe, diffuse peritonitis (Hinchey stages 3–4).

A medical expert (a general surgeon) for the plaintiff expressed the opinion that immediate laparotomy should have been arranged that morning, and delay in waiting for results of a CT scan contributed to the patient's demise. Even after being informed of the results of the CT scan (at 5 p.m.), it was evident that the patient needed an urgent laparotomy. The initial attending surgeon transferred the patient to a tertiary center rather than performing immediate laparotomy, an act also considered negligent, substandard care by the expert witness.

The defendant's excuse for not performing surgery at the rural hospital, as explained at the examination of discovery, was based on the impending unavailability of an anesthesiologist the next day for 48 h; and his limited experience in performing colectomy. The surgeon was a trained urologist but practiced and covered for general surgery for certain procedures.

Medical expert witness and the lawyer for the plaintiffs contend that the explanation by the defendant was unacceptable. First, an anesthesiologist was available on the day of the surgeon's initial assessment and if further surgery for complications arising thereafter was required, then the patient could be transferred to a tertiary care center. Secondly, if the defendant felt incompetent in performing appropriate surgical management, then arrangement should have been made on the morning of the initial assessment to directly transfer the patient for surgery at the university hospital. It was also argued that the defendant should have had the knowledge and skill (based on his training) to perform drainage procedures and colostomy, and colectomy could be deferred to a later date. The delay in transporting the patient back and forth between the two hospitals (for investigation and subsequent management) allowed the infection to progress to a critical stage and led to his demise.

The issues concerning management at the teaching hospital involved failure of the radiologist to refer to the ER for immediate surgery, and delay by the consulting general surgeon to perform immediate laparotomy or transfer. Since the patient was hemodynamically stable at the time of the CT scan, and was referred from another hospital only for imaging and possible percutaneous drainage, there was no medical or legal obligation by the radiologist to refer the patient to the ER of the tertiary care center. However, the patient had to wait almost 10 h before surgery was performed at the teaching hospital with a documented free perforation and diffuse peritonitis. Current guidelines recommend immediate surgery. The second defendant (surgeon) noted that the deceased was hemodynamically stable on arrival to the ER, and his request for an operating room (OR) was only granted for the time scheduled. However, according to the OR nurse manager, the OR time scheduled was based on the classification of the case. The deceased was classified as an urgent/Priority B (implying need for surgery within 8 h), rather than priority A case, which indicates immediate surgery or within 2 h.

Based on our current knowledge of the pathogenesis of severe sepsis, it is reasonable to surmise that surgical intervention on the morning after admission to the rural hospital would likely have improved the outcome. It is less clear whether earlier surgery by 6–8 h at the tertiary center would have affected the outcome. However, the surgical morbidity and mortality are more favorable when the patients have normal vital signs, rather than rushed in when they are hemodynamically unstable.

11.5 General Comments

Medico-legal problems related to cholecystectomy have recently been reviewed by the Canadian Medical Protective Association (CMPA).[27] This review involved 131 legal cases associated with surgical complications from cholecystectomy (both laparoscopy and laparotomy) between 2003 and 2007. The most frequent complications were biliary tract injuries (53%), intestinal injuries (19%), and vascular/hemorrhagic injuries (11%), with other complications (17%), such as wound infection, pulmonary emboli or retained foreign body material.[27] Medical experts review noted that most complications were insidious in onset and often with non-specific symptoms. The onset of symptoms after biliary or intestinal injuries may vary from hours to several days after the procedure, and patients commonly present to the ER or FPs. Major vascular injuries, however, present with rapid clinical deterioration soon after the surgery. In 70 patients suffering biliary tract complications, six died as a result of the complications, 26 patients had permanent disability, and 25 had major, but temporary disability.[27] Fifty-one of these cases were entirely laparoscopic and 13 cases began as laparoscopic procedures but were converted to an open procedure (after bile duct injury, bleeding or adhesions). Thus, 64 of the 70 (91.4%) cases with biliary tract complications were related to laparoscopic procedures. In 58 cases (83%) there was complete ligation or transection of the common bile or hepatic duct, and 17% had lesser injuries leading to leak, strictures, or fistula. In 37 cases (53%), there was mis-identification of the anatomy. The CMPA cases were settled in favor of the plaintiffs in 53% of cholecystectomy cases compared to 32% of all other cases. However, the CMPA paid a settlement to 70% of biliary tract injury cases on behalf of the surgeon.[27]

Criticisms by surgical experts who reviewed the cases include: (1) lack of comprehensive informed consent discussion, including risks of bile duct, intestinal and vascular injuries, (2) failure to take the necessary steps to minimize injury, (3) failure to convert to open approach or perform intraoperative cholangiography when unsure of the anatomy, (4) failure to identify anatomical structures adequately before the surgical clips or ligatures were applied, (5) inadequate operative notes to reflect difficulties encountered during the procedure, (6) inappropriate delay in post-operative investigation and intervention in symptomatic patients, (7) incomplete discharge instructions leading to delay in seeking medical care.

What lessons can physicians learn from these four cases?

- Informed consent discussion should be more comprehensive and not leave the patient with the impression that the procedure is minor (laparoscopy) and without significant complications
- Symptoms of post-operative complications (and even appendicitis) can be non-specific and develop or evolve hours to several days after the procedure (or onset of non-specific symptoms)
- Appropriate discharge instructions should be provided to patients to seek early attention after surgery, or ER visits
- FP and ER physicians will often face the challenges of assessing the patients with early non-specific symptoms of a surgical complication, or appendicitis in the early stages
- Surgeons and other physicians should be cognizant of the complications of *S. aureus* bacteremia (which has become a common complication of CVC in hospital), and if in doubt they should consult a specialist (infectious diseases)
- CT scan is a useful tool for diagnosis, but it is not necessary in cases of peritonitis when the diagnosis is obvious; and it is important to perform urgent surgery rather than delay for the investigation
- Delaying surgery, even for a few hours in cases of diffuse peritonitis or free perforation can be detrimental to the patient.

References

1. Doherty GM (ed.), (2010). Biliary tract. IN Current Diagnosis & Treatment.: Surgery, 13th Edition, Lange McGraw Hill, New York, p544–571.
2. Fletcher DR, Hobbs ST, Tan P, Valinsky LJ, Hockey RL, Pikora TJ, Knuiman MW, Sheiner HJ, Edis A(1999). Complications of cholecystectomy: risks of laparoscopic approach and protective effects of operative cholangiography. A population based study. *Ann Surg* 229: 449–457.
3. Gadaz JR (2000). Update on laparoscopic cholecystectomy including a clinical pathway. *Surg Clin North Am* 80:1127–1149.
4. Johnson AG (2000). Peptic ulcer – stomach and duodenum. In (eds): Morris PJ, Wood WC. Oxford Textbook of Surgery, 2nd Edition, Oxford University Press, Oxford, p1297–1311.
5. Donovan AJ, Berne TV, Donavan JA (1998). Perforated duodenal ulcer. An alternative therapeutic plan. *Arch Surg* 133: 1166–1171.
6. Crofts TJ, Park KG, Steele RJ, Chung SS, Li KK (1989). A randomized trial of nonoperative treatment for perforated peptic ulcer. *N Engl J Med* 320:970–973.
7. Lau WY, Leung KL, Kwong KH, Davey IC, Robertson C, Dawson JJ, Chung SC, Li AK (1996). A randomized study comparing laparoscopic versus open repair of perforated peptic ulcer using suture or sutureless technique. *Ann Surg* 224:131–138.
8. Kaiser AM, Katkhouda N (2002). Laparoscopic management of perforated viscus. *Sem Laparosc Surg* 9:46–53.
9. Casas AT, Gadacz TR (1996). Laparoscopic management of peptic ulcer disease. *Surg Clin N Am* 76:512–522.
10. Flancbaum L, Nosher JL, Brolin RE (1990). Percutaneous drainage of abdominal abscesses associated with perforated viscus. *Am Surg* 56:52–56.

11. Flum DR, Koepse LL T (2002). The clinical and economic correlates of misdiagnosed appendicitis: national analysis. *Arch Surg* 139:799–804.
12. Jaffe BM, Berger DH (2005). The appendix. In (eds): Brunicardi FL, Anderson DL, Bilhar TR, Dunn DL, Hunter JC, Pollvik RE. Schwartz's Principles of Surgery, 8th ed. McGraw Hill, New York p 1119–1137.
13. Doherty GM (ed) (2010) Peritoneal cavity. In: Current Diagnosis & Treatment: Surgery, 13th edition, Lange Medical Text, McGraw Hill, New York, p 464–478.
14. Stone HH, Bourneuf RN, Stinson LD (1985). Reliability of criteria for predicting persistent or recurrent infections. *Arch Surg* 120:17–20.
15. Mermel LA (2000). Prevention of intravascular catheter-related infections. *Ann Intern Med* 132:391–402.
16. Maki DG, Kluger DM, Crnich CJ (2006). The risk of bloodstream infection in adults with different intravascular devices: a systematic review of 200 published prospective studies. *Mayo Clin Proc* 81:1159–1171.
17. Moro ML, Vigano EF, Lepri, The Central Venous Catheter-Related Infectious Study Group (1994). Risk factors for central venous catheter-related infections in surgical and intensive care units. *Infect Control Hosp Epidemiol* 15:253–264.
18. Mermel LA, Allon M, Bouza E, Craven DE, Flynn P, O'Grady NP, Raad II, Rijnders BJ, Sheretz RJ, Warren DK (2009). Clinical practice guidelines for the diagnosis and management of intravascular catheter-related infections: 2009 update by the Infectious Diseases Society of America. *Clin Infect Dis* 49:1–45.
19. Rosen AB, Fowler VG Jr, Corey GR, Woods CW, Downs SM, Biddle AK, Li J, Jollis JG (1999). Cost-effectiveness of transesophageal echocardiography to determine the duration of therapy for intravascular catheter-associated *Staphylococcus aureus* bacteremia. *Ann Intern Med* 130:810–820.
20. Fowler VG Jr, Olsen MK, Corey GR, Woods CW, Cabell CH, Reller LB, Cheng AC, Dudley T, Oddone EZ (2003). Clinical identifiers of complicated *Staphylococcus aureus* bacteremia. *Arch Intern Med* 163:2066–3072.
21. Ammerlaan H, Seifert H, Harbarth S, Brun-Buisson C, Torres A, Antonelli M, Kluytmans J, Bonten M; European Practices of Infections with *Staphylococcus aureus* (SEPIA) Study Group (2009). Adequacy of antimicrobial treatment and outcome of *Staphylococcus aureus* bacteremia in 9 Western European countries. *Clin Infect Dis* 49:999–1005.
22. Falione M, Carfagna P, Cassone M, Pistella E, Pavoni G, Nofroni I, Serra P, Venditti M. (2002). *Staphylococcus aureus* sepsis in hospitalized non-neutropenic patients: retrospective clinical and microbiological analysis. *Ann Ital Med Int* 17:166–172.
23. Bullard KM, Rothenberger DA (2005). Colon, rectum and anus. In (eds): Brunicardi FC, Andersen DK, Billair TR, Dunn DL, Hunter JG, Pollock RE. Schwartz's Principles of Surgery, 8th Edition. McGraw Hill, New York, p1055–1137.
24. Fox JM, Stollman NH (2006). Diverticular disease of the colon. In (eds): Feldman M, Friedman LS, Brandt LJ. Sleisenger & Fordhtran's Gastrointestinal and Liver Disease: Pathophysiology/Diagnosis/Management, 8th Edition. Saunders/Elsevier, Philadelphia, p2613–2632.
25. Krukowski ZH, Matheson NA (1984). Emergency surgery for diverticular disease complicated by generalized and fecal peritonitis: a review. *Br J Surg* 71:921–927.
26. Stapakis J, Thickman D (1992). Diagnosis of pneumoperitoneum: abdominal CT vs upright chest film. *J Comput Assist Tomogr* 16:713–716.
27. Le Grand L, Dunn C, Liu R, Hunava D (2009). Medico-legal problems related to cholecystectomy: biliary tract injuries. CMPA Perspective. *Can Med Protect Assoc* 1:16–19. www.cmpa-acpm.ca.

Chapter 12
Litigations in Plastic Surgery

12.1 Case 1: Infection After Liposuction

A previously healthy, 43-year-old female underwent cosmetic surgery at a private, freestanding center consisting of a tummy tuck and liposuction of the abdomen, hips and thighs. Soon after surgery, the patient complained of shortness of breath, but no specific abnormalities could be found. At the post-operative visit 2 days later, the wounds were found to be in satisfactory condition. Four days after the operation, the patient attended the ER of a university teaching hospital with worsening shortness of breath. She was admitted to hospital and treated for pulmonary emboli, but this diagnosis was never confirmed. On admission, the intern diagnosed an abdominal wound infection, but this was not confirmed by the senior medical resident or the attending medical staff. An abdominal wound swab taken was reported as growing normal skin flora. The wound drain was removed 3 days after admission and the patient was discharged home a week after hospital admission.

The patient was subsequently seen in follow-up by the plastic surgeon about 2 weeks later. At this visit, there was evidence of abdominal wound infection, requiring surgical drainage of approximately 200 mL of pus that grew Pseudomonas aeruginosa (with usual susceptibility). She was treated with intravenous anti-pseudomonal antibiotics for a week and wound debridement. The patient recovered fully from the infection, but was left with residual scarring. Evidence was also provided of a female friend of the patient who underwent similar cosmetic surgery at the same center (on the same day), and her post-operative course was also complicated by a P. aeruginosa wound infection. The bacteria were never retained for further testing for genotype to determine the relationship of the two recovered isolates.

12.1.1 *Medico-legal Issues*

The patient (plaintiff) initiated medico-legal actions against the surgeon and alleged that the infection was the result of contamination at surgery. This was the result of poor aseptic technique and negligence by the cosmetic center. As a result of the

I.W. Fong, *Medico-Legal Issues in Infectious Diseases*, Emerging Infectious Diseases of the 21st Century, DOI 10.1007/978-1-4419-8053-3_12,

complications of the surgery, she suffered physically and mentally, plus the residual scar was unsightly and caused prolonged mental anguish. As a result, financial compensation of a substantial amount was being sought. Interestingly, the surgeon was not sued for the complication of pulmonary thrombo-emboli.

An expert medical witness for the plaintiff was of the opinion that infection by such an uncommon wound pathogen of two women performed on the same day was likely related to break in aseptic technique or defective sterilization of the surgical equipment. He was also critical of the post-operative care, as on the morning of the third post-operative day, the plaintiff phoned the defendant's office complaining of shortness of breath. She was neither examined nor advised to seek medical attention at an ER, but was told that her symptoms would resolve spontaneously. Evidence was brought forth at the examination of discovery from the admitting intern that the patient's wound discharge, on admission for possible pulmonary emboli "smelled of a possible pseudomonas infection" and this was corroborated by the nurse present at the ER at that time.

12.1.2 Medical Aspects

It was estimated that 2.7 million cosmetic procedures were performed in the US in 1998, and over the 7 years preceding 2004, the number of abdominoplasty procedures (for obesity) had increased by 344%.[1] In a survey of 497 plastic surgeons involving 20,029 procedures, 35% were liposuction of the abdomen, 10% were limited abdominoplasties, and 55% were full abdominoplasties.[1] In a recent study of surgical complications in plastic surgery, national (US) databases were analyzed for abdominoplasty and breast augmentation patients.[2] The complication rates were similar for the two procedures. Hematoma occurred in 0.5–0.9%, infection in 0.7–3.5%, and thrombo-emboli in 0.1–0.3% of abdominoplasties.

Freestanding, or office-based outpatient plastic surgery facilities have been found to be safe, as well as convenient. In a retrospective study of 5,316 procedures performed over 6 years, only 35 (0.7%) complications occurred, with no deaths reported.[3] Most complications were secondary to hematoma formation (77%), and post-operative infection requiring surgical intervention occurred only in 0.11%. The seven patients requiring hospitalization were for arrhythmias, angina, and pulmonary emboli. This report was similar to a previous study involving 241 accredited office surgical facilities and 400,075 operative procedures over 5 years.[4] Significant complications (hematoma, hypertensive episode, wound infection, sepsis, hypotension) were infrequent, occurring only in 1 in every 213 cases. The overall risk of complications was similar to a hospital ambulatory surgical facility.

With respect to abdominal contour surgical complications, there are minor differences between the types of procedures for wound infection rates; i.e., 1% for liposuction, 0.02% for limited or mini-abdominoplasties, and 1.1% for full abdominoplasties.[1] There were no significant correlations between years in practice and the total volume of abdominal liposuction and these complications. Risk factors for surgical site infection in cosmetic procedures have been found to include the mean duration of the procedure (>2 h), general anesthesia, and placement of a Blake drain.[5,6] A study from Italy on surgical site infection in plastic and reconstructive surgery reported an overall wound infection rate of 3%, and a multivariate analysis identified diabetes, chronic obstructive lung disease, and surgical drains as significant factors for infections.[7]

P. aeruginosa, although ubiquitous in our environment, is an infrequent cause of surgical site infection in clean or cosmetic surgery. In hospitals, pseudomonas account for 8% of surgical site infection on national surveillances.[8] Pseudomonas species' natural habitats include water, soil, plants, vegetables and fruits, yet community-acquired infections are rare.[9] *P. aeruginosa* is problematic in hospitals and immunocompromised conditions such as cystic fibrosis, burns, cancer, and prolonged catheterization or exposure to broad spectrum antibiotics. The organism can be found in aqueous solutions such as disinfectants, ointments, soaps, irrigation fluids, eye drops, dialysis fluids, and equipment. It is also frequently found in aerators, traps of sinks, whirlpool, hydrotherapy baths, swimming pools, hot tubs, showerheads, cosmetics, contact lens solution, and inner soles of sneakers.[9] *P. aeruginosa* is infrequently present in healthy people, but rates of colonization with low density of pseudomonas in healthy adults can be found in up to 7% on the skin, throat, and nasal mucosa, and up to 20% in the stool.[10] Healthy humans are highly resistant to pseudomonas infection and it is an opportunistic pathogen for damaged tissue and poorly perfused open wounds. On wound dressings or discharge it produces a blue-green color and characteristic odor, described as like "rotten fruit."

In healthy subjects, infection with *P. aeruginosa* in the community is usually rare, localized, or superficial, and associated with contact with contaminated water or solution. These conditions include pseudomonas folliculitis (associated with hot tubs, whirlpools, water slides, swimming pools, contaminated bath sponges), otitis externa ("swimmer's ear"), conjunctivitis/keratitis (from contact lens, eye solutions, minor trauma), osteomyelitis of the foot in children (nail puncture through sneakers), and endocarditis/bacteremia in intravenous drug abusers (IVDA).[9,10]

It is generally considered that the majority of surgical site infection arises from contamination of the open wound at the time of surgery; most frequently from endogenous bacteria colonizing the patient, and occasionally from the operating team or the environment of the facility.[8] If this paradigm were accurate, it is difficult to explain the lack of correlation between pre-operative bacterial colonization of the skin and the microbes causing post-operative wound infections reported in some studies.[11] Inoculation of the wound post-operatively from surrounding contaminating bacteria is considered rare, but it is difficult to prove and maybe underestimated. The notion that 24 h after surgery the wound closure acts as a microbial barrier against infection from without lends credence to the idea that open drains allow direct entry of organisms into the wound. However, conflicting data exists where some prospective studies show greater risk for surgical site infection when wound drains were present, but others did not.[8]

12.1.3 Medico-legal Discussion

The main issues in this case are: (1) When and where was the *P. aeruginosa* introduced into the plaintiff's wound? (2) Was the wound infection due to faulty technique or break in asepsis or contamination of surgical equipment in the office-associated facility? (3) If the latter were correct, does this represent medical negligence?

Introduction of *P. aeruginosa* in the plaintiff's wound at the time of surgery would be a very rare event for an office-associated facility. However, the fact that both the plaintiff and her friend experienced a pseudomonas infection at the same facility on the same day, strongly implicates the cosmetic surgical center. It would also be unlikely that this bacteria was introduced from colonization of the patient's skin. The routine surgical skin preparation technique and method of sterilization of the equipments were reviewed by an independent medical expert, and no defects or deficiencies could be identified. However, an onsite inspection or direct observation of any procedure was never performed, so whether any break in aseptic technique could have occurred cannot be adequately addressed. Theoretically, pseudomonas could have been introduced during liposuction, as large volumes of fluid were used to irrigate and suction the subcutaneous space and fatty layer. However, this was performed as usual, with commercial preparation of sterile saline. If contamination occurred at the source of preparation, this generally involves batch contamination with outbreaks involving multiple patients from different sites. Occasionally, local outbreaks of infection have been traced to connecting suction tubes and bottle containers, if they are not changed between cases. This was never investigated in this case.

Counsel for the defendant with support from other medical expert witnesses contend that the infection was likely introduced post-operatively via the wound drains. This could have occurred at home while taking a bath or shower (common places for pseudomonas to reside). The organism could also have been acquired when the plaintiff was admitted to the hospital for possible pulmonary embolism. There was no real evidence that the plaintiff and her friend were infected by the same organism, as no molecular typing was ever performed. Although the admitting intern described an infected wound with a "pseudomonas odor" on hospital admission, this was not verified by the senior resident and the wound swab culture only reported normal skin flora. The absence of pseudomonas from the wound discharge a few days post-operatively suggest the organism likely gained entry to the wound in hospital via the drain.

12.2 Case 2: Complication After Browlift

A 55-year-old female underwent bilateral blepharoplasty and endoscopy browlift at a private plastic surgery clinic for aesthetic reasons. She was previously well, except for a past history of migraine and allergy to sulfonamides. The procedure was performed under local anesthetic and intravenous sedation without any intraoperative complications. The operation lasted 80 min, and while no perioperative antibiotic was given, an anchor screw was inserted over the right temporal area posterior to the hairline. This was apparently used for stabilization of the browlift. The early postoperative course was uncomplicated, but 6 weeks later, the patient reported some drainage from the lateral temporal region of the wound. Minor surgery was performed 2 weeks later under local anesthetic to remove the suture and anchor screw, but no microbiologic cultures were performed.

A month later (3 months after the initial surgery), she returned to the surgeon's office with an abscess over the right brow. This was drained and debrided by a different physician and the patient was placed on oral cephalexin for 10 days. Two weeks later, there was a flare-up of the local wound and so repeat debridement and drainage were performed. This local procedure was repeated again 3 weeks later when there were recurrent symptoms, and she was empirically treated with ciprofloxacin for 2 weeks. Wound swabs were performed on two occasions after completing oral antibiotics, but failed to grow any pathogen.

About 5 months after the original surgery the patient was referred to another plastic surgeon at a university teaching hospital for consultation. A computerized tomography (CT) revealed erosion of the outer table of the skull with involvement of the diploe, but the inner table was intact. The consulting surgeon recommended discontinuation of the antibiotics with the intention of performing deep bone debridement for a culture in about 2–3 weeks. About 3 weeks after the consultation, (right when the patient was due for pre-operative assessment), she presented to the ER of the hospital with severe headache, fever, photophobia, and impaired concentration ongoing for 5 days. A repeat CT scan of her head revealed progressive osteomyelitis of the skull with development of a subdural empyema compressing the brain.

The patient was started on broad-spectrum intravenous antibiotics and surgery was performed later that day by a neurosurgeon. A right frontal craniotomy was performed with debridement and drainage of the empyema and infected tissue. The pus grew Streptococcus milleri and she was treated with 6 weeks of intravenous penicillin. (Her course was complicated by seizures requiring long-term antiseizure medications). Six months later reconstructive surgery was performed for the skull defect. Two years later, secondary infection and erosion of the cranioplasty site occurred. Further sequestrectomy was done and intravenous vancomycin was given for 6 weeks to treat a staphylococcal infection.

12.2.1 Medico-legal Issues

The patient (plaintiff) hired a lawyer to pursue actions for medical malpractice against both the initial and the consulting plastic surgeon. The legal counsel requested an independent medical review of the case to determine any negligence in the medical care of the plaintiff. The main issues that were to be addressed were as follows:

1. Was the initial plastic surgeon negligent in the care of the plaintiff's wound infection?
2. Does his failure to perform a CT scan earlier in her course or failure to consult a specialist earlier represent substandard care?
3. With respect to the consulting plastic surgeon, was his recommendation to discontinue the antibiotics the direct cause of the progression of the infection?
4. Was his failure to consult an infectious disease specialist before stopping the antibiotics medical negligence?
5. Was his failure to arrange urgent surgery since he knew the plaintiff had evidence of osteomyelitis of the skull, representative of medical malpractice?

12.2.2 Medical Aspects

Blepharoplasty (plastic surgery of the eyelid) is often performed in conjunction with browlift for periorbital aesthetic surgery.[12] These cosmetic procedures are used to preserve upper orbital fullness, define upper lid crease, correct excessive and prominent fat pads, correction of brow position, and for correction of mid-facial descent. Browlift can be used for ptosis, asymmetry between the nasal and lateral brows; and brow elevation and stabilization is used in order to avoid worsening brow ptosis after upper blepharoplasty.[12] Browlift in aesthetic plastic surgery can be performed by the open procedure or endoscopically. There is ongoing debate about the various methods of flap fixation to maintain the elevated brow position (while the soft tissues are re-adhering at a higher position). The methods used medial to the anterior temporal crest include bolster fixation, mattress sutures, control tunnels, Kirshener wire, external screws and fibrin glue.[13] However, it has been proposed that fixation of the elevated forehead flap through an endoscope was not necessary and that maintenance of brow position could be accomplished by adequate release alone.[14] The complications of browlift include: sensory nerve deficit, frontalis muscle paralysis, skin necrosis, alopecia, infection, hematoma and bleeding, asymmetry of eyebrows or eyelids, chronic pain, permanent over-correction, abnormal soft tissue contour, abnormal hair part, and visible scar.[12,13]

Cranial osteomyelitis is a very uncommon condition that can occur spontaneously, post-traumatically or post-surgically (iatrogenic); the latter being most common in my experience, but the least reported in the literature. Spontaneous cranial osteomyelitis may occur in chronic frontal sinusitis (Pott's puffy tumor) as a complication of odontogenic infection, but rarely as a result of malignant otitis externa in diabetics. Frequently, patients present with chronic draining wounds or sinuses or with overlying soft tissue swelling (as in Pott's puffy tumor). CT scan is very accurate in differentiating between soft tissue and bone infection, but the MRI is best for assessment of the calvaria and skull base.[15]

Osteomyelitis of the skull following craniotomy for intracranial pathology is the most frequent form of skull infection. In this situation, appearance of wound infection can occur simultaneously with intracranial complication, i.e., extradural empyema, subdural empyema, and less commonly meningitis and brain abscess.[16] Other rare complications may include intracranial sinus thrombosis or septic venous phlebitis. Infections of the skull from external source (trauma, external fixation with screws) more commonly present with chronic draining sinus with slower progression to the inner table and subsequent extradural or subdural empyema. Cranioplasties following decompressive craniectomy has been reported to be complicated by infection in about 7%, with predominantly *S. aureus* and *Staphylococcal epidermidis* (50%).[17]

Infectious rates of plastic surgery are quite variable depending on the type of procedure; indications for surgery, presence of prosthesis, and duration of the procedure are some of the most important ones. In general, reconstructive surgery for serious deformities and requiring grafts and prosthesis (especially involving the mandible or encroach upon the sinuses) have higher infection rates (8–17%).[18] In these procedures, it has become standard practice to administer perioperative prophylactic antibiotics, and there is some evidence to support this custom. Infections in cranio-facial surgery most commonly are due to *S. aureus* but oral *Streptococcus* species and *Bacteroides* are common with procedures impinging on the oral mucosa.[19] Although infections following cosmetic surgery are very low in healthy subjects, there are several reports of rapidly growing mycobacterial infections (*Mycobacterium abscessus, M. chelonae, M. peregrinum, and M. fortuitum*) after soft tissue augmentation, mesotherapy (microinjection into the dermis) or liposuction, in small outbreaks or isolated cases.[20–23] A characteristic feature of these infections include cutaneous or subcutaneous nodule, with or without draining abscess, and often presents as an insidious late post procedural complication.[24,25] Late infectious, however, after breast implant, can be secondary to *Enterobacter* species whereas *S. aureus* more commonly presents as early infection (within 20 days).[26]

12.2.3 Medico-legal Discussion

There were several aspects of the initial plastic surgeon's management of the plaintiff's infection that could be considered substandard medical care. The appearance of a late draining wound over the temporal area at the site of an inserted screw should have immediately alerted the physician to deep bone infection. It was negligent medical care not to have sent the explanted screw or debrided tissue for microbiological culture at the first post-operative visit with wound infection. The defendant should have known that removal of the foreign body would not be enough to cure the infection, as the screw was anchored in the skull. Despite several repeated visits for recurrent abscess and drainage from the wound, no investigations were done (such as a CT scan) to define the extent of the infection. The surgeon should have consulted a specialist (in infectious diseases, neurosurgery or another plastic surgeon) much earlier when the plaintiff kept returning with relapses of the wound infection. It seemed highly likely that had appropriate investigations and management been instituted within the first month of the plaintiffs presentation with the infection, the complication of subdural empyema, repeated hospital admissions, and need for reconstructive surgeries could have been avoided.

The case against the consulting plastic surgeon at the university teaching hospital was much weaker. Investigations showed that it was appropriate to discontinue the empiric antibiotic before surgery in order to obtain deep tissue specimens for culture in order to identify the etiologic pathogen. A more contentious issue was the duration of time it took (3 weeks) to arrange for surgical debridement and hospital admission. At the time the consulting surgeon diagnosed osteomyelitis of the skull, the involvement was localized and did not appear to extend to the inner table of the skull, nor were there any clinical and imaging evidence of extension to the extradural or subdural space. Thus, there was no need then to arrange for emergency surgery. In hindsight, it would have been more appropriate to have had surgery earlier (within 10 days of the assessment when the diagnosis of osteomyelitis of the skull was made). This may be considered an error in clinical judgment, but was not clear evidence of medical negligence or malpractice. It is possible that a judge or jury may have a different opinion on this viewpoint. Failure of the consulting surgeon to properly counsel the patient and monitor her condition (after withdrawal of the antibiotic) for worsening could be considered negligent care.

12.3 Case 3: Infection After Breast Reconstruction

Three years before, a 42-year-old female had undergone a lumpectomy, adjuvant chemotherapy, and radiotherapy for breast cancer. She consulted a plastic surgeon at a university teaching hospital for right breast reconstruction and left breast reduction in 2002. Besides the breast cancer, she had a history of chronic seizures and depression. The operation itself was performed without any immediate complications. On post-operative day 7, she was assessed by the surgeon in the clinic and there was a red, swollen, circular area of the right medial breast that was painful and tender. Oral ciprofloxacin 500 mg twice daily was started for suspected wound infection, with no cultures obtained or local drainage performed.

A follow-up visit 2 weeks later in the plastic surgery clinic revealed the left breast wound was healed, but the right breast was very tender, with hard-indurated tissue from the sternum to the nipple. The ciprofloxacin was continued until 2 weeks later with no improvement, so surgical debridement and drainage of necrotic tissue was performed. The antibiotic was then changed to intravenous ceftriaxone, and tissue culture grew Peptostreptococcus.

Five days after the debridement, there was still a moderate amount of wound drainage, and the breast was tender and indurated. Intravenous antibiotic was continued for another 2 weeks on home parenteral therapy. About 10 days after the drainage/debridement procedure, there was evidence of extension of the wound infection to the base of the neck, across the sternum and into the left breast, with odious seropurulent drainage. A repeat wound culture grew Streptococcus viridans and mixed anaerobes. Computerized tomography (CT) revealed edema and gas in the tissues, but no abscess or foreign body was reported. Intravenous ceftriaxone was reordered for another 2 weeks.

Six weeks after the second surgery, the patient reported that a 1-inch piece of plastic spontaneously extruded from the right breast wound, with subsequent improvement of the infection. Further surgical debridement was performed, and eventually the wound infection gradually cleared with closure of the wound after 6 months.

12.3.1 Medico-legal Issues

The patient sued the surgeon for medical negligence in leaving a piece of plastic in her breast that directly resulted in a chronic breast infection. Furthermore, the plaintiff claimed inadequate management of her breast infection, which resulted in prolonged unnecessary suffering, repeated surgery, and hospitalization. As a result of the adverse outcome, she was left with a deformed, scarred breast, which indirectly resulted in worsening depression and a feeling of inadequacy.

The defendant countered that the infection and necrosis of the plaintiff's breast was the result of previous radiation. Furthermore, the surgeon could not explain the presence of a piece of plastic in her breast, and was alleging that the foreign body was introduced at her home during the packing and dressing of the wound.

12.3.2 Medical Aspects

The source of the retained piece of plastic in the plaintiff's breast remained unclear, as this was not recognized as representing any part of the surgical implements or paraphernalia used at surgery. The home care nurses in charge of the patient's wound dressing were unable to explain the foreign body as well, as packing of the wound was performed with plain packing strip or ribbon gauze. Although the defendant did not accuse the plaintiff of inserting the piece of plastic into her breast wound directly, this was implied in the defense statement.

Another issue of contention raised by the plaintiff was negligent care of her wound infection by the surgeon. Expert medical witness for the plaintiff identified several areas of concern in the management of the plaintiff's wound infection that was considered substandard care. The expert witness noted that the hard induration of the breast indicated that here was deep soft tissue infection of the breast, and not a simple superficial cellulitis at presentation 1 week post-operatively.

Therefore, prescribing antibiotics alone was inappropriate, as the patient required surgical debridement earlier, once infection was recognized with appropriate tissue cultures. It was surmised that the outcome could have been catastrophic if the infection was caused by a more virulent bacteria – i.e. *Streptococcus pyogenes*. Moreover, ciprofloxacin was not even an appropriate choice of antibiotic for empirical treatment, as the drug should be highly effective against *S. aureus* and *Streptococci*, which are the most likely pathogens in this infection. It was further surmised that if appropriate surgical and medical therapy were instituted promptly, the prolonged illness could have been avoided and the outcome would be better.

12.4 Case 4: Complication of Breast Reduction Mammoplasty

A 38-year-old obese female (BMI = 35) with very large breasts consulted a plastic surgeon for breast reduction mammoplasty as she was suffering from chronic shoulder and upper back pain. There was no significant past illness, and surgery was arranged after explanation of the potential risks of the procedure. The surgery was performed at a community hospital with no intraoperative complications, and lasted 2½ h.

Three days post-operatively she developed pain and discharge from the wound. Prior to hospital discharge, the patient was given a prescription for azithromycin to be filled for signs of infection (redness and tenderness). On day 8 she visited the hospital ER with fever, chills, and progressive cellulitis of both breasts, with an area of necrosis on the right breast. Cultures were taken and the patient was discharged home on cephalexin 500 mg four times a day. Her symptoms progressed and she was readmitted to the ER 2 days later with fever, and bilateral cellulitis of the breast with a leukocyte count of 26,300 cells/uL. Intravenous penicillin and clindamycin were instituted; local debridement was performed in the ER and again the next day in the operating room. She required further extensive debridement 3 days later with evidence of a "gangrenous" loss of the majority of the skin over both breasts, and suppurative inflammation of the fat and soft tissue. The patient required continued debridement and skin grafting and was hospitalized for 3 weeks.

Due to residual deformities of both breasts, she was referred to another surgeon 4 months later for cosmetic surgery. This reconstructive procedure included a split abdominal tram flap with an adjustable volume saline implant, and lasted 4.5 h. Perioperatively, clindamycin 600 mg IV was given for prophylaxis. A week later, she was readmitted to hospital with evidence of infection and necrosis of the tram flap, requiring extensive debridement and drainage of pockets of necrotic fat with abscesses. Tissue cultures grew group B Streptococcus resistant to clindamycin and macrolides and so she was treated with intravenous cefazolin. She required multiple debridements and mesh split-thickness skin grafts to the chest wall and abdominal wound. Eventual healing with multiple scars occurred over 5–6 months.

12.4.1 Medico-legal Issues

Medical malpractice litigation was initiated by the patient against both plastic surgeons. Claims against the first surgeon included the following: (1) failure to take proper precautions to prevent infection, (2) delay in institution of appropriate treatment after development of breast infection, (3) failure to counsel and arrange early post-operative follow-up that would have detected early signs of infection. In the statement of claim, it was eluded that negligence of the surgeon resulted in her breast infection, deformities, pain and surgery, and the need for multiple surgeries.

Accusations against the second surgeon for negligence included: (1) failure to administer the proper prophylaxis to prevent infection, (2) inadequate disclosure about the risk of repeat infection, (3) inadequate close monitoring post-operatively for infection, as she had a history of previous infection, and (4) despite her previous infection, he should have not recommended a high risk procedure for infection with foreign body.

12.4.2 Medical Aspects

In 2000, it was estimated that at least 2 million women in the US had breast implants, and that close to 200,000 would be implanted every year.[27,28] Breast augmentation is the third most common type of plastic surgery done for cosmetic reasons in the US, after nose reshaping and liposuction. Breast reconstruction and implantation are used primarily for breast augmentation for aesthetic reasons, or following mastectomy or breast surgery for cancer.

Infection is the leading complication that occurs after breast augmentation surgery, in about 2.0–2.5%.[29] An international survey of 10,941 patients with breast augmentation reported infections in 2.5%, with 1.7% acute post-operative and 0.8 late infections.[30] Risk factors for infection after cosmetic breast surgery have not been carefully assessed by prospective studies but mainly by retrospective cohorts. The underlying clinical condition and surgical technique are the most important determinants.[29] Precise technique to prevent and eliminate hematoma and tissue ischemia is essential for excellent outcome.

Breast reconstruction after cancer surgery has a tenfold greater risk of infection than augmentation aesthetic procedures. Pre-existing tissue scarring and skin atrophy from previous surgery and tissue ischemia from previous radiation (endarteritis of small blood vessels) can increase the risk of infection to 24–53% in some series of breast implant for reconstruction.[29,31-33] Adjuvant chemotherapy was also associated with a higher rate of infection (10.7% vs. 1.5%) after immediate reconstruction.[34]

There is also evidence that immediate implant placement after mastectomy is associated with a higher risk of infection than delayed placement.[29] This may be related to introduction of endogenous bacteria from initial surgery and insertion of a prosthesis before clearance by the host's immune system.

The source of microbes causing infection after breast implants may include the patient's skin flora or mammary ducts, contaminated implants, contaminated saline, the surgical team or surgical environment, and later hematogenous seeding from a remote infection. Development of contact dermatitis at a surgical site from adhesive bandage, leading to skin and then implant infection by *Enterobacteriaceae* and *Pseudomonas* species have been reported.[35] Rare miscellaneous predisposing factors for breast implant infection include preceding infection, pyoderma, breast trauma or penetrating injury, breast massage and dental surgery.[29]

Most early infections appear within a month after surgery, a median of 10–12 days, but can range from 6 days to 6 weeks, depending on the virulence of the microorganism. The type of implant or surgical procedure does not seem to influence the rate of infection, but most infections for saline implants and breast expanders occur within 5–8 weeks; and with silicone breast implants, over half were reported after 26 weeks.[29] The clinical manifestations, acuteness of illness or tempo of the infection may vary with the organism. For instance, most patients presenting with typical symptoms of infection with wound erythema, tenderness, swelling and draining wound with or without fever, are most commonly due to *S. aureus* or coagulase negative *Staphylococci* (CONS), streptococci and occasionally gram-negative bacilli. However, patients may present very early (<4 days post-operatively) with signs of sepsis, but minimal or no obvious inflammatory changes of breast infection, due to toxin-producing *Streptococcus pyogenes* or *S. aureus*. There have been several cases of early *Staphylococcal* or *Streptococcal* toxic shock syndrome reported after breast implants.[29,36]

Late infection that occurs months or years after breast implantation is rare (0.8%), and usually results from secondary bacteremia and invasive procedures elsewhere. When late infection is defined as after 20 days post-surgery, it is more frequent that earlier infection 3.47% versus 2.08%, in a prospective study of 288 silicone gel implants.[26] The length of time to infection was shown to be bimodal and organism-related with *S. aureus* and streptococci presenting much earlier than *Enterobacteriaceae* species, and rare cases of atypical mycobacteria. These latter organisms are ubiquitous (found in water, soil, hospital water conduits and dust), and present with chronic draining sinus or with large fluid collection around the implant, with negative routine cultures.[24,25,37]

Appropriate management of breast implant infection should include aseptic aspiration of fluid collection under ultrasonography for gram-stain, acid-fast stain and cultures for aerobic, anaerobic and mycobacteria and fungi. Cytology or histology can also be useful to define the inflammatory reaction, such as presence of granulomas. Initial antibiotic should be guided by the gram-stain and suspected likely organism, then modified according to the culture and susceptibility results. Surgical removal of the implant is mandatory in most cases. In the presence of severe sepsis without an obvious source, surgical removal of the implant should be considered.[29]

Breast reduction mammoplasty infectious complication is no greater than any clean surgical procedure ($\leq$2.0%). The indications for mammoplasty are mainly for physical (neck and upper back pain), psychological or aesthetic reasons.[38]

Breast reduction surgery is performed much less frequently than breast augmentation procedure. Besides infection, the complications of mammoplasty include hematoma, nipple-areolar necrosis from reduction of blood supply to the nipple, problems with wound healing, but extensive flap necrois is rare, seroma, under-resection, asymmetry of the breast and ruckers ("dog ears" due to excess skin).[38]

12.4.3 *Discussion of Medico-legal Issues*

In Case 3, the risk of infection would be relatively high due to previous chemotherapy and adjuvant radiotherapy for breast cancer. Although antibiotic prophylaxis for breast surgery is controversial, most surgeons administer perioperative systemic cephalosporins for breast implant. Necrosis of the wound would also be higher from previous radiotherapy. Review of the management, however indicated suboptimal or substandard therapy. At the first sign of infection, a diagnostic aspiration should have been attempted with admission to the hospital for intravenous therapy and early surgical debridement. This approach may have shortened the course of illness and modified the outcome.

Concerning Case 4, there was no clear indication for antibiotic prophylaxis as there was no implant. However, the patient should have been admitted to hospital for intravenous and surgical therapy at the first ER visit for infection, rather than discharged on oral cephalexin. Had an aggressive approach been taken in the management of the infection, it was more likely than not, the infection could have been limited to the right breast, and resulted in an improved outcome.

12.5 General Comments

What can we learn from these four cases? Despite our best efforts, human error will always occur. We (as physicians) should strive to reduce these errors to as close to zero as possible. When error does occur, (i.e., accidental retention of a piece of gauze in a surgical wound) we should be forthright with our patients, apologize for any mistake, and accept responsibility for the adverse outcome. Often frankness and communication with patients or relatives will avert medico-legal lawsuit.

Physicians and surgeons should be more cognizant of the fact that foreign body-related infections can be subtle and insidious, and in the early stages, (when aggressive intervention is most effective) there may not be pus or overt signs of inflammation. Moreover, any drainage of the wound, whether purulent or not, should be sent for bacterial culture. All too often, physicians order tests and do not follow-up on the results. It is most important in patients with prosthesis or wounds overlying bone and major blood vessels that drain fluid (best obtained with cleansing of the skin surface with chlorhexidine), whether serosanguinous, blood or clots, should have stat gram stain, and the results of cultures be checked in 24 h and again in 48–72 h. If the gram stain reveals pus cells with visible bacteria, antibiotics should be started based on the results of gram-stain morphology, and either admit the patient for deeper surgical drainage and debridement, or reassess the patient within 48–72 h. On reassessment, the identity and susceptibility of the bacteria should be available and the need for intravenous or antibiotic modification can be made. The need for further surgical debridement and readmission to hospital should then be assessed.

It should be recognized and appreciated by all physicians that late post-operative wound drainage usually implies deep wound infection and would be rarely be of a superficial origin. Whenever there is an operation on bone or entry into bone by a screw, or even close proximity to bone, osteomyelitis should be excluded. The best imaging techniques for diagnosis and delineation of the extent of infection are MRI and CT scan.

Prompt attention to the patient's post-operative complications are essential, and it is imperative in infections to aggressively utilize combined medical and surgical therapy early to limit the spread and possible metastatic seeding of the infection. Delayed diagnosis and tardy institution of appropriate management are major reasons for bad outcomes and litigations.

References

1. Wysenbeek AJ, Volchek J, Amit M, Robinson D, Boldar I, Nevo Z, (1998). Treatment of staphylococcal septic arthritis in rabbits by systemic antibiotics and intra-articular corticosteroids. *Ann Rheum Dis* 57:687–690.
2. Matarasso A, Swift RW, Rankin M, (2006). Abdominoplasty and abdominal contour surgery: national plastic surgery survey. *Plast Reconstr Surg* 117:1797–1808.
3. Alderman AK, Collins ED, Streu R, Grotting JC, Sulkin AL, Neligan P, Haeck PC, Gutowski KA, (2009). Benchmarking outcomes in plastic surgery: national complication rates of abdominoplasty and breast augmentation. *Plast Reconstr Surg* 124:2127–2133.
4. Byrd HS, Barton FE, Orenstein HH, Rohrich RJ, Burns AJ, Hobar PC, Haydon MS, (2003). Safety and efficacy in an accredited outpatient plastic surgery facility: a review of 5316 consecutive cases. *Plast Reconstr Surg* 112:636–641.
5. Marello DC, Colon GA, Fredricks S, Iverson RE, Singer R, (1997). Patient safety in accredited office surgical facilities. *Plast Reconstr Surg* 99:1496–1500.
6. Fatica CA, Gordon SM, Zins JE, (2002). The role of preoperative antibiotic prophylaxis in cosmetic surgery. *Plast Reconstr Surg* 109:2570–2573.
7. Drapeau CM, D'Aniello C, Brafa A, Nicastri E, Silvestri A, Nisi G, Petrosillo N, Italian Group of Surgical Infections in Plastic Surgery, (2007). Surgical site infections in plastic surgery: an Italian multicenter study. *J Surg Res* 143:393–397.
8. Wong ES, (2004). Surgical site infections. In (ed): Mayhall CG. Hospital Epidemiology and Infection Control, 3rd Edition, Lippincott, Williams &Wilkins, Philadelphia, p287–310.
9. Kiska DL, Gilligan PH, (1999). Pseudomonas. In (eds): Murray PR, Baron MA, Tenover FC, Yolken RH,: Manual of Clinical Microbiology, 7th Edition, *Am Soc Microbiol,* ASM Press, Washington DC, p517–525.
10. Pier GB, Ramphal R, (2005). *Pseudomonas aeruginosa.* In (eds): Mandell GL, Bennett JE, Dolin R. Principles and Practice of Infectious Diseases, 6th Edition, Elsevier/Churchill/ Livingstone, Philadelphia, p2587–2615.

11. Andenaes K, Lingaas E, Amland PF, Giercksky KE, Abyholm F, (1996). Preoperative bacterial colonization and its influence on post-operative wound infections in plastic surgery. *J Hosp Infect* 34:291–299.
12. Codner MA, Ford DT, (2007). Blepharoplasty. In: (eds) Thorne CH, Beasley RW, Aston SJ, Bartlett SP, Gurtner GL, Spear SC: Grabb & Smith's Plastic Surgery, 6th Edition Wolters Kluwer/Lippincott, Williams & Wilkins, Philadelphia, p486–497.
13. Paul M, (2001). The evolution of the browlift in aesthetic plastic surgery. *Plast Reconstr Surg* 108:1409–1244.
14. Oslin B, Cone GB, Vasconez LO, (1995). The biplanar endoscopically assisted forehead lift. *Clin Plast Surg* 22:633–638.
15. Seabold JE, Simonson TM, Weber PC, Thompson BH, Harms KG, Rezai K, Madsen MT, Hoffman HT, (1995). Cranial osteomyelitis: diagnosis and follow-up with In-III white blood cell and Tc-99m methylene diphosphonate bone SPECT, CT, and MR imaging. *Radiol* 196:779–788.
16. Anslow P, (2004). Cranial bacterial infection. *Eur Radiol* 14:E145–E154.
17. Miyata I, Masaoda T, Nishiura T, Harada Y, Ishimitsu H, (1990). Infection after cranioplasty following decompressive craniotomy. IRYO-Japan. *J National Med Services* 44:213–301.
18. Fialkov JA, Holy C, Forrest CR, Phillips JH, Antonyshyn OM, (2001). Post-operative infections in craniofacial reconstructive procedures. *J Craniofacial Surg* 12:362–368.
19. Spacey YJE, Bettens RMA, Mommaerts MY, Adriaen SJ, Van Landugt HM, Abeloos JVS, De Clerq CAS, Lamoral PRB, Neyt LR, (2005). A prospective study of infectious complications in orthognathic surgery. *J Cranio-Maxillofacial Surg* 33:24–29.
20. Toy BR, Frank PJ, (2003). Outbreak of *Mycobacterium abscessus* infection after soft tissue augmentation. *Dermatol Surg* 29:971–973.
21. Nayao M, Sonole M, Bando T. Saito T, Shirano M, Matsushima A, Fujiharn N, Takahura S, Linuma Y, Ichiyama S, (2009). Surgical site infection due to *Mycobacterium peregrinum*: a case report and literature review. *Internat J Infect Dis* 13:209–211.
22. Mauriello JA Jr, Atypical Mycobacteria Study Group, (2003). Atypical mycobacterial infection of the periocular region and periocular facial surgery. *Ophthal Plast Reconstr Surg* 19:182–188.
23. Reynier S, Cambau E, Meningaud JP, Guihot A, Deforges L, Carbonne A, Bricaire F, Caumes E, (2009). Clinical management of rapidly growing mycobacterial cutaneous infections in patients after mesotherapy. *Clin Infect Dis* 49:1358–1364.
24. Sampaio JLM, Chimara E, Ferrazoli L, da Silva Telles MA, Del Guercio VMF, Jerico ZVN, Miyashiro K, Fortalezo CMCB, Padoveze MC, Leao SC, (2006). Application of four molecular typing methods for analysis of *Mycobacterium fortuitum group* strains causing post-mammoplasty infections. *Clin Microbiol Infect* 12:142–149.
25. Leonard AL, Brown LH, (2004). Atypical mycobacterial outbreaks associated with cosmetic surgery and aesthetic procedures. *Cosmet Dermatol* 17:636–640.
26. Basile AR, Basile F, Basile AVD, (2005). Late infection following breast augmentation with textured silicon gel-filled implants. *Aesth Surg J 25*:249–254.
27. Herdman, RC, Fahey TJ Jr, (2001). Silicone breast implants and cancer. *Cancer Invest* 19:821–832.
28. Bondurant S, Ernster V, Herdman RC, (eds) (1999). Safety of silicone breast implants. National Academic Press, Washington DC.
29. Pittet B, Montandon D, Pittet D (2005). Infection in breast implants: *Lancet Infect Dis* 5:94–106
30. De Cholnoky T (1970). Augmentation mammoplasty. Survey of complications in 10941 patients. *Plast Reconstr Surg* 45:573–577.
31. Gabriel SE, Woods JE, O'Fallon WM, Beard CM, Kirland LT, Melton LJ, 3rd (1997). Complications leading to surgery after breast implantation. *N Engl J Med* 336:677–682.
32. Kjøller K, Hölmich LR, Jacobsen PH, Friss S, Fryzek J, McLaughlin JK, Lipworth L, Henriksen TF, Jørgensen S, Bittmann S, Olsen JH (2002). Epidemiological investigation of

local complications after cosmetic breast implant surgery in Denmark. *Ann Plast Surg* 48:229–237.

33. Nahabedian MY, Tsangaris T, Momen B, Manson PN (2003). Infectious complications following breast reconstruction with expanders and implants. *Plast Reconstr Surg* 112:467–476.
34. Vandeweyer, E, Deaemaker R, Nogaret JM, Hertens D (2003). Immediate breast reconstruction with implants and adjuvant chemotherapy: a good option? *Acta Chir Belg* 103:98–101.
35. Brand KG (1993). Infection of mammary prosthesis: a survey and question of prevention. *Ann Plast Surg* 30:289–295.
36. Holm C, Muhlbauer W (1998). Toxic shock syndrome in plastic surgery patients: case report and review of the literature. *Aesthetic Plast Surg* 22:180–184.
37. Haiavy J, Tobin H (2002). *Mycobacterium fortuitum* infection in prosthetic breast implants. *Plast Reconstr Surg* 109:2121–2128.
38. Spear SL (2007). Breast reduction: inverted T-technique. In: (eds) Thorne CH, Beasley RW, Aston SJ, Bartlett SP, Gurtner GC, Spear SL Grabb and Smith's Plastic Surgery, 6th Edition. Walters Kuwer/Lippincott Williams and Wilkins, Philadelphia, p593–615.

Chapter 13
Litigations for HIV Related Complications

13.1 Case 1: I Have HIV Infection

In 1992, a 27-year-old male with same sex exposure requested human immunodeficiency virus (HIV) testing anonymously at a walk-in clinic. He was advised that the test (HIV serology) was positive and he requested a repeat test (anonymously) 1 month later, which was also reported as being positive. About 2 years later, he was assessed by a general practitioner for symptoms of depression and continued medical care. At that time, investigations revealed a CD4 T-cell count of about 700 cells/uL. Sometime in 1996 a repeat blood test revealed a CD4 cell count just <500 cells/uL. No consultation to an infectious diseases specialist or HIV clinic was made. The GP(general practitioner) then initiated a regimen consisting of didanosine, lamivudine, and saquinavir for HIV infection. At that time, testing for HIV viral load was not generally available to the medical community, but became procurable in 1997. Initially, the patient tolerated the regimen well and over the next 3 years his CD4 cell count was maintained above 600–700 cells/uL and the HIV viral load remained undetectable (<50 copies). However, the patient started to show morphologic changes of moderate facial and peripheral lipoatrophy, developed mild sensory peripheral neuropathy, and increased liver enzymes attributable to fatty liver, and elevations of the fasting serum glucose. In the summer of 2000, although the CD4 cell count remained stable, the HIV viral load was reported as being over 7,000 copies/uL. At this time, the patient was referred to a university hospital HIV clinic.

At the HIV clinic, the HIV viral load and serology were repeated (as the GP never had a documented test result). The test results revealed undetectable (<50 copies) HIV-RNA and both the HIV antibody screen (ELISA) and Western blot for HIV-1 and HIV-2 were reported negative by the reference laboratory. The anti-HIV medications were discontinued and the tests repeated a month later with similar results. A special PCR (polymerase chain reaction) was performed for HIV-1 proviral DNA that was undetectable. It was finally concluded that the patient did not suffer from HIV infection, and although there was some improvement in his drug-related complications, after 6–12 months he was left with some residual abnormalities. Further investigation by the laboratory that reported an HIV viral

I.W. Fong, *Medico-Legal Issues in Infectious Diseases*, Emerging Infectious Diseases of the 21st Century, DOI 10.1007/978-1-4419-8053-3_13,

load of >7,000 copies/uL came to the conclusion that there was either a mix-up in the blood specimen samples or error in the labeling or reporting. Efforts to verify or clarify the initial HIV serology were unsuccessful as no permanent records were kept for anonymous HIV serology results.

13.1.1 Medico-legal Issues

The patient (plaintiff) initiated litigation against the GP (defendant) for medical malpractice. Specific charges were: (1) the GP should have repeated the HIV serology to confirm that the plaintiff was HIV infected, (2) the defendant was negligent in starting treatment for HIV infection without proof of disease, (3) the physician lacked knowledge of HIV infection and should have referred the patient to a specialist or HIV clinic, (4) treatment of toxic medications were given for several years without any clear indication, and (5) the GP did not adequately inform the patient on the pros and cons of therapy, nor explain the potential toxicities and side-effects.

Financial compensation by the plaintiff was sought for psychological suffering over the years with the false impression that he was HIV infected, and physical suffering from the side effects of the medications and the need to take unnecessary large amounts of pills for several years. The side effects had affected his social life and left a permanent physical stigma, and also adversely affected his performance at work (due to absenteeism from adverse events). The latter had resulted in his inability to perform at a high level and thus retarded his progress in his career path. All these effects have indirectly affected his earning ability over 3–4 years, and also future earning capacity.

13.1.2 Medical Aspects

The present AIDS pandemic is caused by the HIV-1 strain and HIV-2 is predominantly found in West and Central Africa but is rare in developed nations. Seroconversion after exposure usually occurs within 2 weeks to 3 months, but occasionally may take 12 months or longer.[1] Delayed or protracted time for seroconversion may be seen especially in immunosuppressed subjects.[2] Usually by 6 months after exposure, seroconversion should occur in 95% or more of cases.[3] A period of viremia and antigenemia without detectable antibodies occurs within 4–6 weeks of initial HIV infection. At this phase, high levels of plasma p24 antigen or viral RNA can be detected, and the viremia and antigenemia decline to very low levels coinciding with seroconversion.

Detection of antibodies to HIV remains the most cost-effective and commonly used method to prove HIV infection. Enzyme-linked immunoassay (ELISA) is the most commonly used assay to test for HIV-1 and HIV-2 because of its low cost, standardized procedure, reliability, and rapid turnaround.[1] For experienced laboratories under optimal conditions (commonly licensed kits) the sensitivity and specificity of the ELISA are both 99%.[1] False negative reactions can occur in infected persons early in the course before seroconversion and in immunosuppressed patients. False positive ELISA results can occur for various reasons, including human error, variability in the test kits, hemodialysis, auto-immune disease, multiple myeloma, hemophilia, alcohol hepatitis, positive rapid plasma regain (RPR) test, and for unknown reasons (idiopathic).[1] The ELISA uses HIV antigen to bind IgG HIV antibodies in the test sample.

The Western Blot test (WB) is the most commonly used confirmatory test for the presence of HIV specific antibodies. Compared to ELISA, the WB is more expensive, time-consuming and requires more technical expertise to interpret. False negative WB can also occur in the very early phase of HIV infection before development of antibodies. False positive reactions can occur in auto-immune disorders, polyclonal gammopathies, hyperbilirubinemia, subjects with human leukocyte antigen (HLA) antibodies, and healthy individuals. In low risk populations, the chance of false-positive reaction of ELISA and WB combined is extremely low – 1 in 135,000.[4] The probability of another test being false positive in the same person tested at another time for both tests would be 1:135,000 $\times$ 135,000 or 1 in 18 billion chance.

Although the polymerase chain reaction (PCR) can be used to detect the HIV genome before antibody production, the PCR is highly prone to contamination with nucleic acids, which causes many false-positive reactions and therefore has not been recommended for diagnostic purposes. These PCR tests are thus mainly used for serial measurements of plasma HIV-1 RNA for quantitation over the range of 75–500,000 RNA copies/mL to monitor progress and response to therapy. In high risk populations, detection of HIV-DNA by PCR has been found to have false-positive rates of 2–3.4%.[5,6] Data from Bayer on the versant HIV-1 RNA 3.0 assay (bDNA) found that all 22 of 912 false-positive samples quantitated were 1,000 copies/mL or less (personal communication with Dr. R. Ziermann from Bayer).

13.1.3 Medico-legal Discussion

The main issue in this case is related to acceptance of a patient's history of a serious disease (from a test performed elsewhere) without verifying the results. Although physicians commonly accept the history of a patient's underlying illness as valid, treatment for a disorder with potentially toxic agents should always require verification of the diagnosis. It could therefore be argued that the GP was remiss in instituting a cocktail of medications without having a confirmed copy of the test result for HIV infection. This is particularly damaging for an asymptomatic subject with no history of opportunistic infection or clinical evidence of AIDS complication. Moreover, a CD4 count cannot be used as a surrogate marker for the diagnosis of HIV infection.

Although the $CD4^{+}$ T lymphocyte quantitative count is a very useful and standard test to monitor patients for progression of HIV disease or response to therapy, it can be low in many conditions. The normal $CD4^+$ T lymphocyte count usually averages 8.0–10.5 × 10^8 cells/L (800–1,050/mm^3), but the range of normality (2 standard deviations of the mean) is quite wide (500–1,400 cells/mm^3).[7] About 80% of the normal blood lymphocytes are T lymphocytes and nearly two-thirds of blood T lymphocytes are $CD4^+$ (helper) lymphocytes and most patients with lymphocytopenia have reduction in absolute number of $CD4^+$ T lymphocytes.[8] There are many conditions that can be associated with lymphocytopenia and lower than normal $CD4^+$ lymphocyte count. Although HIV infection is the most common viral infection associated with $CD4^+$ lymphocytopenia, other viral infections can transiently decrease $CD4^+$ cell counts (including measles, corona viruses and others).[8] The list of conditions associated with $CD4^+$ lymphocytopenia (besides viral infections) include bacterial and fungal sepsis (including tuberculosis), major surgery, recent trauma or hemorrhage, malignancy, glucocorticoid use, cytotoxic chemotherapy, radiotherapy, auto-immune diseases, nutritional deficiencies, organ transplantation, acquired common variable immunodeficiency, and idiopathic $CD4^+$ lymphocytopenia.[8] Furthermore, it is common to observe biologic variations in the absolute $CD4^+$ cell count even in HIV infected subjects without any other factors. A healthy adult may have, at some time, transient decrease in $CD4^+$ cell count below 500. Whether the plaintiff's $CD4^+$ cell count decline was due to viral upper respiratory tract infection or other causes was not clear. In HIV-infected patients, the T lymphocytes decline by 4% per year for every $\log_{10}$ HIV RNA copies/mL in the plasma.[7]

Currently, the optimal time to start antiretroviral therapy (ART) for asymptomatic HIV patients is not clear. There is consensus that patients with AIDS complications or symptomatic disease should be started on ART. There is still controversy as to the optimal time to initiate ART in asymptomatic patients. Some guidelines recommend considering starting ART below 350 cells/mm^3 and others recently <500 cells/mm^3, but there are no randomized controlled trials to provide a clear answer.

How can we resolve the issue of two HIV serology tests taken at separate times in the same subject being false positive? There are several possibilities, none of which can be proven in or out of court. It is possible, that since blood samples taken in the clinic were labeled with a code number to provide anonymity, that the samples were mislabeled and originated from a truly HIV infected subject. However, the chance of that occurring twice in a row would be extremely low or unlucky. It is also possible that the plaintiff suffered from a mental disorder or delusion (such as Munchausen's syndrome) and imagined that he had a positive HIV serology. There was no indication of a psychiatric disorder from the GPs office records. Rare false claims of a medical disease (including HIV infection) may be encountered under unusual conditions where the person can expect some form of material gain, i.e., financial, improved living conditions, sympathetic reduction in sentences for criminal offenses, etc. None of these appeared evident from review of the records.

Feigned HIV infection has been reported in malingering patients[9–11] and in young women with psychosocial disorders with history of prolonged sexual, physical and emotional abuse.[12] A retrospective study from an HIV clinic in a municipal hospital identified seven patients with fictitious HIV infections, six of whom had a history of illicit narcotic abuse.[13] A survey of ten other local hospitals found that known cases of alleged (fictitious) HIV infection occurred at eight of the hospitals but only one of the ten hospitals routinely documented HIV infections before initiating care.[13] In a specialist HIV unit in Central London over a 5-year period, 12 patients (1.7% of admissions) with feigned HIV/AIDS were identified.[14]

13.2 Case 2: Missed Opportunities

A young man, aged 36 years, presented to a new family physician (FP) in 1994 with symptoms of 15 lb weight loss, chronic diarrhea for 3 weeks and night sweats. He was found to be unwell, with evidence of significant weight loss from wasting, oral thrush, and oral hairy leukoplakia. An HIV serology performed was positive (screen and confirmatory), and his CD4$^+$ lymphocyte count was 80 cells/mm^3. He was started on antiretrovirals and prophylaxis for pneumocystic pneumonia.

The patient was a practicing homosexual with multiple partners (none of whom were known HIV infected) and he used condoms sometimes, but inconsistently for sexual encounters. He had no known past medical illness and claimed to have previously tested negative for HIV infection in 1988. He claimed to have requested an HIV test in 1990 but there was no record of this in his previous FP records. An HIV serology was performed in the summer of 1993 (which was positive), but the patient was never informed of the results and apparently was lost to follow-up.

The records subsequently in 2000 indicated that the young man was attending an HIV clinic regularly with no opportunistic infection and was clinically stable on a combination of ART with a stable CD4$^+$ lymphocyte count of 160 cells/mm^3 and undetectable HIV (<50 copies).

13.2.1 Medico-legal Issues

The patient in 2000 initiated lawsuit against his original FP for medical malpractice. The charges against the physician were that he failed to perform an HIV test in 1990 despite the plaintiff's request and he was negligent in failing to notify the plaintiff of the result of the HIV serology in 1993. These acts of negligence by the defendant resulted in delay in the diagnosis and treatment, thus allowing his HIV infection to progress to AIDS. Furthermore, failure on the part of the defendant had resulted in missed opportunities to start earlier treatment, and the delay in initiation of ART resulted in a decrease in his expected life span and affected the quality of his life.

The defendant countered that there was no record of the plaintiff ever requesting an HIV test in 1990. Furthermore, the plaintiff never kept the appointment after the positive HIV serology in 1993 to be notified of the result. Moreover, since the plaintiff was subsequently lost to follow-up, he never had a chance to counsel him on HIV disease or institute treatment.

13.2.2 Medical Aspect

Following acute HIV infection, about 50–70% of subjects develop clinical symptoms of variable severity, from mild flu-like illness to aseptic meningitis.[15] There is also evidence that severity and duration of clinical acute illness of a primary infection is of prognostic importance. The risk of developing AIDS within 3 years of seroconversion in subjects who were asymptomatic or had mild illness was only 10% versus 78% (eight times greater) in those with seroconversion illness of at least 14 days.[16] Peak viral replication soon after infection occurs in 2–4 weeks and levels of virus can exceed $>10^7$ copies/mL in plasma. This is associated with a dramatic drop in circulating $CD4^+$ lymphocytes, then a slowing of T-cell loss, and rebound by 9–12 weeks. This rebound of $CD4^+$ cells corresponds to a decline in viral load, which reaches a steady state (set point), which is variable by 9–12 weeks. Clinical progression of HIV disease has been tied to a set point level with lower levels associated with better prognosis.[17] At 1 year after seroconversion, patients demonstrate a fall of about 349 $CD4^+$ cells/mm^3 (mean baseline 999 cells/mm^3), followed by a more gradual decline in $CD4^+$ cells in the later period of infection.[18] There is usually a variable period of 8–10 years span before patients develop AIDS (about 50%). The typical HIV-infected person shows a progressive decline of $CD4^+$ lymphocytes (50–100 cells/year) over time.

Long-term non-progression or elite controllers represent <5% of HIV-infected subjects who maintain relatively normal $CD4^+$ cell count and very low or immeasurable viral load for 8 years to decades without therapy.[19] This is a heterogenous group of elite controllers whose benign course may result from robust immune responses against HIV, or defective poorly replicative virus secondary to deletion of the nef gene.[20,21] There is evidence that host factors that influence the course of HIV disease are correlated to polymorphisms dominating the HLA region, with class I polymorphism dominating the HLA associations.[22] There is also evidence from studies in the United States and Europe that HLA-B57 and HLA B-27 are strongly associated with long term survival or non-progression.[22] Not all long-term non-progressors are "elite or viremic controllers" (patients with undetectable HIV or plasma HIV RNA levels of 50–2,000 copies/mL). These patients with $CD4^+$ cell counts of >500 cells/mm^3 for at least 10 years most often had HIV RNA levels of >2,000 copies/mL, but had significantly lower HIV RNA than subjects with typical progression. Thus plasma set point HIV RNA levels explain <50% of the variability in rates of clinical progression.[24]

The chemokine receptor 5 (CCR5) protein serves as a co-receptor on $CD4^+$ lymphocytes for certain strains of HIV-1. Homozygosity for a 32-base pair deletion allele (CCR5Δ32) protects against HIV infection (1% of Caucasians) and heterozygosity (individuals with one allele) show a decreased progression to AIDS.[25] There is also evidence that co-infection with GB virus (GBV-C), a flavivirus not known to cause disease (in subjects with GBV-C viremia), have slower progression and slower decrease in $CD4^+$ cell counts than those without GBV-C infection.[23] Although the reasons for this protective effect are unclear, there is evidence that GBV-C inhibits HIV replication in peripheral blood mononuclear cells in vitro, and since GBV-C infects $CD4^+$ cells, this may compete with the HIV for target cells for infection.[23]

A minority of patients with HIV infection can rapidly progress to AIDS within 1–3 years of their infection. This may be related to host genetic factors and age at the time of infection and other extrinsic conditions. Older age at the time of infection (>25 years) has been associated with faster progression of the disease in hemophiliacs and older homosexuals.[23] Concomitant co-infection with cytomegalovirus (CMV) has been associated with rapid progression to AIDS in hemophiliacs and others.[26,27] Co-infection with HTLV-I may increase the risk for development of AIDS while HTLV-2 can delay the progression of disease.[23,28,29] Active tuberculosis can also enhance HIV replication and cause rapid progression to AIDS.[30] However, although treatment of active tuberculosis for 6 months is associated with increased $CD4^+$ cell count, it does not markedly affect the HIV viral loads.[31] The role of hepatitis C-virus (HCV) co-infection on the progression of HIV disease has been conflicting, with some studies showing more rapid progression, but others have found no effect on the development of AIDS.[23]

The clade or strain of HIV-1 may play a role in the course of the disease. Clade D of HIV-1 is associated with faster progression to death in Africa than Clade-A and B.[32] Women in Senegal infected with C, D or G HIV Clade were eight times more likely to develop AIDS than those infected with Clade A subtype (the predominant sub-type).[33] Infection with multiple strains of HIV-1 (more common in women) have also been associated with faster disease progression.[34]

Socio-economic factors such as poverty, homelessness, drug and alcohol abuse, and black race play indirect roles in the prognosis and disease progression of HIV infection, primarily through lower access to medical management, delay in instituting antiretrovirals and poorer compliance with medications. Although a previous study found that alcohol and psychoactive drugs did not accelerate HIV disease,[35] there is in vitro evidence that alcohol, cocaine and narcotics can impair the immune response to HIV-1 and allow enhanced replication in peripheral blood mononuclear cells.[23] A case report of rapid progression to AIDS within a year of infection has also been attributed to alcoholism.[36]

13.2.3 Medico-legal Discussions

It would appear that the plaintiff became infected with HIV sometime between 1988 (reported HIV-negative) and 1993 (first noted HIV-positive). However, by 1994 he had progressed to symptomatic AIDS. Thus, his course was more rapid than usual HIV infected individuals were, and especially as no other conditions or factors were recognized that could accelerate his course of disease. However, the failure of the FP to notify the patient of his HIV-positive status before recognition of his condition was only 1 year. Would an earlier diagnosis by 1 year and assuming institution of ART then, affected the outcome as to lifespan and quality of life? Appropriate treatment a year earlier with ART would likely have aborted or ameliorated his symptomatic disease, of weight loss, diarrhea, and malaise. However, it is less clear whether his expected lifespan would be any greater. If we assume that over the preceding year his $CD4^+$ cell count probably declined by 50–100 cells/mm^3, then even at that time he would have already progressed to AIDS ($CD4^+$ cell count <200 cells/mm^3). There is reasonable good cumulated evidence that starting therapy when the $CD4^+$ cell count is very low (<200 cells/mm^3) is associated with less chance of immune reconstitution and greater risk of opportunistic complications than those started on ART when the $CD4^+$ cell count was >200 cells/mm^3. The optimum $CD4^+$ cell count for initiating ART has not been well established although recent large observational cohort studies (retrospective and prospective) and suggests better outcomes for HIV infected patients receiving earlier ART with $CD4^+$ cell count ≥350 cells/mm^3 or >500 cells/mm^3; the data however is flawed and controversial.[37,38] Lack of randomization in these studies could result in significant biases as motivated, health-conscious individuals would likely do better that those less motivated. It is not clear in these studies as to the cause of excess mortality in those not accepting treatment. For instance, it would be expected and predictable to find excess mortality (from any disease) in marginalized people (homeless, alcoholics, drug abusers), who are less likely to start ART, which may be unrelated to HIV complications, such as suicides, homicide, accidents, drug overdose, liver failure or other diseases more prevalent in these groups (diabetes mellitus, cardiovascular disease, chronic lung disease and cancer).

The defendant denied the plaintiffs claim that earlier HIV test (in 1990) was requested. It could be argued, however, that the FP should have been doing regular HIV serology in an individual that belongs to a high-risk group (with the patient's consent). The Center for Disease Control and Prevention (CDC) estimate that nearly 50% of men who have sex with men (MSM) with HIV infection are unaware of their status. The CDC National Behavior Surveillance System of high-risk venue-based recruitment found 25% of MSM tested to be infected with HIV, and nearly 50% of the HIV infected individuals were unaware of their infection.[39] In New York City, the HIV incidence rate among MSM was 2.3%, with 52% of those infected being unaware of their HIV seropositivity.[39] It is estimated that 21% of HIV-infected people in the US who are unaware of their infection may account for up to 52% of new infections. CDC HIV testing guidelines recommend annual testing for high-risk populations (including MSM), and since 2006 have recommended universal opt-out HIV screening in all health care settings.[40] Thus, the plaintiff could argue that the defendant fell below the standard of care by not recommending and performing annual HIV tests. Furthermore, if he were found to be HIV seropositive earlier, (by 1990 or before) with careful monitoring and institution of ART before his $CD4^+$ cell count fell <350 cells/mm^3, his quality of life and life expectancy would be greater.[41]

What is the effect of life expectancy with late treatment initiation for HIV disease? In a recent study using a state-transition model of HIV disease, the projected life expectancy of HIV uninfected and HIV infected persons with similar risk profiles were compared.[41] Those with HIV infection lost 11.92 years of life if they received care concordant with guidelines and late treatment initiation resulted in 2.60 additional years of life lost (greatest for Hispanics [3.90 years]).[41]

13.3 Case 3: Visual Impairment in HIV

An infectious disease (ID) specialist/internist was consulted to assess a 41-year-old male with mild pancytopenia and a past history of bilateral pneumonia the year before. The patient had a history of multiple sexual contacts with prostitutes 5 years prior and had refused HIV testing the year before when he developed pneumonia (which was suspected to be pneumocystic pneumonia [PCP]). At this office visit, he agreed to an HIV test and a $CD4^+$ cell count. The patient was called for a return appointment to discuss the results of the test a month later, but this appointment was cancelled by the patient for personal reasons. The blood test revealed the patient was HIV seropositive with a very low $CD4^+$ cell count of 5 cells/mm^3, but the results were not given over the phone or by mail. Thus, the subject remained unaware of his HIV status and severe immune deficiency.

About 3 months later, the patient attended an optician for blurred vision and he was referred to a hospital ER for an ophthalmologist consultation. He was briefly assessed by the attending ER physician, but due to the long waiting period pending full eye assessment, he left prematurely. The patient arranged an appointment with the ID specialist in the ER of the suburban community hospital. The subject was told of his HIV status and a brief retinal examination (without pupillary dilatation) by the ID physician revealed no abnormality. An appointment was arranged for another office visit to the ID specialist to discuss HIV therapy in 2 weeks. One week later the subject returned to the ER with respiratory symptoms and poor vision. He was admitted as possible PCP under the care of the ID physician, but no eye examination was performed. A week after his admission to hospital, a neurologist who was consulted found very poor vision with light perception only in the right eye and finger counting on the left eye. Fundoscopy revealed bilateral chorioretinitis and ophthalmology consultation was requested, but treatment of CMV retinitis was only instituted 2 days later. His course was complicated by retinal detachment secondary to CMV retinitis with almost complete blindness in the right eye and severe visual impairment of the left eye – legally blind.

13.3.1 Medico-legal Issues

Malpractice litigation was brought by the patient against the ID physician and the admitting ER physician of the hospital. The charges against the ID consultant were: (1) failure to notify the plaintiff of his HIV status and seriousness of his condition,

(2) failure to do a proper eye examination or refer him to an ophthalmologist when he was first seen in the ER a week before his admission, (3) failure to do a fundoscopy or arrange urgent ophthalmology consultation on admission to the hospital, (4) delay in starting appropriate treatment for CMV retinitis even after the neurologist findings were consistent with the diagnosis.

The claim filed against the ER physician was for neglect in performing an eye examination, despite the patient's symptoms of poor vision and failure to require an urgent ophthalmology consultation. That prompt recognition of CMV retinitis and immediate institution of antiviral therapy could have resulted in better visual result. Failure of the ID physician to inform the plaintiff of the seriousness of his condition, even by phone, could have resulted in prevention of visual loss and admission to hospital if treatment with ART and PCP prophylaxis were started 3 months before his hospital admission.

The defendant (ID specialist) countered that it was the plaintiff who canceled the follow-up appointment for counseling on his condition, and it was neither his policy nor the recommended standard to discuss these issues on the phone. Therefore, failure to initiate earlier ART before the AIDS complications was due to the fault of the plaintiff. Furthermore, his eye examination performed at the first ER visit revealed no abnormalities.

13.3.2 Medical Aspects

Visual complaints in HIV infected persons can be unrelated (as in normal people) or related directly to complications of AIDS or indirectly due to medications. Ocular manifestations are common in people with AIDS, and before the advent of highly active ART (HAART), the majority of patients with AIDS developed some ocular involvement at some time.[42] The most frequent ocular abnormality was usually silent or asymptomatic and occurred in nearly 50% of AIDS patients before the era of HAART – HIV microangiopathy, consisting of cotton wool exudates, and less frequently hemorrhages.[42] Occasionally HIV retinopathy could present with visual impairment from larger branch vein or central retinal vein occlusion.

The most dreaded ocular complications of AIDS were from opportunistic ocular infections (CMV retinitis, herpes zoster (VZV) retinitis, or herpes zoster ophthalmicus, toxoplasma retinitis and ocular syphilis), or neoplasm (Kaposi sarcoma of the lids and conjunctivae, and orbital or intraorbital lymphoma).[42]

CMV retinitis is the most frequent sight threatening ocular complication of AIDS, occurring in the late stages when the $CD4^+$ lymphocyte counts <50 cells/μL. In the pre-HAART era CMV retinitis occurred in 30% of patients with AIDS, and the number of new cases has dramatically fallen since widespread use of HAART by 55–95% (average 80%).[42] The incidence of CMV retinitis among patients with $CD4^+$ cell count <100 cells/μL was 10% per year and for many patients with $CD4^+$ cell count <50 cells/μL, it was 20% per year. Symptoms of CMV retinitis include floaters, flashing lights, loss of visual field, or visual loss. In the early stages with small peripheral retinal lesions patients can be asymptomatic and 13–15% of persons with $CD4^+$ cell count ≤50 cells/μL have asymptomatic CMV retinitis.[43] Lesions adjacent to the optic nerve or fovea (posterior pole of the retina or macula) are immediately vision threatening. The retina has been divided into three zones for clinical assessment of risk to vision. Zone 1 lies within 1,500 μm from the edge of the optic nerve, zone 2 extends from the edge of zone 1 to the equator of the eye, and zone 3 extends from the equator to the pars plana (pigmented posterior zone of the ciliary body). See Fig. 13.1 for the schematic diagram of the zones of the retina. Lesions of zone 1 are immediately sight-threatening and require urgent treatment, whereas lesions of zones 2–3 may be observed for short periods of time without risk of loss of visual acuity.[42] The mean time for progression of peripheral lesions without treatment was found to be 22 days (enlargement to uninvolved retina by ≥750 μm in width).[44] The complications of untreated or delayed treatment of CMV retinitis include impaired vision to blindness, secondary to progressive retinitis with hemorrhages, scarring and retinal detachment. In the pre-HAART era, retinal detachment in CMV retinitis occurred in 25% at 6 months and in 50–60% at 1 year.[42]

The diagnosis of CMV retinitis can be made reliably by an experienced ophthalmologist by dilated direct or indirect ophthalmoscopy. Examination of the fundus through an undilated pupil is inadequate to diagnose or exclude CMV retinitis as only 10% of the retina can be evaluated.[42] The aim of treatment with anti-CMV drugs (ganciclovir intravenously or oral valganciclovir) is to arrest progression of the disease, prevent further spread, and preserve vision. Treatment with anti-CMV agents does not eradicate the virus but delays progression and relapse, until

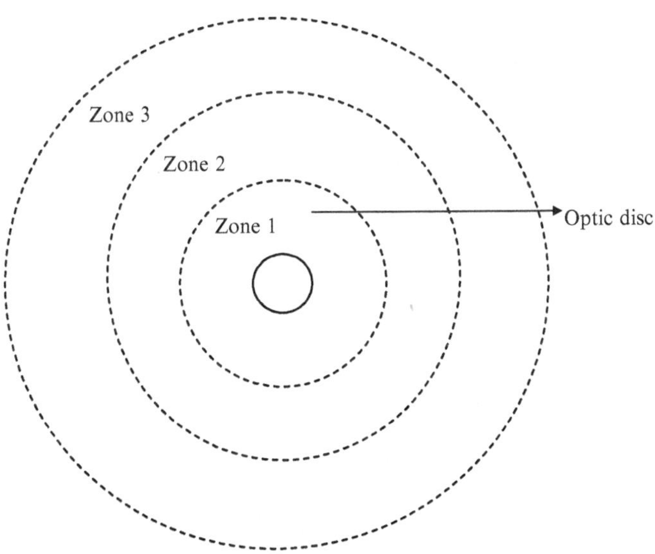

Fig. 13.1 Diagram of the zones of the retina

immunity can be restored by HAART. Anti-CMV therapy, half the dose after induction for 3 weeks, can be discontinued once the $CD4^+$ cell count is >100–150 cells/μL for 6 months. Some experts advocate intravitreal injection or ganciclovir implant in addition to systemic therapy for zone 1 CMV retinitis to avoid loss of vision.[42]

13.3.3 Medico-legal Discussion

For persons recently discovered to be HIV-seropositive, it is ideal to give the results in person confidentially, and at the same time counsel the patient on the disease. However, there are several options available if the individual were reluctant to return for follow-up appointment (or cancels the appointment). The results could be forwarded to the FP and notify him or her of the patient's cancelation plus need for counseling, close monitoring, and for initiating ART or PCP prophylaxis depending on the $CD4^+$ cell count. If the subject has no FP, then the person can be notified of the results by letter or phone, or through the public health department. Although some physicians are reluctant to discuss confidential issues on the phone, there is no edict against this practice. However, confidentiality needs to be maintained and the identification of the person on the phone should be verified. This has become a common practice with financial institutions and even lawyers who discuss medico-legal cases with medical experts on the phone. Since the defendant knew the plaintiff had advanced HIV disease (AIDS) as indicated by the very low $CD4^+$ cell count, it was mandatory that the patient be made aware of the seriousness of his condition as soon as possible by one of the above mechanisms. His failure to impart this information to the plaintiff directly or indirectly could be considered negligence by the court.

Failure of the defendant to perform a dilated ophthalmoscopy or arrange for urgent ophthalmology consultation when the plaintiff initially presented with impaired vision also falls below the standard of practice in an HIV infected individual with a $CD4^+$ cell count of <50 cells/ μL. The physician ought to have known that CMV retinitis was a main concern, and could be sight threatening and that examination by un-dilated fundoscopy would be insensitive and inaccurate. Based on the evidence presented, it could be argued by the plaintiff's lawyer that had the patient been notified earlier of the seriousness of his condition and accepted treatment with HAART 3 months before his hospital admission, it is likely that he would have had a better quality of life and preservation of his vision.

Although counsel for the defendant may counter that the plaintiff should be responsible for his own health (as he canceled the follow-up appointment), there were several avenues available to the defendant to ensure that the patient became aware of his serious illness, and he failed to utilize any of them. Whether or not a court may consider these failures as human errors from oversight in a busy medical practice and not medical malpractice would be difficult to predict.

13.4 General Discussion

What lessons can we learn from these cases?

- All patients with self-reported HIV-seropositive status should be verified by repeating the test.
- Request documentation of HIV serology from the FP or referring physician for documentation.
- Never institute ART without confirmation of HIV-seropositivity.
- Baseline $CD4^{+}$ cell count and HIV viral load with genotype testing for resistant mutations should be performed.
- High-risk people should have annual HIV serology and all health care contacts should be offered the tests.
- A standard practice for reporting HIV-seropositivity should be adopted, and alternative methods for notification of remissive individuals should be a part of the standard protocol.
- HIV infected patients with $CD4^{+}$ cell count <200 cells/μL should be immediately alerted to the seriousness of their condition and the need to start ART and prophylaxis.
- Visual disturbance in AIDS warrants urgent attention. A detailed examination after pupillary dilatation by direct ophthalmoscopy can be done by any physician to determine the presence of any abnormality. However, an urgent ophthalmology consultation is desirable.
- Any evidence of CMV retinitis in zone 1 of the retina should be considered an ophthalmologic emergency, and requires immediate attention.
- Remember, the lack of communication directly or indirectly to our patients is one of the main roots of medico-legal malpractice litigation.
- Physicians should pay more careful attention to patients' symptoms and complaints and act with reasonable promptness.
- Deal with patients' complaints as you would want to be done to yourself or relatives.

References

1. Bylund DJ, Ziegner UHM, Hooper DG (1992). Review of testing for human immunodeficiency virus. *Lab Immunol* II, 12:305–333.
2. Marlink RG, Allen JS, McLane MF, Essex M, Anderson KC, Groopman JE (1986). Low sensitivity of ELISA testing in early HIV infection. *N Engl J Med* 315:1549.
3. Horsburgh Jr CR, Ou LY, Jason J, Holmberg SD, Longini IM Jr, Schable CA, Mayer KH, Lifson AR, Schochetman, G, Ward W, Rutherford GW, Evatt BL, Seage GR III, Jaffe HW (1989). Duration of human immunodeficiency virus infection before detection of antibody. *Lancet* 2:637–639.
4. Burke DS, Brundage JF, Redfield RR, Damato, JJ, Schable CA, Putman P, Visintine R, Kim HI (1988). Measurement of the false positive rate in a screening program for human immunodeficiency virus infection. *N Engl J Med* 319:961–964.

5. Lifson AR, Stanley M, Pane J, O'Malley PM, Wilber JC, Stanley A, Jeffrey B, Rutherford GW, Sohmer PR (1990). Detection of human immunodeficiency virus DNA using polymerase chain reaction in a well-characterized group of homosexual and bisexual men. *J Infect Dis* 161:436–439.
6. Horsburgh CR Jr, Ou CY, Jason J, Holmberg SD, Lifson AR, Ward JW, Seage CR, Mayer KH, Evatt BC (1990). Concordance of polymerase chain reaction with human immunodeficiency virus antibody detection. *J Infect Dis* 162:542–545.
7. Letvin NL (2008). Immunology of HIV infection. IN (ed): Paul, WE. Fundamental Immunology, 6th Edition, Wolter Kluwer/Lippincott, Williams & Wilkins, Philadelphia, p1204–1232.
8. Kipps TJ (2006). Lymphocytes and lymphocytopenia. IN (eds): Licktman MA, Beutter E, Kipps TJ, Seligsohn U, Kaushansky K, Prchal JT. William's Hematology, 7th Edition, Mc-Graw Hill Medical, New York, p1087–1097.
9. Zumwait RE, McFeeley PJ, Maito J (1987). Fraudulent AIDS. *JAMA*, 257:3231.
10. Evans GA, Gill MJ, Gerhart S (1998). Factitious AIDS. *N Engl J Med* 319:1605–1606.
11. Zuger A, O'Dowd MA (1992). The baron has AIDS: a case of fictitious human immunodeficiency virus infection and review. *Clin Infect Dis* 14:211–216.
12. Milano MD, Barnowski C, Fiore T, Gormley J, Rich JD, Emgushove RT, Carpenter CC (2001). Factitious HIV syndrome in young women. *AIDS Read* 11:278–282.
13. Craven DE, Steger KA, La Chapelle R, Allan DM (1994). Factitious HIV infection: the importance of documenting infection. *Ann Intern Med* 121:763–766.
14. Churchill DR, De Cock KM, Miller RF (1994). Feigned HIV infection/AIDS: malingering and Munchausen's syndrome. *Genitourin Med* 70: 314–316.
15. Niu MT, Stein DS, Schnittman SM (1993). Primary human immunodeficiency virus type I infection: review of pathogenesis and early treatment intervention in humans and animal retrovirus infection. *J Infect Dis* 168:1490–1501.
16. Pedersen C, Lindhardt BO, Jensen BL, Lauritzen E, Gerstoft J, Dickmeis E, Gaub J, Scheibel E, Karlsmark T (1989). Clinical course of primary HIV infection . *Br Med J* 299:154–157.
17. Mellors JW, Renaldo CR Jr, Gupta P, White RM, Todd J, Kingsley LA (1996). Prognosis in HIV infection predicted by the quantity of virus in plasma. *Science* 272:1167–1170.
18. Stein DS, Korvick JA, Vermund SH (1992). $CD4^+$ lymphocyte cell enumeration for prediction of clinical course of human immunodeficiency virus disease: *J Infect Dis* 165:352–263.
19. Fauci AS, Pantaleo G, Stanley S, Weissman D (1996). Immunopathogenic mechanisms of HIV infection. *Ann Intern Med* 124:654–663.
20. Learmont JC, Geczy AF, Mills J, Ashton LJ, Raynes-Greennow CH, Garsia RJ, Dyer WB, McIntyre L, Oelrichs RB, Rhodes DI, Deacon NJ, Sullivan JS, For the Sydney Blood Bank Cohort Research Group (1999). Immunologic and virologic status after 14 to 18 years of infection with an attenuated strain of HIV-1. *N Engl J Med* 340:1715–1722.
21. Kirchoff F, Greenough TC, Brettler DB, Sullivan JC, Desrosiers RC (1995). Brief report: absence of intact Nef sequence in a long-term survivor with non-progressive HIV-1 infection. *N Engl J Med* 332:228–232.
22. Keet IPM, Tang J, Klein MR, LeBlanc S, Enger C, Rivers C, Apple RJ, Mann D, Goedert JJ, Miedema F, Kaslow RA (1999). Consistent associations of class I and II and transporter gene products with progression of human immunodeficiency virus type I infection in homosexual men. *J Infect Dis* 180:299–309.
23. Levy JA (ed) (2007). Overall features of HIV pathogenesis prognosis for long term survival. IN: HIV and the Pathogenesis of AIDS, 3rd Edition, ASM Press, Washington DC, p317–361.
24. Hunt PW (2009). Natural control of HIV-1 replication and long-term non-progression: overlapping but distinct phenotypes. *J Infect Dis* 200:1636–1638.
25. Dean M, Carrington M, Winkler C, Huttley GA, Smith MW, Allikmets R, Goedert JJ, Bachbinder SP, Vittinghoff E, Gomperts E, Donfield S, Vlahov D, Kaslow R, Saah A, Rinaldo C, Detels R, O'Brien SJ (1996). Genetic restriction of HIV-1 infection and progression to AIDS by a deletion allele of the CKR5 structural gene. *Science* 273:1856–1862.

26. Webster A, Lee CA, Cook DG, Grundy JE, Emery VC, Kenoff PBA, Griffiths PD (1989). Cytomegalovirus infection and progression towards AIDS in hemophiliacs with human immunodeficiency virus infection. *Lancet* 11:63–66.
27. Robain M, Bonfassa F, Hubert J, Persoz A, Burgard M, Meyer L (2001). Cytomegalovirus seroconversion as a cofactor for progression to AIDS. *AIDS* 15:251–256.
28. Bartholomew C, Blattner W, Cleghorn F (1987). Progression to AIDS in homosexual men coinfected with HIV-1 and HTLV-1 in Trinidad. *Lancet* ii:1469.
29. Page JB, Lai S, Chitwood DD, Klimas NG, Smith PC, Fletcher MA (1990). HTLV-I/II seropositivity and death from AIDS among HIV-I seropositive intravenous drug users. *Lancet*: 335:1439–1441.
30. Wallis RS, Vjecha M, Amir-Tahmasseb M, Okwera F, Byekwaso S, Nyole S, Kabengera S, Mugererwa RD, Ellner JJ (1993). Influence of tuberculosis on human immunodeficiency virus (HIV-I): enhanced cytokine expression and elevated B2-microglobulin in HIV-I associated tuberculosis. *J Infect Dis* 167:43–48.
31. Morris L, Martin DJ, Bredell H, Nyoka SN, Sack SL, Pendle S, Page-Shipp L, Karp CL, Sterling TR, Quinn TC, Chaisson RE (2003). Human immunodeficiency virus-I RNA levels and CD4 lymphocyte counts during treatment for active tuberculosis in South African patients. *J Infect Dis* 187:1967–1971.
32. Vasan A, Renjiifo B, Hertzmark E, Chaplin B, Msamanga G, Essex M, Fawzi W, Hunter D (2006). Different rates of disease progression of HIV type I infection in Tanzania based on infecting subtype. *Clin Infect Dis* 42:843–852.
33. Kanki PJ, Hamel DJ, Sankale JL, Hsieh CC, Thior I, Barin F, Woodcock SA, Grueye-Ndiaye A, Zhang L, Montano M, Siby T, Marlink R, Ndoye I, Essex ME, MBoup S (1999). Human immunodeficiency virus type I subtypes differ in disease progression. *J Infect Dis* 179:68–73.
34. Sugar M, Lawreys L, Baeten JM, Richardson BA, Mandaliya K, Chohan BH, Kreiss JK, Overbaugh J (2003). Infection with multiple human immunodeficiency virus type I variant is associated with faster disease progression. *J Virol* 77:12921–12926.
35. Kaslow RA, Blackwelder WE, Ostrow DG, Yerg D, Palenicek J, Coulson AH, Valdiserri RO (1989). No evidence for a role of alcohol or other psychoactive drugs in accelerating immunodeficiency in HIV-I positive individuals. *JAMA* 261:3424–3427.
36. Fong IW, Read S, Wainberg M, Chia WK, Major C (1994). Alcoholism and rapid progression to AIDS after seroconversion. *Clin Infect Dis* 19:337–338.
37. Kitahata MM, Gange SJ, Abraham AG, Merriman, B, Saag MS, Justice AG, Hogg RS, Deeks SG, et al. For the NA-ACCORD Investigators (2009). Effect of early versus deferred antiretroviral therapy for HIV on survival. *N Engl J Med* 360:1815–1826.
38. When to Start Consortium (2009). Timing of initiation of antiretroviral therapy in AIDS-free HIV-I infected patients: a collaborative analysis of 18 HIV cohort studies. *Lancet* 373:1352–1362.
39. Centers for Disease Control and Prevention (2005). HIV prevalence, unrecognized infection and HIV testing among men who have sex with men – five US cities, June 2004-April 2005. *MMWR Morb and Mort Wkly Rep* 54:597–601.
40. Centers for Disease Control and Prevention (2006). Sexually transmitted diseases treatment guidelines 2006. *MMWR Morb and Mort Wkly Rep* 55:1–100.
41. Severe P, Juste MA, Ambroise A, Eliacin L, Marchand C, Appollon S, Edwards A, Bang H, Nicotera J, Godfrey C, Gulick RM, Johnson Jr WD, Pape JW, Fitzgerald DW (2010). Early versus standard antiretroviral therapy for HIV-infected adults in Haiti. *N Eng J Med* 363:259–265.
42. Jabs DA, Thorne JE (2008) Ophthalmologic Disease. IN (eds) Dolin R, Masur H, Saag M. AIDS Therapy, 3rd Edition, Churchill Livingstone/Elsevier, Philadelphia p1169–1186.
43. Baldassano V, Dunn JP, Feinberg J, Jabs DA (1995). Cytomegalovirus retinitis and low $CD4^+$ T-lymphocyte counts. *N Engl J Med* 333:670.
44. Lalezari JP, Stagg RJ, Kuppermann BD, Holland GN, Kramer F, Ives DV, Youle M, Robinson MR, Drew WL, Jaffe HS (1997). Intravenous cidofovir for peripheral cytomegalovirus retinitis in patients with AIDS. *Ann Intern Med* 126:257–263.

Chapter 14
Overview of Medico-legal Issues

14.1 Identification of the Problem

Medical malpractice litigation is usually the result of an unexpected and undesirable outcome in our customers (patients) and when the affected individual or family perceives the outcome as being due to carelessness or poor clinical care. In the present era of modern medicine, the management of any given patient involves many layers of physicians and health care professionals from encounter to completion of care, for any specific condition. With this complexity of medical care, there is an increased risk of some error along the route towards the end result of cure or achieving relief and patient satisfaction. As a result, each health professional owes a separate duty to the patient within his or her area of practice. Although an individual health professional is not directly liable for the mistakes of others, in complex cases with multiple encounters by several health professionals, it is often difficult to determine the individual responsible for the outcome. However, in many of the cases, despite checks along the way involving multiple layers of health care workers, errors are often overlooked and missed by several health professionals.

The three main areas of concern and complaint by plaintiffs that leads to medical litigation are: (1) diagnostic errors, (2) errors in management, and (3) adverse events of treatment that are often related to poor counseling. A new territory for medico-legal action, which is of major concern in Canada, but less relevant in the US, is the wait times for investigations and procedures that may affect the outcome of patients. The individual physician's influence on wait times is limited, and this is more related to the accountability of the health care system and institutions. Benchmarks have been set in emergency departments for triaging patients for assessment according to the severity of illness, and breakdown in this system, which leads to poor outcomes, have resulted in medico-legal litigations.

I.W. Fong, *Medico-Legal Issues in Infectious Diseases*, Emerging Infectious Diseases of the 21st Century, DOI 10.1007/978-1-4419-8053-3_14,

14.2 Diagnostic Errors

Medical diagnoses that are wrong, missed, or delayed make up a large fraction of medical errors that cause substantial injury and suffering. Before the advent of modern imaging and diagnostic techniques, studies found the clinical diagnosis was correct in less than half the patients dying in hospitals. Nowadays, the vast majority of patients are diagnosed correctly, but diagnostic errors continue to be frequent and important, and represent an underemphasized and understudied area of patient safety. A Harris poll commissioned by the National Safety Foundation found that one in six people had personally experienced a medical error related to misdiagnosis.[2] A recent review of 53 autopsy series found an average rate of 23.5% major missed diagnoses (range 4.1–49.8%).[3] Despite decreases of major diagnostic errors observed from 1966 to 2002, it is estimated that contemporary US institutions could observe major diagnostic error rates of 8.4–24.4%.[3]

Although much of the medical reports and lay press have highlighted medication errors and adverse events as major concern for patient safety, recent studies of malpractice claims revealed that diagnostic errors greatly exceed medication errors, 26–32% versus 8–12%.[4,5]

In a recent review of 100 cases of diagnostic errors involving internists, 90 cases involved injury with 33 deaths.[6] The investigators classified diagnostic errors as follows:

- No fault errors – masked or unusual presentation of disease or patient-related error (uncooperative, deceptive)
- System-related errors – technical failure, equipment problems and organizational flaws
- Cognitive errors – faulty knowledge, faulty data gathering, and faulty synthesis.

Overall, the investigators identified 228 system-related factors and 320 cognitive factors, averaging 5.9 per case.[6] In 46% of cases, both system-related and cognitive factors contributed to diagnostic error. Error due only to cognitive factors (28%) or only to system-related factors (19%) or solely to no fault factors (7%) were less common.[6] Thus, diagnostic error is commonly multifactorial in origin, involving system-related and cognitive factors.

A recent survey of 2,201 adults in the US found that 35% of subjects, families, or friends experienced medical mistakes in the past 5 years, half of which were considered diagnostic errors.[7] Misdiagnosis is the greatest concern of respondents visiting outpatient physicians (55%), but also of significant concern in the hospital setting (23%). In the emergency department visits, 38% of patients are concerned about medical errors and the most common worry was misdiagnosis (22%).[8] Diagnostic errors are encountered in every specialty and are usually lowest in the two perceptual specialties, radiology and pathology (error rate of 2.5%), which rely on visual interpretation of objective data. Error rates in clinical specialties are higher, and the estimated rate of diagnostic error in clinical medicine has been about 15%.[9]

Where does the diagnostic process fail? Analysis of diagnostic errors collected over a 3-year period (weekly case conferences) by a survey of physicians identified errors in various stages of the diagnostic process. These include: (1) access and presentation, (2) history taking/collection, (3) the physical examination, (4) testing, (5) assessment, (6) referral, and (7) follow-up. The most common system-related factors involved problems with policies and procedures, inefficient processes, teamwork, and communication. The most common cognitive problems involved faulty synthesis, and premature closures (or failure to consider alternates after an initial diagnosis) was the single most common cause of diagnostic error by internists.[6] Other common causes included faulty context generation (to put together circumstances that are relevant to an event), misjudging the salience of findings, faulty perception and errors from heuristic problem solving.[6] Faulty or inadequate knowledge was uncommon, but overconfidence (miscalculation of one's own sense of accuracy) may be a significant cause of diagnostic error in medicine.[10]

Schiff et al.[1] have identified four key challenges in assessing potential diagnostic error cases: (1) uncertainties about diagnosis and findings, (2) the relationship between diagnosis failure and adverse outcomes, (3) challenges in reconstructing clinician assessment of the patient and clinician actions, and (4) global assessment of improvement. What are the solutions to the problem? There is no easy answer, but several steps have been suggested to minimize diagnostic errors. Strategies to decrease diagnostic errors have been proposed.[10,11]

- Clinicians and health care organizations need to take active steps to discover, analyze, and prevent diagnostic errors.
- Diagnostic errors should be included in quality assurance surveillance and review.
- Elements of the system prone to error should be identified and addressed, such as issues related to diagnostic testing and interpretation.
- Health organizations could minimize errors by system-level interventions to aid the clinician such as rapid or second readings of key diagnostic tests and provide resources for clinical decision support (computer-based).
- Standard protocols for follow-up of abnormal test results, with prompt reading and reporting of x-ray and laboratory tests.
- Identifying "red flag" and "don't miss" diagnoses and use checklists (automated or manual).
- Provide feedback to physicians on missed or delayed diagnoses.
- Safety nets to mitigate harm from diagnostic uncertainty and error. These include observation and follow-up venues, follow-up calls or email within 48 h and automated 2 week phone follow-up to ensure tests obtained for high-risk and uncertain diagnosis.
- Educating patients on importance of follow-up care or advice defining and specifying particular warning symptoms.

Measures to reduce cognitive errors are more complex and difficult to implement. It has been suggested that compiling a complete differential diagnosis in each case may be helpful, and elimination by signs and laboratory results can be useful to prevent premature closure ("jumping to the wrong conclusion"). Educational approaches include strategies to improve physicians' overall knowledge, case rounds (clinico-pathological conferences), medico-legal rounds on missed diagnoses, and approach to "situational awareness and cognitive forcing strategies" to correct the tendency to cognitive errors.[11] This involves training to recognize any specific pitfalls in diagnosis of patients physicians commonly see. However, it also may be useful to make physicians aware of common conditions where delayed or misdiagnoses frequently results in litigations, and the errors that result in these mishaps.

Physicians were taught during medical school and residency training to generate a comprehensive differential diagnosis or a problem-solving list when faced with a complex case (or straightforward case). However, it seems that this practice has largely been abandoned once physicians are in a busy practice, and they rely more on pattern-recognition approach based on past experiences.[10] This is especially evident on review of medico-legal files with records from physicians' offices, clinics, and emergency departments. It is very unlikely that physicians in practice will pursue any special training to improve cognitive processing (synthesis of data collected) to any large degree as suggested by some experts, such as meta cognitive training and reflective practice using the technique of "prospective hindsight."[10,11] That is, to consciously analyze a proposed diagnosis as if it were incorrect and thus forced to consider alternative diagnoses. Many errors in diagnostic thinking can be attributed to shortcuts in reasoning (reduced time spent for assessment), and in the majority of cases they are typically correct and produce a satisfactory result with a minimum of delay, cost, and anxiety. Although these shortcuts are useful and practical, physicians should be aware of the potential diagnostic danger from their reliance on shortcuts in reasoning and use them only when appropriate. An important message that can be readily conveyed and easily remembered by all physicians is to consider the opposite: "What is the diagnosis that I don't want to miss?"[12] This is particularly relevant in the emergency department where most diagnostic errors made in the course of clinical decision-making are due to cognitive erratum.

14.3 Adverse Events in Hospital

Adverse events, defined as unintended injuries or complications caused by medical health care management, resulting in death, disability or prolonged hospitalization, are significant occurrences in many industrialized countries. In the Harvard Medical Practice Study 30,121 randomly selected records from 51 acute care hospitalizations found adverse events occurred in 3.7% of hospitalizations, and 27.6% of these adverse events were due to negligence.[13] A similar study in Canadian acute care hospitals found the adverse event rate was 7.5% of hospital admissions, of which about 37% were preventable.[14] In other countries, even higher rates of adverse events in hospitals were reported: in Australia 16.6% of admissions resulted in adverse events (51% were preventable),[15] in New Zealand the incidence rate was 11.2%,[16] in British hospitals 10.8% of patients experienced adverse events,[17] and in Denmark, 9.0% of all admissions (40% preventable).[18] Of the patients suffering adverse events, the mortality varied from 4.9% to 20.8%, and permanent disability from 2.6% to just under 15%. Adverse events increase with age and were more common among elderly patients, attributed to complexity of their care.[13,19] Preventable adverse events related to medical procedures and falls were especially common in elderly patients.

In the Canadian study,[14] adverse events were more frequent in the teaching hospitals than in large community or small hospitals. The most common types of adverse events were related to surgical procedures, followed by drug, or fluid-related events. In the Utah/Colorado Study, operative procedures accounted for 44.9% of all adverse events (16.9% were negligent and 16.6% resulted in permanent disability), and adverse drug events were the next most frequent (19.3%), with 35.1% being negligent, and 16.6% resulted in permanent disability.[20] In the Harvard Medical Practice Study II,[21] drug complications were the most common type of adverse events (19%), followed by wound infections (14%), and technical complications (13%). Although nearly half the adverse events (48%) were associated with an operation, only 17% were caused by negligence compared to37% for nonsurgical ones. The highest proportion of adverse events due to negligence was greatest for diagnostic errors (75%); non-invasive therapeutic errors of omission (77%), and mishaps in the emergency department (70%).[21] In a prospective study in a tertiary care teaching hospital in British Columbia over a 12-week period, there were 122 patient visits to the emergency department for drug-related problems.[22] Adverse drug events accounted for 39% of the visits, and use of the wrong or suboptimal drug in 11.5%, and 68% of the visits were deemed preventable.[22]

Much less is known about the incidence or prevalence of adverse events in primary care, as few studies have been reported. Preventable adverse events in primary care practice include incidents related to diagnosis, treatment, and preventive care. Review of incident reports entered by eight primary care clinics into a risk management database found the prevalence of adverse events was 3.7/100,000 clinic visits, of which 83% were preventable.[23] Diagnostic errors (26%) and treatment errors accounted for most of the adverse events. Review of the literature on medical errors in primary care concluded that the key safety issues are in the spheres of diagnosis, prescribing, communication, and organizational change.[24,25] Analysis of malpractice claims in the US from primary care (68% in outpatient settings) assessed 23% as due to negligence.[5] Diagnostic error accounted for one third of the claims. A list of categories for underlying causes of malpractice claims included: diagnostic error, wrong patient or body part, medication errors, improper performance, failure to instruct or communicate with the patient, procedure performed when not indicated or contraindicated, delay in performance, retained foreign body after surgical procedure, failure to supervise care, failure to recognize a complication of treatment, failure or delay in admission to hospital, failure/delay in referral to a consultant, improper supervision of residents or staff personnel, failure to properly respond, contraindication of procedure, and not or improperly performing resuscitation.[5]

14.4 Medication Errors and Adverse Drug Events

Medication errors are very common in hospitals but fortunately, most of the errors are minor and do not result in injury. Medication error can be defined as a deviation from the proper written prescription or error at any stage of the medication use process.[26] An adverse drug event is defined as an injury from a drug-related intervention; these may be preventable or non-preventable, and can result from errors in prescribing, dispensing, and administration. Medication errors in hospital occur at a rate of two per patient day[26] and account for 10–25% of all errors.[27] The most frequent errors by category were wrong time (43%), omission (30%), wrong dose (17%), and unauthorized drug (4%).[28]

The causes of medication errors can be summarized as related to various factors indicated below:

1. Lack of information about the patient
2. Lack of information about the drug
3. Communication and teamwork failures
4. Look-alike drug labels and packaging, and confusing, sound-alike drug names or look-alike pills/capsules
5. Unsafe drug standardization, storage and distribution
6. Difficult to read prescriptions, or prescription errors
7. Factors and staffing patterns that do not support safety
8. Inadequate or unsafe medication delivery system
9. Inadequate staff orientation, education and supervision
10. Inadequate patient education on medications and errors
11. Failure of feedback and supportive safety network or absent error-reduction strategies.

Methods to improve or reduce medication errors include establishing a system for identifying, reporting, analyzing, and correcting medication errors.[28] Safety should be promoted by encouraging frank disclosures of errors and near-misses, productive discussions and institution of effective system-based solutions. Regular and assessable quality control checks are necessary.

The US National Academy of Sciences Institute of Medicine (IOM) in its 2000 report,[29] responded to the alarming numbers of medical errors for immediate systematic attention on a large-scale societal level. Medication errors are preventable events that can be related to professional practice, health care products, procedures, and systems, including prescribing order communication, product labeling, packaging, and nomenclature, preparation, dispensing , distribution, administration, education, monitoring, and use.[30] Drug-related problems have been categorized under the following: (1) untreated indication (failure to provide indicated drug for a medical condition requiring drug therapy), (2) improper drug selection (inappropriate drug), (3) sub-therapeutic dosage, (4) failure to receive drug (pharmaceutical, psychological, sociological or economic reasons) (5) overdose, (6) adverse drug reactions, (7) drug interaction (drug-drug, drug-food or drug-test agent interaction), and (8) drug use without any medically valid indication.[31]

In a recent Canadian study,[32] the two main factors contributing to preventable adverse events were communication problems and poor clinical judgment. Communication problems (native foreign language, deafness, and psychiatric disorder) increase the risk threefold of preventable adverse events, and preventive measures will require extra effort and scrutiny.

14.4.1 Adverse Drug Reactions

Every known drug or medication can produce some form of side effect or adverse reaction, however, most of these reactions are minor or easily tolerated, and more severe reactions are uncommon or rare. In the Harvard Medical Practice Study,[21] drug complications were the most common type of adverse event (19%). The types of errors related to drug treatment included: error in dose or method of use (42%), failure to recognize possible drug-drug interactions (8%), inadequate follow-up of therapy (45%), use of inappropriate drug (22%), avoidable delay in treatment (14%), professional practice outside area of expertise (5%), and others (9%). It was judged that 52.8% of the drug treatment errors were negligent.[21] In a review of malpractice claims in primary care 8% were related to medication errors.[5]

Adverse reactions can be preventable or non-preventable and most are reversible on withdrawal of the drug, but some can produce long-standing or permanent disability, and sometimes fatal reaction. Drug interactions can produce predictable side effects (toxic reactions which are dose or cumulative-dose related), or idiosyncratic, unpredictable reactions, or partly predictable hypersensitivity reactions. Some classes of drug have a high risk for toxic reactions and are administered by a specialist in a hospital setting, i.e., cancer chemotherapeutic agents and amphotericin B, that require close clinical and laboratory monitoring.

There are many factors that may influence the risk of adverse drug reactions and these can be divided into host factors and extrinsic factors or a combination of both. A common host factors include aging, as the elderly are more susceptible to many drug reactions. This may be related to the risk of multiple chronic illnesses or altered liver and kidney clearance, and possible low body mass. Others include comorbid illnesses (chronic kidney and liver disease), atopy and allergic predisposition and genetic predisposition (single nucleotide polymorphisms of genes that predispose to specific toxicity of certain drugs). Extrinsic factors may include the properties of the drug itself, dose-related events, concomitant medications, drug-drug interaction, alcohol intake, food-related events, and even psychological factors.

The WHO definition of adverse drug reaction (ADR) in use for >35 years, is "a response to a drug that is noxious (harmful or injurious) and unintended, and occurs at doses normally used in man for the prophylaxis, diagnosis or therapy of disease, or for modification of physiological function."[33] Some authors opine that this definition has several limitations and does not include minor unwanted reactions or adverse effects caused by error or reaction caused by ingredients other than the active agent.[34]

ADRs are common in medical practice and a meta-analysis of 39 prospective studies in the US between 1966 and 1996 reported ADRs (excluding medication errors) of 10.9% in hospitalized patients.[35] Serious ADR accounted for 6.7% of all hospital admissions and lethal reactions occurred in 0.32% of all hospital admissions. These statistics are not confined to the US but similar incidence of ADRs have been reported in European countries and Australia.[35] Worldwide, the frequency of ADRs within various healthcare systems account for 3.3–7.5% of all hospital admissions and about 0.3–1.4% of hospital fatalities.[36]

The economic impact of ADRs is huge, not only on the healthcare system, but in the public sector as well, due to absenteeism from work. In the US, an adverse drug event in hospitalized patients increases the average length of stay by 2.2–3.2 days, with associated increase cost of $3,244–4,355 per case.[36] ADRs resulting in hospital admission in Europe led to a mean length of stay of 9 days and patients had to stop work for an average of 20 days.[37] In France, 5–9% of hospital costs related to medical admission were due to ADRs.[37] It has been estimated that 32% of ADRs are preventable,[37] and that 20% of hospital costs are avoidable. In a US study, it was found that avoidable drug events were relatively more serious, caused longer hospital length of stay, and incurred greater hospitalization costs than unavoidable ones.[38,39]

ADRs have been classified under the following categories[36]:

A. Augmented – dose-related and predictable toxic events (warfarin, digoxin, aminoglycosides, etc.)
B. Bizarre – unpredictable, idiosyncratic reaction (aplastic anemia to chloramphenicol, interstitial nephritis to penicillin, etc.)
C. Chronic – related to long term use +/– dose (i.e., steroid induced immune suppression, Cushing's syndrome, etc.)
D. Delayed – chronic use after sometime on treatment, dose-effect (carcinogenesis with cytotoxic agent, lipodystrophy on antiretrovirals)
E. End of Treatment – after withdrawal of drug (opiate withdrawal syndrome, cardiac ischemia after ß-blockers, adrenal insufficiency after chronic steroids)
F. Failure – unsuspected failure due to drug-drug interaction causing inadequate levels.

This classification, however, does not include a common adverse reaction to antimicrobials which can occur at any time during or after treatment, and is unpredictable – i.e., *Clostridium difficile* colitis. Superinfection with broad-spectrum antimicrobials could be included under D classification.

14.4.2 Management and Prevention

The management of any ADR will depend on the seriousness and nature of the event. Immediate life threatening reaction (i.e., anaphylactic shock) requires immediate rapid action with emergency measures and withdrawal of the offending agent. Less dramatic, but potentially serious reactions should also require discontinuation of all implicated drugs as soon as possible. After resolution of the reaction, cautious introduction of essential medicines may be considered. If several agents could be responsible for the ADE, non-essential drugs may be withdrawn first, preferably one at a time depending on the severity of the reaction. If the offending agent is fairly clear then a benefit-risk decision with input from the patient about the need for the drug, or possible substitution with another medicine. Lower doses of the drug may be used when the adverse event appears to be dose-related. In many cases, it is prudent to reassess the need and indication for the various medicines. Use of several medicines together (polypharmacy), although sometimes necessary, not only increases the risk of additive side effects and drug intolerance, but also may cause toxic reaction through drug interactions.

When a patient cannot manage without a drug responsible for the ADE and there is no suitable alternative, or the alternatives have similar side-effects, then symptomatic therapy of the reaction while continuing treatment with the same agent may be undertaken (for non-severe reaction with close observation and monitoring).

Various strategies have been proposed to improve detection, prevention, and management of ADRs.[34,36] These measures can be subdivided under the following headings: (1) Infrastructure – universal improved use of information technology, (2) Structure – development of local ADR teams, policies and monitoring schemes and improved regulatory and control systems, (3) Staff – improve staffing for support, training, continued education and monitoring, with less individual blame, and encouragement of ADR reporting.

A recent study on the effects of incorporating bar-code verification technology within an electronic medication system found a substantial reduced rate of errors in order transcription and in medication administration, as well as potential adverse drug events (50.8% relative reduction).[40] Although these strategies may be more suitable for inpatient care in healthcare institutions, some aspects such as bar-code technology can be easily adopted for outpatient settings and independent pharmacies. The challenges may be greater for office practices and private pharmacies interactions.

The basic tenet of the Hippocratic Oath that physicians should always keep in mind is "first to do no harm" to our patients. Before instituting or recommending any form of therapy, physicians should ponder for a moment on the risk-benefit aspects. The questions that we as physicians should ask ourselves before starting treatment can be quickly reviewed: (1) Is this therapeutic intervention necessary? (2) Do I have a clear therapeutic objective in mind? (3) Are there potential adverse reactions with the risk of treatment? (4) Do I need to treat as long or can the course be shortened? (5) Can I simplify the management? (6) Is the patient aware of and do they agree with the management plan, risks and benefits? If all physicians take a few minutes to contemplate these questions before ordering treatment on such patient, it is likely that many unnecessary and potentially toxic medicines or interactions could be avoided.

14.5 Review of Medico-legal Experience

In the past 15 years, I have reviewed and provided opinions as an independent medical expert on over a 100 medico-legal cases involving some form of infectious disease. These reviews were requested by lawyers representing the plaintiffs or defendants (physicians) with no selection bias. An analysis of 91 cases will be provided where there was sufficient data remaining in retained files.

There were similar numbers of cases reviewed for both the defendants' and plaintiffs' lawyers (see Table 14.1). Over half of the cases resulted in death, loss of limb, or permanent disability, and most of the remaining cases had prolonged illnesses and hospitalization, and required repeat or multiple surgical interventions. Emergency physicians and family physicians were the most frequent disciplines involved in these legal actions, followed by general, orthopedic and plastic surgeons, and internists. However, in nearly 50% of the cases, multiple disciplines or physicians were involved who were accused of negligence (range 2–5 physicians). Based on my opinion, it was felt that 60% of the adverse reactions could have been prevented or ameliorated.

The most common reason for the adverse events leading to litigation was missed or delayed diagnosis in 47.2%. This was often combined with poor medical notes, poor communication with the patient and family, and pursuit of a single wrong

Table 14.1 Medico-legal cases related to infectious diseases

• Total no. cases reviewed	91
• Reviewed for the defendant(s)	44
• Reviewed for the plaintiff(s)	47
• Resulted in death	27 (29.6%)
• Resulted in permanent disability or loss of limb	24 (26.3%)
• Resulted in prolonged illness, multiple procedures or cosmetic disfigurement	28 (30.7%)
• Preventable outcomes	55 (60.4%)
Highly likely	50
Probable	5
• Multiple physicians or disciplines accused of negligence	41 (45%)
• Lawsuits mainly against the institutions	14 (15.3%)
• University tertiary care centers involved in litigation	14.3 (15.3%)
• *Medical disciplines involved*	
Emergency physicians	25
Family physicians	24
General surgery	12
Internist	11
Plastic/cosmetic surgery	11
Orthopedic surgery	11
Obstetrics/Gynecology	5
Neurosurgery	3
Others (ENT, ID, CVS, Vasc.)	8

Abbreviations: *ENT* ear, nose and throat, *ID* Infectious Disease, *CVS* Cardiovascular surgery, *Vasc* Vascular surgery working diagnosis (without generation of differential diagnosis) which was perpetuated by other physicians in the same or different specialty. Unnecessary treatment or overtreatment was responsible for only 13 (14.2%) of the cases, and this resulted primarily in toxic adverse events which could have been avoided. The majority of the drug toxicities were related to gentamicin; prolonged unnecessary use (6 cases), and all vestibular toxicity. In 5/6 cases, gentamicin was not needed and could have been discontinued within 3 days after culture results failed to show any gram-negative coliforms. In all the cases reviewed, there was either no or inadequate counseling provided to patients on the potential side effects or risk-benefit aspect. Surprisingly, several of the physicians and nurses were not aware that nausea, vomiting, and dizziness were the earliest manifestation of aminoglycoside toxicity (thus often continued), and that vestibular toxicity was more frequent than auditory toxicity.

Inadequate management (other than delayed treatment for missed diagnosis) such as an inadequate course of antibiotics or surgical procedure and failure to give prophylaxis for surgical infection (when indicated) accounted for 12 cases (13.2%). The conditions most commonly associated with delayed or missed diagnosis include: necrotizing fasciitis/myonecrosis – 12 cases (5 due to group A *Streptococcus* and 4 to *Clostridia* species.), intraabdominal and pelvic infections – 10 cases (4 due to group A *Streptococcus* with toxic shock syndrome), bacterial endocarditis (7 cases), septic arthritis (6 cases), osteomyelitis (5 cases), bacterial meningitis (4 cases), and spinal epidural abscess (3 cases).

14.5.1 Lessons to be Learned from Case Review

The review of medico-legal cases has been a learning experience, which has broadened my outlook and improved both my knowledge of medico-legal aspects and of medicine in general. It gives one the opportunity to understand and appreciate patients' feelings, frustrations, and complaints about the medical system and their perceptions of physicians' lack of empathy and failure to understand their sufferings. This experience has certainly provided insight about physicians' failings in general, and some ways and means that all clinicians can adopt to better serve their customers (patients).

First, it is important to review common themes among medical practice suits that most lawyers in the field are quite familiar with. These are outlined below:

- Poor medical notes, often illegible, too abbreviated without any details of the patient's symptoms or clinical findings. This is characteristic of the medical notes from ER physicians, general practitioners, and surgical disciplines in the medical malpractice cases reviewed.
- "Tunnel vision" or failure to generate a differential diagnosis is another common feature of these cases. It is surprising that pursuit of the same wrong diagnosis is often followed by other consultants that assess the patients.

- Inability to recognize the serious nature of the patient's illness. This failure is most commonly seen in the ER visits, often even on repeat visits, but has been observed in other specialties as well. This is of particular concern for young children and adults who present to the ER in the wee hours in the morning, to be dismissed as some "viral illness" with no details provided or important negative findings. It is highly unlikely that a parent will attend the ER after 11 p.m. to 6 a.m. for a child with a mild illness, and this alone should raise concerns of a serious disorder. Parents are most familiar with their children and usually know when they are seriously ill.
- Belittling of patients' complaints as due to "histrionic's personality," or attributing their complaints such as severe muscle/soft tissue pain to recent trivial accidents. This is a common feature in necrotizing fasciitis/myonecrosis where the pain is usually out of proportion to the physical signs, which can be minimal and subtle in the early stages.
- Lack of a clear plan for further investigations and management.
- Lack of communication with patients and families with respect to primary diagnosis and differential diagnoses.
- Failure to counsel patients on potential side effects of medications, follow-up monitoring, and what to do in case of worsening of their symptoms.
- Failure of health care personnel to routinely question patients and monitor to detect early signs of drug toxicity.

To reduce unwarranted adverse events and minimize the risk of missing or delaying the diagnosis of serious medical illness, physicians need to take all symptoms seriously, and attempt to communicate more openly with patients by outlining risk-benefit aspects of treatment, and assure proper understanding of the plan of action. Before dismissing patients' symptoms as trivial due to a benign minor condition, we should ask ourselves:

Have I excluded the most serious conditions that I cannot afford to miss? What would I do or recommend if this patient were my child or parent? Does the patient understand the nature of his or her illness? Is the patient aware of the nature of treatment, other alternatives, and the risk of medication or procedure? Have I spent enough time to discuss all the issues and concerns that the patient and family express? Would this problem best be served by another specialist or consultant?

Physicians should be candid with their patients and acknowledge as soon as possible any errors or mistakes that were made. If we erred in making a diagnosis or gave inappropriate treatment, then open discussion, take acceptance of the blame, and give apologies where warranted. Often this simple act of acknowledgement and apologies may avert litigation, whereas denial and neglect will more often encourage atonement through a malpractice lawsuit. Several hospitals, healthcare organizations, and legislators in the United States and other countries are developing or setting standards to encourage open communication with patients after harmful errors have occurred.[41] Implementation of bold disclosure policies have been reported to reduce potential malpractice litigation.[42]

14.6 Addendum

Two recent publications are worthwhile reviewing as a final conclusion of this chapter. It has been 10 years since publication of the Institute of Medicine's report[29] that medical errors cause up to 98,000 deaths and more than 1 million injuries each year in the US. A recent review from 2002 to 2007 of 10 hospitals in North Carolina found no significant changes in the overall rate of harms per 1000 patient days or the rate of preventable harms.[43] Thus healthcare related harms remain common with little evidence of widespread improvement over the past decade. Therefore, stronger efforts are needed by healthcare institutions and physician to enforce and monitor safety interventions into routine practice. This report is another belated "wake-up" call for physicians and hospital administrators.

An encouraging recent report from the University of Michigan Health System (UMHS) medical error disclosure program,[44] maybe a stimulus for other healthcare institutions to adopt. Since 2001, UMHS began implementation of a disclosure-with-offer program, by responding to all open and new malpractice claims by admitting fault and offering compensation when internal investigation reveals medical error. As a result of this program the average monthly cost rates decreased for total liability, patient compensation and non-compensation related legal costs.[44] This novel model may start a new approach to medico-legal compensation that may dramatically result in huge cost savings if widely adopted.

References

1. Schiff GD, Kim S, Abrams R, Cosby K, Lambert B, Elstein AS, Hasler S, Krosnjar N, Odwazny R, Wisniewski MF, McNutt RA (2003). Diagnosing diagnostic errors: lessons in patient safety: From Research to Implementation, vol. 2. Rockville, MD. Agency for Health Care Research and Quality p255–278. AHRQ Publication No. 050021.
2. Golodner L (2004). How the public perceives patient safety. Newsletter of the National Patient Safety Foundation. 1997:1–6.
3. Shojania KG, Burton EC, McDonald KM, Goldman L (2003). Changes in rates of autopsy-detected diagnostic errors over time. A systematic review. *JAMA* 289:2849–2856.
4. Sato L (2001). Evidence-based patient safety and risk management technology. *J Qual Improv* 27:435.
5. Phillips R, Bartholomew L, Dovey S, Fryer GE Jr, Miyoshi TJ, Green LA (2004). Learning from malpractice claims about negligent adverse events in primary care in the United States. *Qual Saf Health Care* 13:121–126.
6. Graber ML, Franklin N, Gordon R (2005). Diagnostic error in internal medicine. *Arch Intern Med* 165:1493–1499.
7. You Gov Survey of Medical Misdiagnosis (2005). Isabel Healthcare – Clinical Decision Support System. Available at: http://www.isabel-healthcare.com
8. Burroughs TE, Waterman AD, Gallagher TH, Waterman B, Adams D, Jeffe DB, Dunagan WC, Garbutt J, Cohen MM, Cira J, Inguanzo J, Fraser VJ (2005). Patient concern about medical errors in emergency departments. *Acad Emerg Med* 23:57–64.
9. Elstein AS (1995). Clinical reasoning in medicine. IN: Higgs JJM (ed). Clinical reasoning in the Health Professions. Butterworth-Heineman Ltd, Oxford, England, p49–59.

10. Berner ES, Graber ML (2008). Overconfidence as a cause of diagnostic error in medicine. *Am J Med* 121: S21–S23.
11. Croskerry P (2003). Cognitive forcing strategies in clinical decision making. *Ann Emerg Med* 142:110–120.
12. Redelmeier DA (2005). The cognitive psychology of missed diagnosis. *Ann Intern Med* 142:115–120.
13. Brennan TA, Leape LL, Laird NM, Herbert L, Localio AR, Lawthers AG, Newhouse JP, Weiler PC, Hiatt HH, (1991). Incidence of adverse events and negligence in hospitalized patients. Results of the Harvard Medical Practice Study I. *N Engl J Med* 324:370–376.
14. Baker GR, Norton PG, Flintoft V, Blais R, Brown A, Cox J, Etchells E, Ghali WA, Hėbert P, Majumdar SR, O'Beirne M, Palacios-Derflingher L, Reid RJ, Sheps S, Tamblyn R (2004). The Canadian Adverse Events Study: the incidence of adverse events among hospital patients in Canada. *CMAJ* 170:1678–1686.
15. Wilson RM, Runciman WB, Gibberd RW, Harrison BT, Newby L, Hamilton JD (1995). The quality in Australian Health Care Study. *Med J Aust* 163:458–471.
16. Davis P, Lay-Yee R, Briant R, Ali W, Scott A, Schug S (2002). Adverse events in New Zealand public hospitals 1: occurrence and impact. *N Z Med J* 115:U271 (1167).
17. Vincent C, Neale G, Woloshynowych M (2001). Adverse events in British Hospitals: preliminary retrospective chart review. *BMJ* 322:517–519.
18. Schiøler, T, Lipczak H, Pedersen BL, Mogensen, TS, Bech RB, Stockmarr A, Svenning AR, Frolich A (2001). Danish Adverse Event Study. Incidence of adverse events in hospitals. A retrospective study of medical records. *Ugeskr Laeger* 163:5370–5378.
19. Thomas EJ, Brennan TA (2000). Incidence and types of preventable adverse events in elderly patients: population based review of medical records. *BMJ* 320:741–744.
20. Thomas EJ, Studdert DM, Burstin HR, Orav EJ, Zeena T, William EJ, Howard KM, Weiler PC, Brennan TA (2000). Incidence and types of adverse events and negligent care in Utah and Colorado. *Med Care* 38:261–271.
21. Zed PJ, Abu-Laban RB, Balen RM, Loewen PS, Hohl CM, Brubucher JR, Wilbur K, Wiens MO, Samoy LJ, Lacaria K, Pursell RA (2008). Incidence, severity and preventability of medication-related visits to the emergency department: a prospective study. *CMAJ* 178:1563–1569.
22. Leape L, Brennan TA, Laird N, Lawthers AG, Localio R, Barnes BA, Hebert L, Newhouse JP, Weiler PC, Hiatt H (1991). The nature of adverse events in hospitalized patients. Results of the Harvard Medical Practice Study II. *N Engl J Med* 324:377–384.
23. Fisher G, Fetters MD, Munro AP, Goldman EB (1997). Adverse events in primary care identified from a risk-management database. *J Fam Pract* 45:40–46.
24. Wilson T, Sheikh A (2002). Enhancing public safety in primary care. *BMJ* 324:584–587.
25. Elder NC, Dovey SM (2002). Classification of medical errors and preventable adverse events in primary care: a synthesis of the literature. *J Fam Pract* 51:927–932.
26. Flynn EA, Barker KN (2007). Research on errors in dispensing and medication administration. IN: Cohen MR (ed). Medication Errors, 2nd Edition, American Pharmacists Association: Washington DC, p15–41.
27. Leape LL (2007). Systems analysis and redesign: the foundation of medical error prevention. IN: Cohen MR (ed). Medication Errors, 2nd Edition, American Pharmacists Association: Washington DC, p3–14.
28. Cohen MR (2007). Causes of medication errors. IN: Cohen MR (ed). Medication Errors, 2nd Edition, American Pharmacists Association: Washington DC, p55–66.
29. Kohn LT, Corrigan JM, Donalson MS (eds) (2000). To Err is Human: Building a Safer Health System. National Academics Press, Washington DC.
30. Teagarden JR (2007). The clinical bioethics of safe medication practice. IN: Cohen MR (ed). Medication Errors, 2nd Edition, American Pharmacists Association: Washington DC, p587–604.

31. Hepler CD, Strand LM (1990). Opportunities and responsibilities in pharmaceutical care. *Am J Hosp Pharm* 47:533–543.
32. Bartlett G, Blais R, Tamblyn R, Clermont RJ, MacGibbon B (2008). Impact of patient communication problems on the risk of preventable adverse events in acute care settings. *CMAJ* 198:1555–1562.
33. WHO (1972). International drug monitoring: the role of national centers. *Tech Rep Ser. WHO no.* 498.
34. Edwards IR, Aronson JK (2000). Adverse drug reactions: definition, diagnosis and management. *Lancet* 356:1255–1259.
35. Lazarou J, Pomeranz BH, Corey PN (1998). Incidence of adverse drug reactions in hospitalized patients. *JAMA* 279:1200–1205.
36. Khong TK, Singer DRJ (2002). Adverse drug reactions: current issues and strategies for prevention and management. *Expert Opin Pharmacother* 3:1289–1300.
37. Wasserfallen J, Livio F, Buclin T, Tillet L, Yersin B, Biolaz J (2001). Rate, type and cost of adverse drug reactions in emergency department admissions. *Eur J Int Med* 12:442–447.
38. Bates DW, Spell N, Cullen DJ, Burdick E, Laird N, Petersen LA, Small SD, Sweitzer BJ, Leape LL (1997). The costs of adverse drug events in hospitalized patients. *JAMA* 277:307–311.
39. Classen DC, Pestotnik S, Evans RS, Lloyd JF, Burke JP (1997). Adverse drug events in hospitalized patients. Excess length of stay, extra costs, and attributable mortality. *JAMA* 277:301–306.
40. Poon EG, Keohane CA, Yoon CS, Ditmore M, Bane A, Levtzion-Korach O, Moniz T, Rothschild JM, Kachalia AB, Hayes J, Churchill WW, Lipsitz S, Whittemore AD, Bates DW, Gandhi TK (2010). Effect of bar-code technology on the safety of medication administration. *N Engl J Med* 362:1698–1707.
41. Gallaher TH, Studdert DM, Levinson W (2007). Disclosing harmful medical errors to patients. *N Engl J Med* 356:2713–2719.
42. Kraman SS, Hamm G (1999). Risk management: extreme honesty may be the best policy. *Ann Intern Med* 131:963–987.
43. Landrigan CP, Parry GJ, Bones CB, Hackbarth AD, Goldman DA, Sharek PJ (2010). Temporal trends in rates of patient harm resulting from medical care. *N Engl J Med* 363:2124–2134.
44. Kachalia A, Kaufman SR, Boothman R, Anderson S, Welch K, Saint S, Rogers MA (2010). Liability claims and costs before and after implementation of a medical error disclosure program. *Ann Intern Med* 153:213–221.

Author Index

I.W. Fong, *Medico-Legal Issues in Infectious Diseases*, Emerging Infectious Diseases of the 21st Century, DOI 10.1007/978-1-4419-8053-3,

D

N

Subject Index

R

S

FSC
www.fsc.org
MIX
Papier | Fördert gute Waldnutzung
FSC® C083411

Kolibri 360 GmbH
Ferdinand-Jühlke-Straße 7
99095 Erfurt, Deutschland
produktsicherheit@kolibri360.de